Huntsville's Technological Evolution

A Technical History of Greater Huntsville, Alabama from 1800 to the Present

by

Raymond C. Watson, Jr., Ph.D., P.E.

Order this book online at www.trafford.com
or email orders@trafford.com

Most Trafford titles are also available at major online book retailers.

Print information available on the last page.

ISBN: 978-1-4907-6553-2 (sc)
ISBN: 978-1-4907-6552-5 (hc)
ISBN: 978-1-4907-6554-9 (e)

Library of Congress Control Number: 2015915885

Trafford rev. 11/12/2015

 www.trafford.com

North America & international
toll-free: 1 888 232 4444 (USA & Canada)
fax: 812 355 4082

CONTENTS

PROLOGUE

In the beginning God created the heaven and the earth. . . . And God said, "Let us make man in our image, after our likeness: and let them have dominion . . . over all the earth and subdue it." Genesis 1:1-28

While the human race grew and dominated the earth, contributions to the arts, science, and technology evolved through man's God-given talents. As expressed in this book, the technological history of what would eventually become Greater Huntsville is taken up as the 18th century closed. During the following two centuries, activities in this very limited geographical area certainly affected the overall technological growth of the entire United States of America. The objective of this book is to trace the many and divergent technological accomplishments within Greater Huntsville, show their cohesive relationships, and give credit to the personnel that led in these endeavors.

Representative Examples

When the 21st century began, the region of Greater Huntsville had experienced close to 200 years of technological evolution. James (John) Ditto had come to the area in 1802, soon opening a trading post on the north bank of the Tennessee River. To improve local travel, he built a ferry across the river in 1807 – the first recorded technological advancement in the area – allowing improved transportation over a fraction of a mile. Leap ahead 200 years. Transportation technology in Greater Huntsville has evolved to where people working just a short distance from Ditto's Landing are developing rockets that might safely propel future explorers over 35,000,000 miles to Mars.

Another spectacular change is found in communications. The early settlers to this area had no means of exchanging messages with the outside world other than hand-written letters that were carried from post to post. Today, fiber-optic networks routinely carry information between Huntsville and many areas of the word at gigabit rates (1,000,000,000 electrical impulses per second) – sufficient to transmit over 3,000 average-sized (50,000 words) books per second. It should be noted that the foundational technology for such communications was electronics, based on the radio vacuum tube patented in 1912 by Lee de Forest, who was raised in Talladega, Alabama.

Huntsville has witnessed, and been a part of, revolutionary changes in military technology. In 1813, Colonel Andrew Jackson assembled at Huntsville an army of frontiersmen from across the region to put down a rebellion of Creek Indians in the southern part of the territory. Among the troops was a company of Tennessee Riflemen (including Davy Crockett and Sam Houston); armed with flintlock-action longrifles, these sharpshooters could hit a moving target at a range of 200 yards. In 1984, local Army and industry engineers conducted an experiment demonstrating the equivalent of hitting a bullet with a bullet; in this, a missile intercepted a small target at an altitude of over 100 miles and moving with a closing speed of 13,600 miles per hour.

Revolutionary evolution in local science is shown in astronomy. There is no record of early settlers in this area using a telescope for viewing celestial bodies, but low-powered instruments were available and might have been used. The Russian Academy of Sciences – one of the world's most renowned scientific establishment composed of some 500 institutions and over 50 thousand scientific researchers – was recently planning a satellite carrying a highly accurate X-ray telescope. For the design and testing of the telescope's collecting mirrors, they turned to scientists at the local NASA Marshall Space Flight Center.

While widely known for Government rocketry, Greater Huntsville has had many lesser-known commercial activities, some successful and others not so. It is likely that few readers are aware that a competitor to Disney World almost opened a giant amusement park nearby. An automobile manufacturer planned to make Huntsville the Detroit of the South, and for many years the largest plant for automotive electronics was here. Until losing to labor unions, a telephone manufacturer produced near 2.5 million sets per year, and a tire manufacturer had one of the largest plants in the Nation. Presently, Greater Huntsville is headquarters for the largest poultry breeding firm in the world, and a local pharmaceutical firm ranks near the top in producing generic drugs. Two of America's most successful telecommunications hardware and software firms were founded and continue operations in Huntsville. All of these, and more, are described in this book.

In 2013, as much cotton lint was produced in Madison County as there was in the year just before the Civil War. Amazingly, this was grown in fields whose total acreage was only seven percent of the acreage in 1860! The yield (pounds per acre) had increased by a factor of about 14 times; technology improvement played a major role.

Jascha Heifetz was considered by many to be the greatest performer of the modern recording age; he was one of the very few, if not the only, player who could hit the high note at the end of Tchaikovsky's Violin Concerto and give it vibrato in the fifth of a second or so of its playing time. Once following a great performance, a woman caught him as he returned to his dressing room. "Mr. Heifetz, your violin sounded the best that I've ever heard," she gushed. Heifetz turned his violin toward his ear and softly replied, "Strange, now it's not making a sound."

This brief venture into the arts is to introduce the absolute importance of the individuals involved in the technological evolution of Greater Huntsville – take them out, and nothing would have happened. Regardless of the financial investments, the shops and laboratories, the tools and power sources, and all of the other materiel factors, it was the people – from single individuals to large groups – that made Greater Huntsville what it is today.

Why This Book?

There have been many books written about the historical development of Huntsville and Madison County. Most of these have centered on the cultural aspects, and there has certainly been much of importance from this standpoint. Others have diverse emphases; one masterpiece of the early 1970s – *A Dream to Remember*, in multiple volumes by James Record – is a highly comprehensive history of Madison County government. Record also provided an excellent early history of the region. However, it is the author's very biased opinion that the technological evolution of Greater Huntsville provided the foundation on which this great region stands – and, until now, this has not been given an end-to-end treatment.

In addition to documenting in this book the technological accomplishments in a series of eras, a major attempt was made to cite the individuals and groups that made it happen. Greater Huntsville – a term for the city and much of the surrounding Madison County that was adopted when the city limits were not enlarged as needed – is primarily a Federal Government metropolis. Correctly, the political aspects of this are acknowledged, and detailed information about the many agencies and associated contracting firms are included, but the people deserve and are given full credit.

I came to Huntsville in 1960, hired by Milton Cummings (founder of Cummings Research Park) to establish the Research Laboratories of Brown Engineering Company (BECO) – the first independent high-technology firm in the region. I was also on the faculty of the Huntsville Center of the University of Alabama, and have continuously been involved with local academics. BECO later became Teledyne Brown Engineering (TBE), and the Research Laboratories evolved to the 1,000-employee Advanced Science and Technologies (AS&T) Group. For many years, I have also operated R.C. Watson & Associates, providing engineering and management consulting.

Through the above-noted activities, I have had a personal involvement in many of the major technical endeavors in this region during the past five-plus decades. I also have an intense interest in history and for years have researched the earlier technical aspects of Huntsville and the region. (It is unusual for an engineer, but I have undergraduate minors in both history and literature.) Consequently, as I approach the ninetieth decade of life, I feel an obligation to document my knowledge of the technological evolution of Greater Huntsville.

I might note that in 2007, my book *Solving the Naval Radar Crisis* was published by Trafford. This also filled a gap in technical history, documenting the need and development during World War II of what was likely the most intense and challenging technical training program ever given by the U.S. Navy (see "Electronics Training Program" on *Wikipeda*). This book was followed by *Radar Origins Worldwide*, 2009, published by Trafford and also gap-filling in context.

Whence Came the Information?

Typically, a book concerning history is filled with numbered reference notes at the end of many sentences and paragraphs. These are intended to convey to the reader the specific source of the information. This is important for journal papers, encyclopedia articles, and many types of books, any of which might in turn be chain-cited by other authors. This present book was mainly written for what might be called "recreational readers"; few of these are interested in concise references.

Such readers might include individuals who themselves had an involvement with the technical activities described and could use the book as a substitute for their lacking diary – "Oh yes, I remember the XXX project (or Dr. YYY)!" Others might be newcomers to the area and would like to have a general knowledge of the past persons, organizations, and technical activities. Most recreational readers might eventually treat the book like a "coffee table" document – browsing it in a leisurely manner; the material in each section is mainly complete in itself, and, although small in size, the many pictures make perusing easy.

As previously described, I had a personal involvement in many of the activities of the past half century; for such, the source is often my memory and a large collection of working documents in both hard copies and digital files.

Consequently, there are no directly listed references throughout the text. Each of the five chapters ends with several pages of what is called Selected Bibliography / Additional Information. This is mainly to aim the reader to an enlargement of the information contained in the book. Many of the references are from the on-line encyclopedia *Wikipedia* with the URL given. I am a contributor and strong supporter of this encyclopedia, and am the originator or editor of many of the listed articles.

It has become a requirement of publishers for the author to show the source of each graphic included in a book. This has been included herein as an appendix. Many of the graphics have been collected over the years, and it has been a major task to trace the origin. Most of the pictures are in the public domain and thus unprotected; protected intellectual property is the exception.

Acknowledgments

Over the five years of preparing this book, many different people were contacted – far too many to list. Their contributions were invaluable – from the spelling of a name to lengthy descriptions of organizations and activities. With inputs from different directions, conflicting material was inevitable, leading to more contacts and searches. If this were a true history book, such conflicts would be directly acknowledged; herein, however, I simply made biased selections. Beyond simple contacts and answering of question, a number of individuals who had been personally involved with some of the activities were personally interviewed.

The Web Sites and digital archives of both the Army Missile Command and the Marshall Space Flight Center were extensively used. Here it is noted that, for security reasons, the responsible agencies have recently made many of these government Sites unavailable to the public.

Listings and descriptions of non-government technical firms and their key personnel have been given extensive space. Much of the current information came from Web sites of the firms, but the periodical publications through the years of the *Industrial Directory* of the Huntsville/Madison County Chamber of Commerce has been invaluable. The Twenty-fifth Edition (2014) is the most recent source of much of the cited information concerning leaders and numbers of employees in firms.

The Heritage Room of the Huntsville-Madison County Public Library was a major source of photographs, brochures, newspaper clippings, and other materials from their extensive archives; special thanks are given to Susanna Leberman.

My friend and neighbor, Robert Simpson, was highly helpful as a reader of drafts. Wally Kirkpatrick, Michael Kirkpatrick, Ray Sells, and others at DESE Research are especially recognized for their support and encouragement to me throughout the book-preparation process.

Thank God for the Internet, the World Wide Web, e-mail, and Microsoft Word, and for His allowing me to complete this endeavor.

Raymond C. Watson, Jr., Ph.D., P.E.
Huntsville, Alabama; RCW-Assoc@comcast.net

Chapter I

ORIGINS and MATURING

The 19th century – 1800 through 1899 – is the originating and maturing period of Huntsville and Madison County. To cover the history – particularly the technological evolution – of this area in this extended period, the information has been divided into the four segments: Early Times, King Cotton, Transportation and Utilities, and Maturing Times. To an extent, these are also time periods, but there is considerable overlapping.

EARLY TIMES

The area eventually containing Madison County was in the region called *Ah-la-bama* by the native Muscogee (also called Creek) Indians who occupied the lower portion of the region; in their language, Mvskoke, this was a phrase meaning, "We will rest here." There is little recorded as to the first exploration of this area – some historians believe this was by Spanish explorer Hernando de Soto in 1540. French maps from the late 17th century show a large traversing river looping south and then north around the area – later called the "Great Bend" – certainly indicating early French explorations (the town of Mobile was formed by the French government in 1602). The name Tennessee River first appears on maps of the late 18th century; it is believed to have come from Tanasi, a Cherokee Indian village.

Early Explorer

NATIVE AMERICANS

The area in North Alabama adjacent to the Tennessee River is commonly called the Tennessee Valley, herein simply "the Valley". The Cherokee Indians were the first well-identified inhabitants spread across the Valley, but in about 1650, they withdrew from the region to an area in the mountains to the northeast, reserving the flat portions of the Valley as a large hunting ground. Shawnee Indians then moved southward from around the Cumberland River and occupied land in the Valley. This led to many years of warfare between the

Cherokees and intruding Shawnees. The Chickasaw Indians from the western portion of the Valley eventually teamed with the Cherokees, and by the early 1720s, the Shawnees had been driven northward into the Ohio River area. Then for almost half a century, the Tennessee Valley was without permanent occupancy.

In about 1765, some of the Chickasaws moved into an area near the Tennessee River in what is now the southern portion of Huntsville, and formed a large settlement. This was challenged by the Cherokees, and they attacked their former allies. In 1769, there was a major battle at the Chickasaw settlement; the Chickasaws won, but at such a great loss that they withdrew from the settlement. Thereafter, the area of the abandoned settlement was known as the Chickasaw Old Fields (a square with about three-mile sides); this became a benchmark for future divisions of the land.

Through the following three decades, both tribes claimed the land on both sides of the river, westward to the Buffalo River, and eastward to the great ridge dividing the waters of the Tennessee and Cumberland rivers. The United States recognized both claims.

As the overall territory developed, the Chickasaw nation became in debt to the White traders and merchants, and also needed funds for local improvements. In the Chickasaw Treaty, signed 23 July 1805, the land between the east boundary and a direct line running at about 45 degrees northwest from the Old Fields to the ridge near the main source of the Buffalo River was ceded to the United States. For this, the Chickasaw nation was paid $20,000, the debt of $2,000 was settled, and the Chickasaw king, Chinubbee Mingo, was to be paid an annual annuity of $100.

Similarly, in the Cherokee treaty of 7 January 1806, all their territory north of the Tennessee River and west of a line drawn from the upper part of the Chickasaw Old Fields northerly to the Elk River, was ceded to the United States. For this, the Cherokee nation was paid $10,000 and the Cherokee chief, known as Black Hawk, was to be paid an annual annuity of $100. In addition, a grist mill would be built in Cherokee country, and a machine for cleaning cotton (a hand-powered cotton gin) would be provided; these showed that many of the Cherokees had become farmers.

The triangular tract of country acquired by these two treaties became the original Madison County of 1808. For the next two decades, the Indians lived peacefully in the land adjacent to Madison County. Then, under the Indian Removal Act passed by Congress on 28 May 1830, Indians from five tribes in the southeastern United States, including the Cherokees and the Chickasaws, were forced to move to the Indian Territory (Oklahoma). The Cherokee people called this journey the "Trail of Tears."

8

Overall, the Cherokees were the most advanced of the native Indian nations across the Southeast. Even before the arrival of European explorers, their  villages often contained full houses. As described later, Cherokee warriors joined Andrew Jackson and U.S. troops in defeating the Creek Indians in 1814. The Cherokees had a written language (a syllabary) developed by George Guess/Gist (Chief Sequoyah) between 1809 and 1824. By 1830, when forced from their homelands,

Early Cherokee House   most were literate in their own language –
Holy Bible translations, other books, and newspapers were published using the syllabary.

THE LAND

Before 1800, there was confusion as to which State or Territory owned the land in the region containing the Great Bend of the Tennessee River. The first officially recorded exploration of the area by Americans was in 1777. North Carolina had given "bounty land" to its Revolutionary War veterans and, believing that the region was their territory, sent a group to examine land along the river up to what is now Muscle Shoals. The group, however, was driven out of the area by Chickasaw Indians who felt that they were losing land already agreed for them.

In 1780, a band of some 160 persons, led by John Donelson in 30 flat-bottom boats, came by the area without incident on the Tennessee River. They passed through the shoals, eventually joined the Cumberland River, and then went upstream to settle Nashborough (later named Nashville) in 1784. Earlier, Donelson and James Robertson had travelled overland from Watauga (the first settlement in what is now Northeastern Tennessee) to establish a stockade on the banks of the Cumberland River in 1779; Donelson then returned to Watauga and formed the pioneering party. At that time, the land around present-day Nashville was in North Carolina (it became Tennessee in 1796).

The Georgia Legislature declared the region that is now Alabama and Mississippi to be in their Territory in 1783. For the next two decades, speculators formed companies for dividing and selling this land. One such group, in what was later called the Great Yazoo Fraud, gained the blessing of the Georgia Legislature to pay the State $500,000 for over 21 million acres (approximately the size of England, Scotland, and Wales). The Tennessee Land

Company, owned by this group, started selling thousands of acres in this area to buyers who had never even seen the land.

By the Treaty of Madrid, in 1795, Spain ceded to the United States the lands east of the Mississippi between 31°N and 32°28'N. In 1798, Congress organized this district as the Mississippi Territory, with the Territorial Governor's office at St. Stephens, a few miles north of Mobile

Georgia eventually surrendered its claim in 1802, transferring to the United States an area that later became Alabama and Mississippi. The Mississippi Territory extended its boundaries in 1804, taking in the eastern portion of this new land, including that along the Tennessee River. The previous land sales, however, were considered legal; recorded in a deed dated 11 July 1808, Martin Beatty of Lee County, Virginia, had paid $1,000 for 1,000 sight-unseen acres that included all of the land of the early settlement around the big spring that is at the heart of present-day Huntsville. Such land sales remained on the books until Congress passed an 1814-Act appropriating $5 million to settle all Yazoo-related claims. Beatty relinquished his title to the big-spring land at that time.

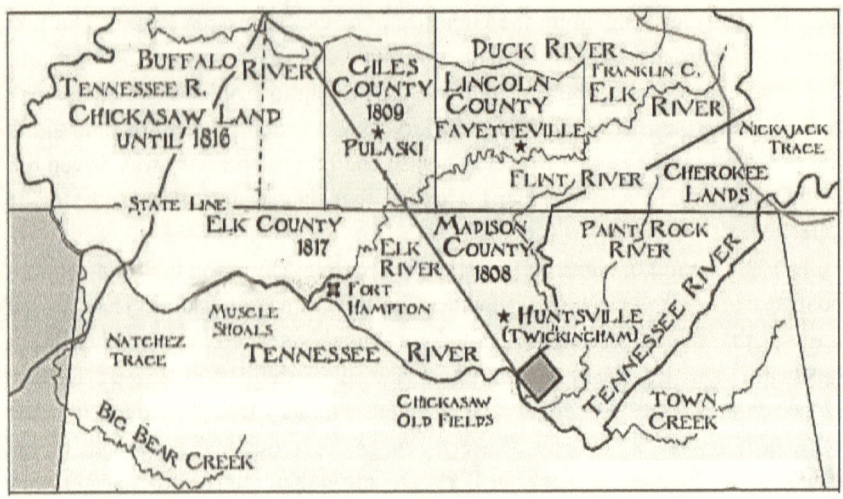

Upper Mississippi Territory and lower Tennessee in Early 1800s

Mapping

Following the Revolutionary War, President Thomas Jefferson proposed a system for use in the United States in mapping and controlling the land. All areas not specifically under a warrant to an individual or other private entity was designated Public Land, owned by the United States and potentially available for sale. To handle this, the Public Land Survey System (PLSS) was devised to divide all of the land (public and private) into townships, primarily squares six miles on the side. Townships, in turn, are divided into 36 square sections, each being one mile on the side and containing 640 acres. Sections, in turn, are usually divided into quarter-sections of 160 acres, these into quarter-quarter sections of 40 acres, and finally into lots of various size and shape.

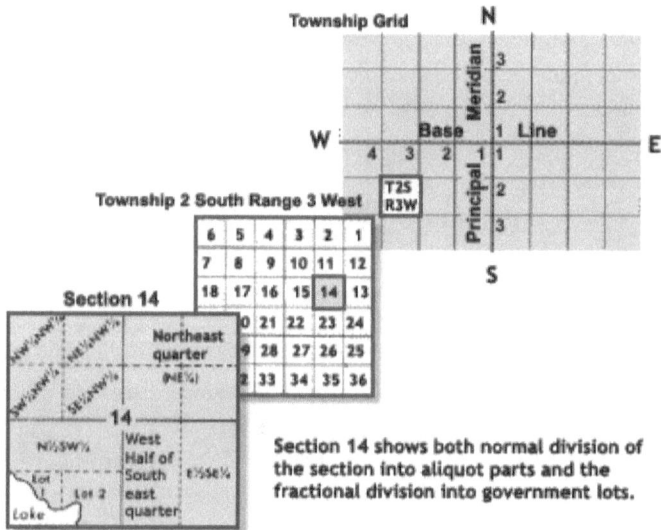

Section 14 shows both normal division of the section into aliquot parts and the fractional division into government lots.

Public Land Survey System

The selection of 640-acre townships as a standard was to allow up to seven divisions by two (halvings) but still retain a whole number of acres. A half of a quarter-quarter-quarter section is five acres, which is a readily surveyed 50-square-chain area. (A chain is a measuring tool 66-feet long, then 80 chains equals one mile.) To identify the location of townships, sections, and their subsections, rectangular grids are used. For these grids, horizontal baselines and vertical meridians were established, with the designated Center Point being the crossing of the Principal Meridian and a Base Line.

After the Mississippi Territory expanded, the Federal Government used the PLSS to map the southern half of this region in 1805, and then followed with the northern half in 1807. For the northern half, the Base Line was the Tennessee border at Latitude 34°-59'-27", and the Principal Meridian – later called the Huntsville Meridian – along Longitude 86°-34'-16". All land in Madison County is mapped into townships, sections, and quarter-sections measured from this 1807 Center Point. Settlers mainly came into the area along a southward path that was near the same as the Principal Meridian.

When the U.S. Congress established Tennessee as a State in 1796, they set the southern border to be exactly the 35-degree north latitude. The first official mapping (in 1818) used multiple observations with a marine sextant and published tables called *ephemerides* to place a border benchmark at a location near a place called Nickajack Cave. Thereafter, this benchmark was used to locate all borders in the region. With improvements in measurement instruments and processes, this benchmark, and thus the border, was eventually settled to actually be at the above-cited 34°-59'-27" – 33" or about a mile south of the intended 35°! Through the years, the band generated by this difference has been a disputed area for taxes, voting, water rights, and other matters. The decisive fact, however, is not where surveyors meant to draw the line – it is where people have accepted the line to be over time.

Madison County

Madison County, in the Mississippi Territory, was created by the Territorial Governor on 13 December 1808. It was named for James Madison, then Secretary of State and President-elect of the United States. Both Cherokee and Chickasaw Indian tribes had originally claimed the area as hunting grounds, but any villages were gone by this time. The initial area – defined by the Chickasaw (1805) and Cherokee (1806) Treaties – was about 25 miles wide at the top along the Tennessee border; some 30 miles in straight-line length, but tapering down on each side following the original Cherokee and Chickasaw boundary lines (to the east and west, respectively); and ended at the bottom at the Chickasaw Old Field with about three miles along the Tennessee River. This encompassed roughly 540 square miles (near 345,000 acres). Land areas were added several times, and then redefined through 1887; finally giving Madison County a total of 806 square miles (515,840 acres) and its present shape.

A geographical survey of Madison County was made for the Federal Land Office and conducted by the official territorial surveyor, Thomas Freeman of Nashville; this was completed in May 1809. Freeman, who had learned surveying as an officer in the U.S. Army, had earlier surveyed the Meridian Line running through Madison County.

The established practice of surveyors at that time involved observing the pole star (Polaris) to find true north at the Center Point, then using a vernier compass (a magnetic compass with two vertical sights)

Surveying with Vernier Compass

12

mounted on a tripod and a Gunter's chain (a 66-foot-long metal chain of 100 links; 80 chains then equals 5,280 feet or a mile). Through this practice, a surveyor could find the desired path and set up markers at corners of townships and sections.

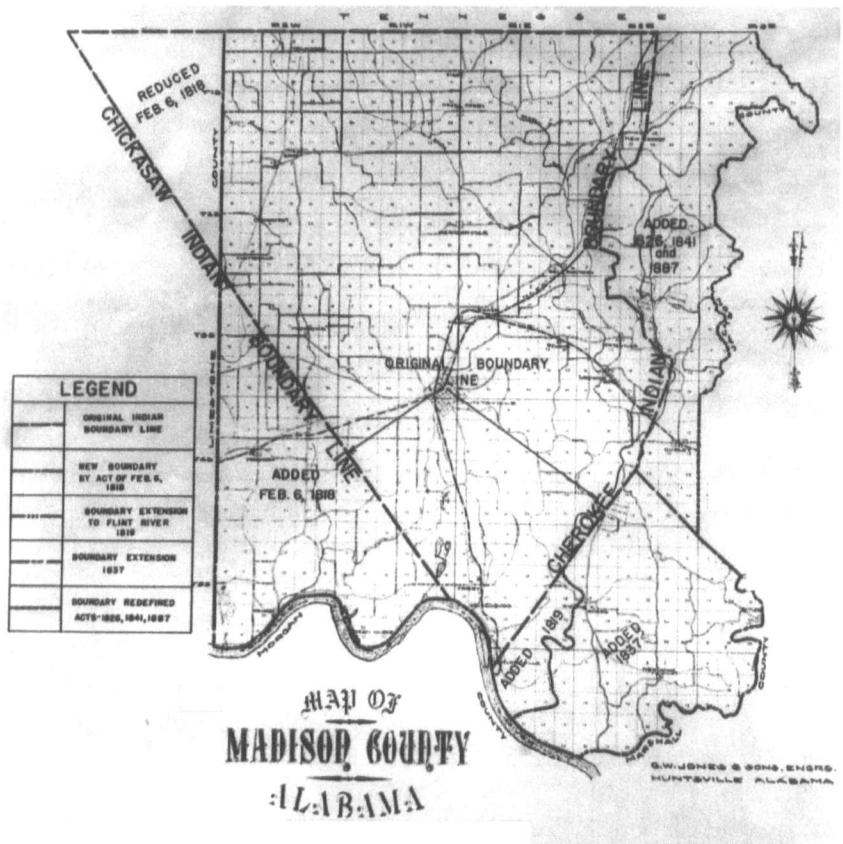

Madison County Showing the Various Changes in Area Through 1887
Prepared by G.W. Jones & Sons, Engineers, in 1934

The basic art and science of surveying is ancient; it was known to the early Greeks and Romans, as shown by their roads and aqueducts. Because of the fundamental value of land, surveying was one of the first occupations in most states to require official recognition. At age 17, before the Nation was formed, George Washington became a recognized surveyor, serving for some time as the official surveyor for Culpepper County, Virginia. Washington, who had no formal higher education, learned surveying from the book, *The Young Man's Companion: Or Arithmetick Made Easy*, by William Mather, 1737.

Theodolite

It is noted that the technology used in surveying greatly changed in the early 19th century. A telescope was added above the vernier compass, and a finely divided circular scale allowed accurate horizontal angular measurements – with other minor improvements, this became the theodolite. Although this instrument was relatively expensive, it was quickly adopted by professional surveyors. It is possible that Freeman used a theodolite in his geographical survey of Madison County.

While making the land survey, Freeman also took the first census of Madison County; released in January 1809, it showed the total population to be 353 heads of families, with 1,150 free White males. 723 White females, and 332 slaves. Many people had come into the new land, although orders were to wait for the public land sale.

THE PIONEERS

As previously noted, at the start of the 19th century there were no Indians permanently living in the area that eventually became Madison County. White settlers began to arrive at this time, all intent on obtaining rights to land that had previously been Indian territory.

It is believed that James ("John") Ditto (1743-1828) was the first settler in the area that is now lower Madison County. Ancestry documents indicate that he arrived at Chickasaw Island (now Hobbs Island) on the Tennessee River in 1802. Some believe that Ditto had come down overland through the Big Spring area, but it is more likely that he had drifted down the river on a small flatboat, possibly accompanied by his wife and several of his grown sons and carrying goods for trading with the Indians.

Unverified documents indicate that Ditto opened a trading post beside the river in 1805, and then started a ferry service for crossing the river in 1807. This river-bank site was later called Ditto's Landing. Ditto also started a boatyard, building flatboats with a shallow draw for carrying goods down the Tennessee River and past the shoals.

In 1804, Isaac Criner (1783-1876), his uncle (or brother?) Joseph Criner (1767-1843), and Thomas McBroom (1784-1843), all from East Central Tennessee, followed a trail that would later be the Winchester Road to explore the area just below the Tennessee border – a distance of at least 175 miles. Explorers such as these were normally on foot, carrying a minimum load for survival in the wilderness. They certainly carried a flint-lock musket and likely

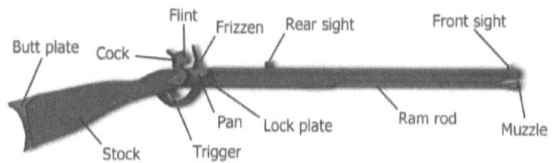

Flint Frizzen   Rear sight   Front sight
Butt plate Cock
Pan   Lock plate   Ram rod   Muzzle
Stock   Trigger

Smooth-Bore Flint-Lock Musket

had a pocket compass – a very common and relatively inexpensive device – and used this to maintain direction and

identify the return path. The musket and compass then might be considered the first new technologies introduced to this area.

After returning to East Tennessee for their families, the Criners and McBroom came back in early 1805 – likely with their possessions in a horse-

Pocket Compass

or mule-drawn wagon. The Criners built log cabins near what is now Mountain Fork of the Flint River; later settlers named this community New Market, considered to be the earliest town in Madison County. McBroom travelled further southeast a distance of about 20 miles, settling his family near what is now Gurley.

In the spring of 1805, John Hunt and Andrew Bean, exploring from their homes near Tazewell in East Tennessee (about 35 miles northeast of present-day Knoxville) crossed into the new territory and found the Criners – this would have been a total trek of about 270 miles. There they were told about a large spring with an abundance of game further south that had been reported by an earlier explorer, Samuel Davis; Indians called this *Waiki Lako* – "Big Spring."

After staying overnight with Joseph Criner, Hunt and Bean continued exploring on a southwest path of about 15 miles and soon found this spring. The spring was semi-circular, some 100 feet in diameter, and flowed from beneath a

Big Spring and Bluff

high bluff. Nearby was the start of a crude cabin (possibly made by Samuel Davis); Hunt and Bean built a cabin on this site.

Bean decided to return to Tennessee, but Hunt, who was then 55 years old and had earlier been a sheriff and a militia captain, thought that building a dwelling was the way to establish first rights to purchase the land when it was opened for sale. Hunt returned to Tennessee, settled his affairs in Tazewell, and returned to his cabin with his wife and three of his sons in the late summer of 1805.

While in Tennessee, Hunt had given glowing descriptions of the land that he had found, and was soon followed by other families; this was the beginning of a small community that they called Hunt's Station.

SETTLERS AND SETTLEMENTS

The early settlers mainly reached this area along natural paths from Tennessee. The first roads through the county followed these paths, leading northwest toward Nashville or northeast toward Winchester in Lincoln County. Along either road, there were no large streams or steep mountains, so transportation using carts and wagons to these destinations soon became common. The two paths came together just above the Tennessee border, and this southward route was selected when placing the Principal Meridian. Some of the settlers possibly reached this area from upstream on the Tennessee River using flatboats that simply drifted with the current.

Flatboat Transportation

Squatters

The pioneering period for this area was relatively brief, basically between 1805 and 1810. In this time, many families seeking new land and fresh opportunities trekked into the territory. In addition to New Market and Huntsville, early towns included Hazel Green, Meridianville, Maysville, and Scotts Mill (now Brownsboro). The first settlers in these communities were true pioneers, living under the dangers, toils, and privations of pioneer life; however, unlike in many other pioneering areas, there were no problems with the Indians. The settlers were known as "squatters," building on government-owned land with the hope of buying their homestead when the land was sold.

Most of the land was covered in trees and thick shrubs, and had to be cleared. The logs were used in constructing the first houses, barns, and other buildings. The food supply initially came from the land: wild turkeys, bears, deer, and many species of birds were abundant; there was honey and a variety of wild berries; and the clear streams were filled with fish. Some of the settlers brought with them several chickens and pigs – these quickly multiplied; others likely brought a cow and a few sheep and goats.

Cottage industries soon provided items such as gun powder, nails, fittings, lamp oil, utensils, and barrels, but salt, flour, musket lead, raw iron, and certain other critical supplies had to be brought in on pack animals. Gun powder, vital for the pioneers, was made of 75 percent potassium nitrate (saltpeter, that was readily available from local caves), 15 percent charcoal (made from several types of softwood), and 10 percent sulfur (obtained by sun-evaporating water from so-called sulfur springs – there is a large such spring a few miles east of Hazel Green).

The first crop grown was likely corn; for conversion to meal, this had to be ground by hand or carried to a grist mill that was some 40 miles away in an area that is now Franklin County, Tennessee. There were also patches of potatoes, tobacco, and

Typical Early Log House Interior

cotton, and families also raised hogs, chickens, sheep, and goats. There were not many cows; bulls were dangerous to be brought in, so cows had to be taken some distance to be bred. Some of the families had spinning wheels and looms to make cotton and wool cloth for their own use.

Land Sale

In 1808, about 300 of the squatters partitioned the Federal Government to hold a public land sale. After Madison County was created as part of the Mississippi Territory and the land surveyed, a Public Land Office was opened at Nashville. On 25 August 1809, the land sale began. Persons from Madison County had to make the trip to Nashville – averaging some 100 miles and over 20 hours on horseback – and wait until their section came up for bid. Most of the bidding was for farm land, which sold for about $1.75 per acre; occupied land, however sold for much more. Buyers had 10 years to make full payment. Eventually, less than half of the squatters were able to buy their land. Overall, the sale was so successful – over 53,000 acres in Madison County had been sold by late 1810 – that the Land Office was transferred to Huntsville in 1811.

At the initial land sale, Hunt found that he had failed to pay a fee required to register as a squatter, meaning that he had to bid against all other interested persons. LeRoy Pope – a wealthy tobacco planter and lawyer from Petersburg, Georgia – with William Anderson and James Jackson – both land speculators from Nashville – as partners, successfully bid for the quarter section (160 acres) that contained Big Spring and Hunt's home; Pope paid $23.52 per acre, an exorbitant price at that time. Unable to match Pope's bid, Hunt successfully bid on a quarter-quarter section (40 acres) located about three miles south of Big Spring.

For several years following his land purchase, John Hunt survived by farming; his daughter and her family had bought an adjacent farm. With failing health, Hunt could not continue payments on his land, lost it, and then lived with his daughter. The pioneer settler of Huntsville died in 1822, and it is believed that he was buried in an unmarked grave at the Acklin (later Sively) graveyard; this site is long lost, but, tragically, thought by some to be where the Huntsville trash dump is now located. This, as well as the absence of information on James Ditto, shows an early shortcoming of the city in not accurately and completely maintaining historical information.

Since LeRoy Pope was the owner of the most of the land being developed around the big spring, he influenced the Territorial Legislature to have the village named Twickenham – a name taken from a place in England that he associated with some of his ancestors. This name shows on documents for several years; in November 1811, however, the Territorial Legislature changed the name to Huntsville, giving credit to Hunt for its start. After some protest, Pope consented; on 9 December 1811, Huntsville became the territory's first incorporated town.

In 1814, Pope built a brick mansion on what was called The Hill (now Echols Hill), then the highest point in the town; this is now the oldest documented brick residence in Alabama. Pope went on to become a leading figure in local political and business activities.

Pope's Mansion on The Hill

The land sale had brought many new families into Madison County. Included were a number of wealthy slave-owners who carved out large plantations. They mainly came from Georgia, North Carolina, and Virginia; many were well educated and quickly became leaders in the communities. Some had earlier owned cotton farms, and soon recognized that the local climate and soil were excellent for growing cotton.

West of Madison County and between the Tennessee Border and the Tennessee River was Chickasaw Land, and many early settlers coming into the region attempted to become squatters there. In 1810, the U.S. Government built Fort Hampton – a garrison of log buildings – near where the Elk River empties into the Tennessee River. Two companies of soldiers were stationed there to keep intruders off the land; the soldiers also built the first roads in that area, including the one between Athens and Huntsville. The fort was abandoned when the land was ceded by the Chickasaw nation to the United States in 1816. This was possibly the only fort built specifically to protect Indian land from White settlers.

EARLY ORGANIZATION

With an influx of settlers following the land sale, businesses in the town of Twickenham began development. Early businesses included a grist mill and a hand-operated cotton gin owned by David Moore near the Big Spring, and a general store, also started by David Moore in a log building above the bluff. John Bunch built a relatively large frame building for the town's first tavern, and this was also used for meetings.

Organizing

By 1808, Madison County had grown to have 4,016 Whites with 624 families, but only 177 owned their land. There were 519 Black slaves owned by 20 percent of the White families. About 200 persons lived in Twickingham.

Acts of Congress had allowed the Governor of the Mississippi Territory to appoint local civil officers and let the county residents vote on Territorial representatives and a U.S. Congressman. In November 1808, Territorial Governor Robert Williams received word that U.S. President Thomas Jefferson had said that it was "high time" for the Governor to exercise his authority and get Madison County organized. The Governor soon got this underway; on 19 December 1808, Williams appointed Stephen Neal as Sheriff – the first official in Madison County.

In early 1809, Governor Williams made several more appointments for Madison County, These included an Attorney General, a local Resident Judge for the State Superior Court, Clerks for the Circuit and County Courts, a Deputy Territorial Treasurer for to serve the as County, and a County Treasurer.

During 1809, other appointments included Hugh McVay as County Surveyor, and John Martin as Road Apportioner – the official to determine road needs and authority to exercise eminent domain. It appears that John W. Leake was already recognized locally as the County Surveyor, and McVay only served in this position until Leake took over. In any event, appointments as Surveyor and Road Apportioner were certainly steps in the area's technological evolution.

On 7 November 1809, Madison County had its first governing body; appointed by Governor Williams, this was the Court of Justices of Peace and of the Quorum. There were five Justices with indefinite terms; William Dickson was the first Chief Justice, serving 1809-1811, followed by LeRoy Pope, 1811-1819. In 1819, this became the Inferior Court, then in 1821, the Court of County Commissioners of Roads and Revenues, which continued until 1850. After that, the overall governing body was the County Commissioners.

Another major step in organizing the town and county took place in December 1809: the Governor appointed a commission to select a site for the county seat and to acquire 30 to 100 acres of land. The land would be divided into half-acre lots with an area reserved for public buildings. Lots would then be sold with the proceeds used for constructing the public buildings. The selected site was on the bluff above Big Spring. This included 30 acres that were bought from LeRoy Pope (he had bought out his land partners) by the commissioners for $750; the deed, however, was not recorded until 1815.

The 30-acre area included a central three and one-half acre square where a court house was to be built. The remaining 26.5 acres surrounding the square were readily sold in half-acre lots at a price between $200 and $500. About $10,000 was raised and applied to public buildings.

The initial town is believed to have been laid out by John W. Leake in 1810; it covered about 60 acres in a grid of 20 square blocks. The original plan was not recorded and is non-extant; the plat often called the original was actually drawn by Hunter Peel in 1821. Peel's drawing shows the then-existing boundary streets named Holmes, Lincoln, Williams, and Henry (east side of Big Spring) and Gallatin (west side). The streets were not due north-south or east-west as is customary, but aligned with the bluff above the Big Spring at about 34 degrees north of west. The square was about a half-mile west of the Huntsville Meridian, which passes just west of the present Maple Hill Cemetery. In later surveys, the original 20 blocks were retained, but most new streets were oriented north-south or east-west.

In May 1811, the first countywide election was held for three Territorial representatives and one delegate to Congress – extending suffrage to Madison County citizens. Another census was taken in May 1812; this showed about 1,400 families in the county, some 38 percent holding near 3,600 slaves, giving an average of near seven slaves per slave-holding household.

While in Madison County and under the county's officials, the town of Huntsville had its first separate governing body in 1812; this was known as the Board of Trustees with four members, including a President, appointed by the Governor. The first President, who served between 1812 and 1816, is not known. This form served until 1828, when it changed to an elected Mayor-Alderman form; this continued until 1844, and then changed to an all-elected President/Council system.

Town Building

There was no shortage of construction materials. The area around the village had groves of pine, oak, hickory, cedar, walnut, and other trees. A number of settlers made their skills available for cutting timber for beams, siding, flooring, and roofing. Others were finished carpenters, but many of the early frame houses were built by the owners themselves. The first full sawmill and lumber yard in the area appears to have been operated by Henry Turner a few miles west near the present-day Athens. A variety of clay in the area was suitable for molding bricks, which were then usually baked in the sun. There was no local source for window glass; this had to be shipped in. Although materials and talents were available, when LeRoy Pope started his mansion, bricks and other materials were brought in from Virginia on flatboats and the skilled builders (most likely slaves) were likewise from Virginia.

A blacksmith shop was a very important part of any pioneer village. This would consist of a room with a large fireplace (a "forge") that burned charcoal to heat the metal to be worked. An accordion-like blower, operated by hand, served to increase the intensity of the fire. The raw metal was heated in the forge to a desired temperature (as indicated by the color) and was shaped, cut, or punched using a heavy

Blacksmith Shop

hammer and an anvil. Next to the anvil was a bucket of water used to cool ("temper") the metal. The first blacksmith in early Huntsville has not been identified, but a man named Claybrook advertised his shop in 1816.

First Madison County Courthouse

Thomas and William Brandon, skilled stone masons, did much in transforming the log-house community into a brick and mortar metropolis. In 1817, John Hickman completed the first Court House; by this time, the square already had many well-designed brick store buildings constructed by Hickman and William Watkins, and elegant homes were being built across the town, especially to the east between Big Springs and Pope's Hill.

Early Engineers

Hunter Peel was born at Lancashire, England, in 1785, and came to Huntsville in 1816. Having earlier served as an engineer in the British Army, he was soon appointed as the Madison County engineer and was responsible for

21

much of the later surveying in this area. A skilled draftsman, Peel drew many of the area's early maps. At that time, there was almost no formal educational training for engineers in England; thus, it is likely that his engineering knowledge came from reading and apprenticeship.

In 1802, the Military Academy at West Point had been formed by the Army Corps of Engineers, and topics such as surveying, structures, and mechanisms were included; this was, however, for military, not civil, applications. It was the only American school teaching engineering for the first quarter of the century; thus, educated persons serving in the civil sector of early days were likely former Army officers. Although West Point was fully recognized as an institution of higher education, it did not grant degrees until 1933.

Persons engaged in mechanical engineering, such as designing mechanical devices, were usually called practical mechanics. It was only in the latter half of the 19th century that more than a few academic institutions taught engineering. A survey in 1890 showed that only four percent of practicing engineers or surveyors in America had a college education; most had gained their knowledge through self-study and/or as an apprentice.

Military Activities

In those days, individual survival and homeland security depended, to a large measure, on each man being skilled in using a muzzle-loading, flint-lock rifle. Territorial law required every adult White male between ages 18 and 45 to be in a militia. By 1809, there were about 700 men in eight companies in the county. These became the Alabama Seventh Regiment, commanded by Lt. Colonel Nicholas Perkins, and held their first countywide muster on 29 October 1810; Governor David Holms, successor to Governor Williams, attended the muster. During the War of 1812 between England and the United States, two of these companies were called into service; however, they did not participate in actual combat.

In 1813, there was an uprising of the Red Sticks tribe of the Creek Indians at Fort Mims about 30 north of Mobile; some 500 residents were slaughtered – the worst massacre by Indians in American history. Using arms supplied by the British, who were at war with the United States, the Red Sticks uprising soon spread over the lower part of the Mississippi Territory. The Government in Washington made a plea to Tennessee for Colonel Andrew Jackson to help in putting down the rebellion. It was rumored that the Creeks were preparing to attack Huntsville, and Jackson asked the various militias of the region to meet in Huntsville and form a defending army. This occurred on 11 October 1813, starting with a 32-mile forced march from Fayetteville, Tennessee, to Huntsville in five hours.

Companies of frontiersmen from across the region joined Jackson's Tennessee Riflemen at Camp Beaty, a large staging area with a good water supply located near the present-day Brahan Spring Park. There were four companies from Huntsville and one from Hazel Green. Davy Crockett and Sam Houston were among the Tennessee troops.

After remaining at Camp Beaty for some time, Jackson's rugged army crossed the Tennessee River by Ditto's ferry, then moved south. They destroyed the Red Sticks stronghold at Talladega on 9 November; two men from Madison County were among those killed.

Jackson was promoted to Brigadier General, and his army was joined by a regiment of U.S. Army infantrymen, as well as several hundred Cherokee and Choctaw allies. They continued south to decisively defeat the Red Sticks at Horseshoe Bend – a section of the Tallapoosa River in the center of the Territory – on 27 March 1814.

The Red Sticks lost an estimated 850 men, and Jackson lost 47, two being from Madison County; 23 Indian allies also died.

Battle of Horseshoe Bend

Jackson came back through Huntsville for a major celebration in May 1814; LeRoy Pope and his wife Judith entertained Jackson and his staff at the Pope's mansion that was still being completed on The Hill.

Alabama State Formation

The Mississippi Territory was divided on 3 March 1817. The western portion became the State of Mississippi, and the eastern portion became the Alabama Territory, with St. Stephens, a few miles north of Mobile on the Tombigbee River, as the temporary seat of government. On 19 January 1818, the first session of the Alabama Territorial Assembly met at St. Stephens and created 14 new counties to accommodate the increasing numbers of settlers. U.S. President James Monroe signed the enabling act for statehood on 2 March 1819.

With Madison County having the largest population in the Alabama Territory, Huntsville claimed the title of provisional capital and invited delegates of the other 21 then-existing counties to meet in Huntsville and write the State's first constitution. On 5 July 1819, forty-four delegates (two from each of the 22 counties), met in a building on the corner of Gates and Franklin Streets. (The present Constitution Village in downtown Huntsville is a recreation of the central buildings of that time.)   John W. Walker,

Present Constitution Village

one of the delegates from Madison County, served as the convention president. By 2 August the draft constitution was completed. On 14 December 1819, the U.S. Congress converted the Alabama Territory into the State of Alabama – the nation's 22nd state.

A temporary Alabama State Capital was set up in Huntsville, and William Wyatt Bibb, who had previously served as the Governor of the Territory, was elected as the first State Governor. It was soon realized that the Capital should be more centrally located, and in 1820, it was relocated to Cahawba in Dallas County. The State Capital was moved to Tuscaloosa in 1826, and finally to Montgomery in 1840.

A Growing Huntsville

In only a few years after the first settler arrived, the town of Huntsville experienced considerable expansion; in 1812, a weekly newspaper – the

First Library Building

Madison Gazette – was started, a bank and stores were built around the square, and the first Masonic Lodge in the Territory was chartered. Green Academy, the first school chartered by the Territory, was also established in 1812. (Several other private schools were already operating throughout the county.)   A private art gallery valued at $75,000 opened in 1817, bringing

formal culture to the city. A public library opened in 1818. Theater enthusiasts organized a Thespian Society, and music lovers started a Haydn Society.

The Government Land Office had another major sale in 1818. With Madison County already established, the bidding was high, often $50 or more per acre. The minimum sale was 160 acres, and one-fourth payment was required at the time of sale; thus, most of the buyers were wealthy. Some land was purchased by families already settled, but much was bought by people from all areas of the South. The 1820 census showed 1,364 residents in Huntsville and a total of

would usually be three pickings, with a full team of hands at the first, then reduced at the subsequent pickings. There would also be several house hands working in the mansion, grounds, and quarters; a total of near 20 hands would be about the minimum to justify the cost of an overseer.

Other than land and workers, little else was needed for farming. For the representative mid-sized plantation, about eight horses or mules would be needed for the field work, and a pair of oxen for the heavy work. There were no machines for field work. All preparation of the new areas involved the use of oxen- or horse-drawn turnover plows, then use of drawn cultivators to stir and pulverize the soil, and finally harrows to prepare the seed bed. All planting, chopping, hoeing, and picking was strictly by hand, and using, where necessary, implements that were many years old in design. Initially, the plows were wooden with metal attachments for cutting into the soil; in 1819, the cast-iron plow with interchangeable parts was patented and became available. The reaper, invented in 1834, was the first successful mechanized farming equipment, but it was primarily for harvesting grain and thus had little use by Alabama farmers.

The U.S. Census of 1840 shows the following as agricultural production in Madison County, Alabama (the figures would have been for the previous year – 1839):

Bushels of Indian corn	1,357,803
Bushels of potatoes	46,371
Pounds of tobacco	15,655
Pounds of cotton gathered	10,358,897

Since the cotton is indicated as "gathered," it is assumed that this is the non-ginned amount; the lint amount would be about one-third of this figure. The Census also lists 20,726 pounds of wool as produced, showing that the farmers also raised sheep. The production of pork was not shown, but it is known that many farmers also raised hogs.

From the time of first planting, the land in Madison County was, for the most part, used year after year for the same crop – cotton. Local farmers evidently did not pay attention to the admonition in the *Holy Bible* (Leviticus 25:1-6) that land should be left fallow every seven years. Artificial fertilizer had not yet been developed. Although it was known that cow manure was beneficial, it was not available in sufficient quantities to be of much help. As a consequence, by mid-century much of the land had decreased significantly in productivity.

Early Cotton Mills

A cotton mill is a factory for spinning and weaving, converting cotton fiber (lint) into yarn and then into fabrics. The equipment for such factories was invented in Great Britain during the late 1600s, and mills producing thread and cloth operated in England during the 1770s. Although originally closely held, the technology became known in America, and the first mill opened in Massachusetts in 1814.

With the extensive cotton-growing in Madison County and the cost of transporting the ginned bales to cotton mills in England or the northeast, it is logical that such mills would be started in the county. There are references in documents to three attempts at such mills, all on the Flint River no more than 20 miles from Huntsville: Charles Cabiness having a spinning mill at his Barren Fork cotton gin in 1819; Joseph Harding, in 1924, erecting a similar facility in Ford's mill at Mountain Fork; and, also in 1824, Horatio Jones building a yarn-producing and weaving mill with 5 carding engines and 431 spindles at Three Forks. However, even if this is factually correct, the operation and fate of these endeavors does not appear to have been recorded.

The Bell Factory, opened in 1832, is generally regarded as the first cotton

The Bell Factory

mill in Alabama; it was located about 10 miles northeast of Huntsville near Three Forks on the Flint River. Initially operated by Patton, Donegan and Company, the factory name came from a large bell that summoned workers in the morning. The mill was powered by water, and initially had 2,323 spindles and 52 looms; its primary product was bedding.

The original building burned in 1841, but was soon rebuilt with an increased capability. The 1860 Census showed that the company owned 147 slaves. Following the Civil War, the slaves were released; some were hired back as free Blacks and others were replaced by White workers. In 1874, the mill was acquired by J. R. Stevens and William H. Echols. Without the low cost of slave labor, the Bell Factory was not financially successful but remained in operation until 1885.

SLAVERY

Slavery – the buying and selling of human beings, the highest of God's creations – was absolutely essential to the formation and sustainment of King Cotton. When Alabama became a state in 1819, more than 30 percent of the

128,000 inhabitants were slaves, and by 1860, they numbered 435,000 and represented 45 percent of the total population. Evil as it was, this labor system was practiced, condoned, or ignored by many White people who otherwise had Judo-Christian beliefs.

This abominable system did not originate in the South, but migrated from the northeast states where slaves had been brought in to unsuccessfully work in the emerging industries. Most of these had come by way of the West Indies, where the British slave traders of the 16th and 17th centuries had taken them after they had been wrenched from their homes in Africa. It is estimated that, over several centuries some 12 million native Africans were brought in bondage to the Americas. Of these, about 645,000 were eventually shipped to what is now the United States.

By the early 1800s, all northern states (those above the Mason-Dixon Line – the border separating Pennsylvania and Maryland) had either passed laws abolishing slavery or the gradual elimination thereof. The United States passed a law prohibiting international slave trade in 1808, but internal trading from the North to the South greatly accelerated into coastal farming states such as Virginia, North and South Carolina, and Georgia.

Many of the settlers of Madison County came from these coastal states and brought slaves with them. At the time of the first land sale in 1809, the total squatter population was 2,545 persons, including 322 slaves. As cotton farming increased, additional slaves were bought, many through auction houses in Mobile and Montgomery; by 1812, there were some 3,600 slaves in the county. In the census of 1820, the county population was 19,565, including 9,323 slaves.

The 1860 Census for Madison County showed a total population of 26,450, made up of 11,685 Whites, 14,573 Black slaves, and 192 free Blacks. This census also showed 5,836 slaves held by 114 slaveholders with 30 or more slaves – an average of 51 slave each; and 1,003 slaveholders with 8,737 slaves – an average of 8.7 slaves each. The largest slaveholder in the county was Patton, Donegan & Company (operators of the Bell Factory) with 147 slaves.

Most of the slaves in Madison County were born in the United States and had never known freedom. Also, children born to slaves were considered chattel, and thus were slaves by birth. Slaves usually sold for a few hundred dollars each, but strong men who could work the fields often sold for a thousand dollars or more. In each of these censuses, there were a small number of free Blacks included in the count.

The work and treatment of slaves in Madison County seems to have varied broadly. Most worked on farms, but a good number were found as household servants in Huntsville; a few with specials skills were carpenters, masons, and in other trades. On small farms, slaves usually worked directly for the owners and were often treated as fellow workers. On large farms (mainly plantations with 20 or more hands), there would usually be a White overseer, and much of the bad treatment of slaves – perhaps unknown to the plantation owner – could often be attributed to the overseer. While individual slaves would occasionally seek freedom by running away, there is no record of any uprising of slaves in Madison County or elsewhere in Alabama.

Most slaves worked from dawn to dusk, Monday through Friday, and a half day on Saturday. Sundays were usually free of work, and many attended religious services. The spread of Christianity among slaves was influenced by ministers who warned slaveholders to do so or risk damnation. In 1820, slave William Harris formed the first Black Primitive Baptist Church in Alabama, and it had more than 400 members by 1849.

The 1852 Code of Alabama – the official State Law – gave very specific master-slave relationships, describing specific offenses of slave and their subsequent punishment. Some people interpreted one of the clauses as prohibiting slaves to be taught to read and write. It did, however, have a few protective measurems, such as requiring owners to provide their slaves adequate food, clothing, and medical treatment; it also suggested – but not required – that families should be held together when the father was sold, and elderly slaves should have continuing care (the 1860 Census showed four male slaves with ages between 100 and 110 years).

The 1820s and 1830s were in a national period called "the great awakening," during which there was much attention given to spiritual matters. In the North, many church ministers and other leaders became activists in the abolition of slavery. To a degree, this was also taken up in the South. Many newspapers, including one in Huntsville, ran articles condemning slavery.

James G. Birney, a Huntsville alderman and mayor, who also served in the legislature and as a University of Alabama trustee, became an outspoken abolitionist; after leaving the city, he was the 1844 candidate for the U.S. Presidency on the abolitionist (Liberty Party) ticket. Thomas Fearn, a prominent Huntsville physician and businessman, at one time owned as many as 80 slaves but later called slavery "that foulest blot in our national character; that damning curse entailed on us by our forefathers"; he promoted a plan for gradually emancipating the slaves and setting up a new nation in Africa for them.

The matter of slavery was a major issue in States Rights when the Confederacy was formed. In 1863, President Abraham Lincoln issued the Emancipation Proclamation freeing slaves in the Confederate States. Since this was issued under Lincoln's temporary war powers, the following year he obtained passage of the 13th Amendment to the U.S. Constitution outlawing slavery throughout the nation. The Confederacy disregarded both the

Proclamation and the 13th Amendment, but had to conform after surrendering in 1865.

TRANSPORTATION AND UTILITIES

Two services were highly important in the evolution of Greater Huntsville: transportation, for both the city and county, and utilities for the city. Examined here is the transportation infrastructure, allowing the movement of people, materiel, and goods, and thus enabling trade and expanding civilization. The utilities are those technologies and functions that provide services that are used by the public. Both transportation and utilities are continuously evolving; the following considers their initiation and progress through the early 1860s.

TRANSPORTATION

As settlers developed Huntsville and Madison County, there was a need to bring in supplies and equipment; initially, this was done almost exclusively by horse-drawn wagons, but soon some small amount was also floated down the Tennessee River to Ditto's Landing. As time went by, the larger creeks within the county were sometimes used, and attention was also given to down-stream transportation on the Tennessee River

Roads

The first road into Huntsville evolved from the trail coming down from Tennessee. With the first official land survey of 1807, the Principal Meridian (later called the Huntsville Meridian) generally coincided with this trail; therefore, the trail became known as Meridian Road. As the road neared the village, it diverted slightly to the west; then after passing the Big Spring area, it returned to the east side and ran due south 10 miles to Ditto's Landing. Along this portion, the road to the river is about a half-mile east of the Principal Meridian.

Even before 1810, settlers in the county were producing cash crops – mainly cotton – and there was a need to transport this to outside the area. This was by wagon, primarily along a north-west path to the Tennessee border and then on to Nashville. After delivering the cotton, the wagons would return to Huntsville loaded with food and other supplies. As the land-speculation died down, the road system of Madison County evolved.

The building of roads at that time should be noted; it involved simple technology and heavy labor. The land was surveyed to determine the most efficient route between primary points. Changes in elevation were very important; a standard was that a wagon pulled by horses could only traverse a grade of five degrees or less (a rise of no more than 452 feet per mile).

The building started by clearing the path of brush, boulders, trees, and stumps, normally by hand with help from horses (the stump-puller apparatus did not become available until about 1820). Smoothing involved use of hand-held rakes, hoes, and sometimes horse-drawn scrapers; smoothing was the mark of an improved road, separating it from paths for foot or animal travel. A final task involved ditching on the sides to stay the erosion process and avoid wheel ruts. The better, more permanent roads were then spread with small stones and hard clay (called Macadam road construction).

Macadam Road Building

By 1820, there was an established network of roads connecting all of the major areas as well as leading to population centers in adjacent Alabama and Tennessee counties. Some roads were designated by the Federal Post Office Department as mail routes, and therefore became principal roads between population centers. Postings by the Postmaster General in the *Alabama Republican* newspaper 22 September 1820, included the following:

Route No. 304. From Huntsville to Triana, Mooresville, Cottonport, Melton's Bluff or Marethon, and Courtland to Russellville [Alabama], three times a week, seventy-seven miles. Leaves Huntsville every Monday, Wednesday and Friday, at one P.M., and arrives at Russellville on Tuesdays, Thursdays and Saturdays, at eleven A.M.; and vice versa from Russellville to Huntsville.

Route No. 305. Leaves Columbus [Mississippi] Tuesdays, Thursdays and Saturdays at one P.M., and arrives at Huntsville, Wednesdays, Fridays and Saturdays at 11 A.M.[?] [Distance not given.]

Route No. 308. From Huntsville by Hillsboro and Hickory Flat [Tennessee], to Winchester, Tennessee, once a week, fifty-two miles. [Days and time were not given.]

Route No. 309. From Huntsville to Jackson Court House [Scottsboro, Alabama] once in two weeks, forty-six miles. [Days and time were not given.]

Route No. 310. From Huntsville to Gaines Meads, Blount Court House, Jonesboro, etc., to Centerville [Jefferson County, Alabama], one hundred and three miles. [Days and time were not given.]

These destinations were all county seats, indicating the importance of communications between court houses.

An 1825 map of the Huntsville area by Hunter Peel shows the following highways connecting with the town: Athens Road, Ditto's Landing Road, Pulaski Road, New Market Road, and Meridian Road (these Roads were sometimes called Pikes).

Rivers

Alabama has an abundance of waterways, actually more than any other state. From the first settlement of this area, the Tennessee River, forming the southern border of Madison County, was looked upon with great potential for transportation, but it would be many years before this potential was fully attained. Although some limited traffic on this river from Knoxville (about 225 miles) existed during the early 1800s, most of the goods brought in to the area was by wagons from Nashville (about 110 miles) over relatively crude roads.

In 1816, James Crump, a Huntsville merchant, received two wagon-loads of food that originated in Mobile, over 350 miles to the south. For this, a 35-foot keel-boat carried the supplies up a waterways route using the Alabama, Tombigbee, and Black Warrior Rivers. At the fall-line a few miles west of the present Birmingham, the supplies were loaded onto wagons and driven overland to Huntsville. The river route took 20 days, and the wagon portion 3 more. Crump declared the venture profitable, and a newspaper later stated that considerable merchandise was being delivered to Huntsville via this route.

The river route to Mobile was only one-way. It was more economical to carry cotton on wagons to Nashville and return with the wagons filled with goods and supplies. However, as more cotton was grown and bailed, there was too much for wagon transport to Nashville. It then began to be brought to Huntsville, where it was graded and sold to cotton merchants. The merchants accumulated the bales of cotton in warehouses, and then mainly had it taken to the Ditto's Landing on the Tennessee River, where it was loaded on flatboats for carrying to the Mississippi River and transported to New Orleans. There were textile mills in New Orleans, but much of the cotton was transferred to deep-water sailing ships for transport to New England or overseas for British mills.

Flatboats

During the early 19th century, flatboats (often called flat-bottom boats), were the most-used type of cargo transportation on rivers and larger creeks in the Nation. Jacob Yoder is credited with building the first such boat in 1782, and used it to transport flour from southwest Pennsylvania, to New Orleans. (Interestingly, his route started on Redstone Creek.) Flatboats were rectangular in cross-section (having no keel) and tapered upward at the front to reduce drag.

Flatboats' most important characteristic was having a small draw (depth in the water), allowing passage over shoals and shallow rocks. They normally had no oars, and simply drifted with the current; consequently, they were usually intended for one-way travel, being disassembled for the timber at voyage end. For orientation with the current, rudders were on the stern and both sides. To stop at night or at their destination, an anchor was necessary; a loud bell was used for signaling. At one time, Abraham Lincoln worked on flatboats, and made two trips from Illinois to New Orleans; on one of these trips, pirates unsuccessfully tried to kill the captain and take the cargo.

The load-carrying potential of a boat is given by Archimedes' principle: it is buoyed up by a force equal to the weight of displaced water. For a flatboat 10-by 20-feet in size and a draw of 12 inches, this would be over 12,000 pounds; for a large, long-distance boat 20-by 80-feet (sometimes called a New Orleans boat), this would be about 100,000 pounds. At that time, a bale of ginned and compressed cotton usually weighed around 375 pounds; thus, allowing for the weight of the boat itself, a small flatboat might carry about 30 bales, and a large New-Orleans boat could transport 300 or more bales.

Starting a few miles downstream from Decatur, the Tennessee River has a 38-mile stretch of shoals and shallow rocks ending near Tuscumbia. The shoals cause the river at points to expand to as wide as three miles, but with little depth; the shoals stretch divided the river into upper (up-stream) and lower (down-stream) segments. In the early years, this had limited navigation between the two segments to canoes, rafts, and small boats.

Cotton to Market

In about 1815, flatboats started to be used to carry bailed cotton from Ditto's Landing and, somewhat later Triana, downstream to New Orleans. Once there and the cotton delivered, the flatboat would be sold for salvage and the pilot and boat hands would need to make the long walk, or horseback ride, through Choctaw and Chickasaw country back to Huntsville.

New Orleans Harbor

For most of the river trip, the boats floated peacefully with the current, but getting through the shoals was dangerous, shooting rapidly between the rocks and dodging the sand bars, the course guided by oars dipping left and then right. As flatboat traffic increased, licensed pilots who knew safe paths through the shoals would be hired as pilots for the stretch between landings at Decatur and Eastport, where the river returns into the State of Tennessee.

Memphis, on the Mississippi River, was founded in 1814, and later became another option for selling the cotton; however, once the bales were loaded onto flatboats, it was usually more economical to float on to New Orleans. Sometimes a large flatboat would be held at Ditto's Landing or Triana until the spring freshets increased the river depth, making it easier to traverse the shoals. Limestone County, directly west of Madison County, was another major cotton-growing area.

The village of Mooresville, located a few miles west of the county line, was settled in 1815, and a port was developed about a mile away on the Tennessee River. This port enlarged in importance as cotton started to be shipped by flatboats down the lower Tennessee; it also handled some of the cotton from western Madison County. A village called Cottonport developed and was incorporated in 1824. It had a town square, a number of houses, several stores, and a large warehouse; Cottonport was anticipated to be a regional shipping center when the Tennessee River became opened to steamboats. However, major epidemics of malaria caused many of the inhabitants to flee, steamboat traffic was not forthcoming, and the town was abandoned. Within a few years, river floods wiped out all traces of the town.

Canals

To reach the Tennessee River from Huntsville, an early road had been built to Ditto's Landing. It was expensive to transport cotton this way – about one-third of the cost for river shipping all the way to New Orleans. In the western portion of the county, the Indian Creek ran to join the Tennessee River near the village Triana; the upper part of the Indian Creek was Prices Fork, and this gave a passage for small boats from well up in the county down to the river.

On the eastern side of the County, the Flint River emptied into the Tennessee River a few miles upstream from Ditto's Landing. Both of these waterways – Indian Creek and Flint River – had been sufficiently cleared of obstacles (mainly fallen trees) to allow the passage of small keel and flat-bottom boats. Keel boats, although more difficult and expensive to build, used oars and could thus make two-way trips.

In Huntsville, the branch from Big Spring ran into Spring Creek, and this, in turn, ran into the Indian Creek. As more bailed cotton came to the buyers at Huntsville, the Indian Creek Navigation Company was formed with the aim of opening a waterway for larger boats from a wharf near Big Spring all the way to Triana. This would require making a canal with locks for a portion of this route.

Although canals had been used for interconnecting waterways in Holland and England for many years, they only came of interest in America during the late 1790s. People such as Benjamin Franklin and George Washington promoted the building of canals; there was, however, a lack of elementary engineering knowledge for making long-distance levels, removing underwater stumps, disposing of earth, creating locks, and similar tasks. It was not until the building of the Erie Canal in New York (started in 1817 and opened in 1825) that canal building in America came into its own.

On 21 December 1820, the Indian Creek Navigation Company was chartered by the Alabama Legislature to build, with power of eminent domain, and operate it charging a toll fee of $2 for each ton carried. Stock was sold to the public to finance the project – the first canal in Alabama. Many prominent men invested in the stock, including LeRoy Pope and Thomas Fearn, who was the primary promoter and was elected president.

Thomas Fearn (1789-1863) was a well-known local physician; he was nationally credited with discovering the nature of quinine, with the consequent cure of malaria. Fearn was also Chief Surgeon on Andrew Jackson's staff during the War of 1812 and in putting down the 1814 Creek Indian rebellion. He is recognized for forming Viduta ("Life"), a health colony atop Monte Sano Mountain in 1827.

Thomas Fearn

Work on what came to be commonly called Fearn's Canal was started in 1821. Essentially all of the engineering work had to be devised locally; unfortunately, the name of the responsible engineer is not on any known documents. It is likely that Fearn personally made many contributions; while a successful physician, he was also accomplished in technical matters. He later made significant contributions in improving Huntsville's early water works.

In January 1827, it was advertised that the canal was not completely finished, but was starting the passage of boats. The canal was not fully opened until April 1831; at that time boats carrying up to 100 bales of cotton and 50 passengers could make the complete route. Loads shipped down the upper Tennessee and coming into downtown Huntsville via the canal included columns and copper roofing for the architectural masterworks being built in the area.

While the canal was being constructed, the town of Triana flourished. Founded in 1819, Triana was the second town incorporated in Madison County. A wide range of businesses were built there with the expectation that it would become the major transport point for cotton and goods. This, however, would not occur. It was found that the maintenance on large sections of Fearn's Canal was a continuing process and highly expensive; this cost had to be raised by increased selling of stock. Many of the stockholders were nearly bankrupt before the enterprise closed in 1841.

The Flint River Navigation Company was formed in December 1821, with the intent of opening this stream between Scott's Mill and the Tennessee River. It was planned that boats using this would go down to Ditto's Landing or even cross the shoals to carry cotton to further ports. This, however, was a complete failure; the first necessary clearing of trees from the banks clogged the stream almost beyond repair and stopped further progress.

Steamboats

Work on boats powered by steam had been conducted in Europe and the United States since the 1790s, but Robert Fulton demonstrated the first fully useful steamboat in 1807, taking the *Clermont* on a round trip between New York and Albany. The *Clermont* was 133-feet long, 18-feet beam, and 7-feet depth of hold; it averaged about five miles per hour on its first voyage. Within a few years, steamboats were initiated on many rivers; in 1817, the steamboat *Osage* reached Sheffield on the lower Tennessee, carrying a cargo of groceries and hardware from New Orleans.

The first steamboat – interestingly named *Rocket* – to cross the shoals from down-river on the Tennessee was in 1821; it sustained significant, but repairable, damage. The *Rocket* did not attempt a reverse voyage but did begin limited steamboat traffic along the upper part of the river. The next crossing of the shoals by a steamboat was the *Atlas* in 1828; after reaching Knoxville and receiving a royal welcome, it discharged its cargo and steamed back to the lower river, never returning to the upper river again.

In addition to the shoals, there was another problem in steamboat transportation into the upper Tennessee. There was a bend in the river just upstream of South Pittsburg in Tennessee where several geological structures (commonly called the suck, the skillet, or the boiling pot) made navigation by the longer and wider steamboats difficult under the best of conditions and even impossible in certain times. (By mid-century, this obstacle had been largely cleared.)

There was great interest in connecting the lower and upper Tennessee to steamboat traffic. In 1830, the Alabama Government requested the National Congress for funds to build a canal around the worst of the shoals. Congress responded with a grant of 400,000 acres of Federal land, the sale of which could be used for the Muscle Shoals Canal Project. During 1831-1836, a 12-mile long and 60-feet wide canal was built; this had 17 locks each 120-feet long, 32-feet wide, and an average lift of 5-feet. This, however, had a major design flaw: the locks could not accommodate steamboats, which were already coming into use. Also, the shoals beyond each end were not cleared. Therefore, although the canal was completed, it was essentially never used; in a few years, the wooden gates decayed and the entire canal was abandoned.

Around 1835, a small steamboat, *Huntsville*, made it through the shoals (but with major damages), and began runs between Knoxville and Huntsville, and to nearby Decatur.

In 1873, the Muscle Shoals Canal project was revived by Congress; George W. Goethals, who later was the overseer of constructing the Panama Canal, supervised the revised effort. A 14.4-mile canal with 9 locks was completed in 1890; the locks were each 60-feet wide and gave a total lift of 85 feet. Under

Decatur Cotton Port

the same project, a canal around Elk River a few miles upstream was constructed; it was 1.5-miles long with 2 locks the same size as on the Shoals Canal. With these two canals, larger steamboats began limited operations throughout the Tennessee River. It was 1924, however, and the completion of the Wilson Dam near Florence, before the river was fully navigational.

Railroads

Although not in Madison County, a railroad built in 1832 between Decatur and Tuscumbia via Courtland was important to the Huntsville area. Covering 42 miles and initially horse-drawn, the Decatur, Courtland, and Tuscumbia (DC&T) was the first full railroad in Alabama; it acquired a British-build steam locomotive in 1834. Decatur and Tuscumbia are both on the south bank of the Tennessee River, close to the ends of the almost impassable shoals stretch; the

DC&T Railroad was built to allow transportation around the shoals. At Tuscumbia, it connected with a two-mile track built earlier going to the river port at Sheffield. Primarily to access the DC&T, a stage line from Huntsville to Decatur was established, crossing the river on a ferry at Mellon's Bluff (also called Marathon).

The Memphis and Charleston Railroad (M&CR) – forerunner of the Southern Railway – was completed from Memphis to Stevenson, Alabama, in 1855, and passed through Huntsville. This was of great value as an alternative to the Tennessee River for cotton transportation from Huntsville to Memphis and the Mississippi River. The city invested $50,000 in M&CR stock to ensure this route and the inclusion in Huntsville of passenger and freight depots, a large roundhouse, and major shops for the repair of the rolling stock. Coming into Alabama from Memphis, at Tuscumbia it joined the route of the DC&T Railroad; at Decatur, it crossed the Tennessee River and came on to Huntsville. When it reached Stevenson, the M&CR connected with the Nashville and Chattanooga (N&C) Railway, allowing transfers to these cities.

M&CR Locomotive

At Chattanooga, there were rail connections to Charleston, South Carolina, and to Augusta, Georgia; this network became the first rail line connecting Atlantic Ports with the Mississippi River. The M&CR used a 5-foot gauge (rail spacing), and so did the N&C Railway. Unfortunately, gauge had not been standardized, and some other tracks were less than 5 feet and others more, making interchange traffic almost impossible.

UTILITIES

Whether private or public, the utilities of a city or area provide a strong indication of the community's evolution. From its beginnings, Huntsville has been a utilities leader in Alabama and the Nation.

Waterworks

In 1823, Hunter Peel was given a contract to develop a basic waterworks for Huntsville – the first in Alabama and one of the first in the Nation. The flow from Big Spring was so large that a small amount could be diverted for the town's waterworks and hardly affect the downstream flow. (Big Spring has an average flow of about 14 million gallons per day, the third largest of Alabama's springs.) LeRoy Pope, then owner of Big Spring, gave permission for a small dam to be placed on the out-flowing stream, creating a holding pond and giving head (water energy) for a water wheel.

To build the distribution equipment, Peel formed a partnership with practical machinist (mechanical engineer) James Barclay. He built the equipment totally from wood; it consisted of a breast-shot water wheel driving a reciprocating pump to force water from the pond through hollowed-out cedar logs to a

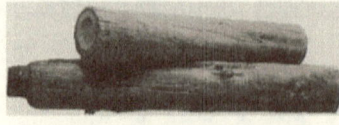

wooden reservoir on the 60-foot bluff over the spring. The reservoir was specified to hold 7,500 gallons (1,000 cubic feet) of water; it was likely cylindrical in shape, possibly 12 feet in diameter and 9 feet in

Hollow Cedar Log Pipes

depth. From the reservoir, which was beside the Court House, underground log pipes ran to hydrants in the yards of customers along the streets. With a 9-foot deep reservoir, the pump would need to provide at least a 69-foot head.

When the contract was given and permission granted for damming the stream, the understanding was that the system would reach as far as Pope's home on The Hill; this height, however, was found to be 96 feet higher than the surface of the pond, much more than the head available from the pump. The waterworks was completed without the Pope segment, and Peel was held responsible for not fulfilling the contract. The origin of this discrepancy led to the breakup of the partnership between Peel and Barclay in 1825, and the operating franchise was bought by Thomas A. Roberts of New York City in 1826.

From the start, there was dissatisfaction with the waterworks. In addition to not extending to the Pope home, there was major leakage and the reservoir could not be kept filled. Anticipating the waterworks, the town had organized a Fire Engine Company, purchased a portable hand-powered pump, and erected a firemen's building adjacent to the Court House. Testing showed that the existing system was inadequate.

A major fire occurred on the west side of the square in 1829; the Fire Company could not obtain sufficient water from the reservoir, so most of the buildings were destroyed. Demands were made for waterworks improvements, and Roberts had Sam D. Morgan supervise rebuilding of the system. This included a higher dam (about nine feet), a nine-inch metal pump (still driven by a water wheel), and new cedar-log pipes that were larger in bore and joined together with iron bands to reduce leakage. The wooden reservoir beside the court house was also rebuilt, becoming 18-feet square and 10-feet high, holding 24,300 gallons.

In 1836, Thomas Fearn and his brother George Fearn acquired the waterworks, and, over the next five years, made further major improvements.

The feeder lines were replaced by five-inch cast-iron pipes, and a 70-feet diameter, 287,500-gallon reservoir was built a few blocks east of the original site. The pump raised the water 96 feet above the level of the spring, and fire plugs were placed around the courthouse square. The Pope family deeded Big Spring and the surrounding area to the city for one dollar in 1843. At that time, there were 111 customers being served.

Huntsville Waterworks 1855

The waterworks was acquired by the City in 1858. John W. Weaver was appointed the Water Superintendent and several fire plugs were added; two years later a new building for the pump was constructed. With water-main extensions and minor pump improvements, the city-operated waterworks remained in use during the Civil War and almost to the end of the 19th century.

Gas Supply

A product called manufactured gas had been used for lighting in Great Britain since the early 1800s, and manufacturing plants at large cities of the United States began in the 1820s. By the 1850s, entrepreneurs in smaller eastern and mid-western cities were opening gas-manufacturing plants. The manufacturing process involved the gasification of combustible materials, most often coal but also oil and wood.

Francis H. Newman, a Huntsville physician and part-owner of a drug and chemical supply store, experimented with various materials for producing this gas, and was successful with rosin. In 1856, Newman formed the Huntsville Gas Light Company to build and operate a gasworks plant; the city bought stock in the firm. A plant using wood as the fuel had been built in Atlanta in 1855, possibly inspiring the effort in Huntsville. The gasworks in Huntsville burned in 1858, but was quickly rebuilt.

The development of the gasworks was the first activity in Huntsville that, in the future, might be called chemical engineering. The gasworks was divided into sections for the production, purification, and storage of manufactured gas. Distillation of resin (sap) from pine trees produced rosin and turpentine; this was done elsewhere and the Huntsville firm purchased the needed rosin.

At the gas plant, the rosin was superheated in a low-oxygen retort, boiling off lighter constituents. The separated gasses then passed on to a condenser – a bank of air-cooled gas pipes – where heavier components were removed. It then bubbled through a sealed tank containing water, removing undesirable lighter compounds. The basic equipment for these processes was obtained from a firm in New York, A pump increased the pressure of the remaining gas and forced it into a storage tank; this tank had a moving top that would rise and fall to maintain even pressure in the distribution pipes.

The plant was located near Big Spring, and pipes were initially run up the bluff to eight gas lamps around the city square. In a short time, the lines were extended throughout much of the residential area. There was no metering of the gas supplied. The city paid the Gas Light Company for each street light installed, and Aaron Franks was hired as a "lamp lighter," turning the gas lamps on and off each day. Similarly, private customers simply paid a flat fee for the gas connection. For several years, the company operated the gasworks with only minor changes, mainly for increasing the output.

Gas Street Light

Telegraph

Credit for the first practical electrical telegraph apparatus in America is usually given to Samuel Morse for a development in 1836, although another American, physicist Joseph Henry, demonstrated a similar device in the same year. Morse's apparatus initially functioned over short distances; his assistant, Alfred Vail, improved the receiver making distant communications possible. Vail also developed the Morse Code for converting between electric pulses and alphanumeric characters. This was the beginning of electrical communication evolution that is still continuing to this day.

The first major demonstration by Morse was from Washington to Baltimore on 24 May 1844; over this, the first message started, "What hath God wrought." Initial lines were strung along railroad right-of-ways, and railroads began using the telegraph for dispatching trains in 1851. Early telegraph systems in America used the simplex circuit.

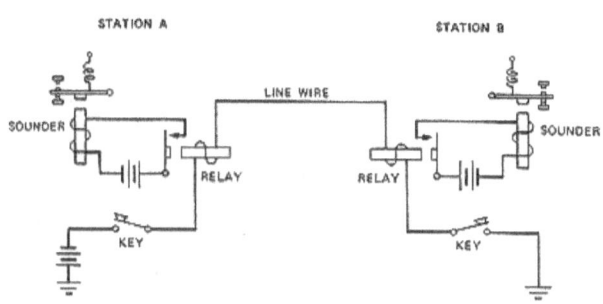

Simplex Telegraph Circuit

As shown in the diagram, basic components were the key, the battery, the line (the ground was normally used as the electrical path to complete the circuit), and the sounder. Only one station at a time could send messages, and the key at the receiving station had to be closed. Vale's contribution was a sensitive relay and a separate battery, making it possible for a weak signal on the line to strongly actuate the sounder.

In 1858, the North Alabama Telegraph Company of New York opened an office in Huntsville; John Fackler was named president of the company. The line ran along existing railroad tracks from Huntsville to Stevenson, Alabama, where it joined lines to Memphis and Montgomery; the local operator was at the Huntsville Depot. Lines to Nashville and Chattanooga were soon installed along the railroad tracks from Stevenson. Initially, messages sent to stations not directly connected required that they be physically relayed (copied and resent).

Western Union started telegraph exchanges in 1861, allowing switching between different systems and bringing widespread communications. Multiple local circuits using separate power sources could be relayed from a single long line. Western Union soon dominated the telegraph business by acquiring small local firms, including the one in Huntsville.

MATURING TIMES

Huntsville was a thriving community by 1860; it had three weekly newspapers, three hotels, seven schools, two stage lines, two saloons, one flour mill, one cotton mill, three brick producers, and a bank with capital assets of $500,000. (The bank building was completed in 1840, and is still in use.) It was served by a railroad that connected from the Atlantic coast to the Mississippi River and north to major cities, and cargo ports on the Mississippi River were only a few miles away at Whitesburg and Triana. A water system served most of the residents, and the downtown was lighted by gas lamps. The professional community boasted four architects, twenty-four lawyers, two physicians, two civil engineers, and a number of clergymen. The Huntsville city limits extended three-quarters of a mile from the Square; the population was near 3,600, and Madison County had over 26,000.

Cotton production was giving the county an excellent economy, and, as the center of trade, Huntsville's businesses were booming. It was also a transportation center, and the prospects for industry were good. The four decades between 1860 and 1900 were maturing times for the greater Huntsville community.

THE CIVIL WAR

For a number of years, there had been harsh debates in Congress concerning States Rights. Central to this was the question of the right of a state to secede from the Union, this being the last resort of southern states in defending their sovereignty; within this sovereignty was the allowance of slavery. A constitutional convention was held in Montgomery, and on 11 January 1861, the majority of the delegates voted to declare Alabama's immediate independence from the United States.

North Alabama voted with the minority who did not want to secede, and three of the five non-secessionists leaders were delegates from Huntsville. One month later, delegates from six other seceded states met in Montgomery to create the new government of the Confederate States of America (the CSA). Eventually, the CSA was composed of 11 states: Alabama, Arkansas, Florida, Georgia, Louisiana, Mississippi, North Carolina, South Carolina, Tennessee, Texas, and Virginia. In addition to the right to succeed and the associated slavery issue, there were economic disputes between the agrarian South and the increasing industrialized North.

The first hostilities occurred on 12 April 1861, when CSA artillery fired upon Fort Sumpter in South Carolina; this was ordered by Leroy Pope Walker of Huntsville, the CSA Secretary of War. The war ended when General Robert E. Lee surrendered to General Ulysses S. Grant at Appomattox, Virginia on 9 April 1865. With some 365,000 total military dead and 275,000 wounded, the Civil War was the bloodiest conflict in America's history.

Capture of Huntsville

On 6 February 1862, three heavily armed Union side-wheeled gunboats steamed up the Tennessee River and began patrolling runs between Paducah, Kentucky, and Florence, Alabama. From that time onward, no cotton from up river could be shipped to New Orleans. Coming southward from Kentucky, the

Union Army of Ohio assembled at Nashville, preparing for a thrust intro Alabama.

Brigadier General Ormsby M. Mitchell, who had previously served as the chief engineer for the Ohio and Mississippi Railroad, was ordered to advance with 8,000 troops to Huntsville and sever the Memphis and Charlestown Railroad as well as control the local telegraph operation. On 10 April 1862, Mitchell and his troops reached about eight miles north of Huntsville and awaited nightfall. Arousing at 2 a.m. the next morning, they entered the city outskirts without being detected, and by dawn they were at the railroad depot.

There was no warning to Huntsville of their approach (the telegraph operators were employees from the North).   The troops captured 18 locomotives, many passenger and freight cars, and the railroad shops; the locomotives had been sent to Huntsville pending their movement to a more secure location.   They also chased and captured a recently-departed train carrying mainly wounded Confederate soldiers, impounding the men in the upper floor of the depot.

Huntsville Occupation

Except for a 10-month period (August 1862 to July 1863) when the troops were sent elsewhere, Huntsville was occupied until the end of the war. During this, there was little property destruction or harsh control of the population.

Two major reasons the city was spared were that it was used as a communications center, and was also a bedroom for Union forces, particularly high-ranking officers. In addition to the downtown square, temporary camps were set up at a number of locations across the city. Only one large building was constructed – a hospital for injured

Union Troops Camped Around Courthouse

Union troops; a number of soldiers died there, and a portion of the local Maple Hill cemetery was purchased for their burial.

Commercial businesses in and around Huntsville were allowed to continue operating, but only if they took an oath of loyalty to the occupying government. Starting in November 1864, citizens were required to obtain a permit to make purchases costing over one dollar. Most of the population resorted to individual gardens, and many kept cows and pigs. The City had the only known Southern-based store that sold goods for tokens that were issued to the Union troops.

As the CSA was being formed, Josiah Gorgas, the CSA's Chief of Ordnance, urgently organized the factories in the South to produce weapons. Gorgas, through Young & Company, arranged for Madison Iron Foundry in Huntsville to turn its full capabilities to casting field artillery. An agreement was made to build howitzers for delivery to Richmond, Virginia, as soon as possible.

Although the work got underway, no deliveries were made; as the Union Army came toward Huntsville early in the war, the foundry was dismantled and shipped to Selma, a safer city in the center of the State. By 1863, Gorgas had created through central Alabama an industrial corridor of iron furnaces, rolling mills, gun –powder plants, arms factories, and arsenals. The complex in Selma ultimately had 3,000 men; it is likely that a major portion of these workers were slaves.

In Madison County, essentially all able-bodied White men were away from their farms and businesses during the war; farming was carried on by women, older men, and remaining slaves. (It might be noted that although logically opposed to slavery, many of the Negro men felt an allegiance with their owners and the land, and continued to work on the farms. In addition to the loss of manpower, many farms had horses taken for the Union troops; the output of the farms greatly decreased.

Military Action

While there was no major military action in Madison County, there were skirmishes and bush-whacking; some of the plantations and towns and essentially all of the school buildings – including the Green Academy – were consequently put to the torch by the occupying Union Army. Nearby villages of Whitesburg and Vienna were burned. In adjacent counties, Guntersville, 30 miles to the southeast, and Decatur, 20 miles to the southwest, were almost totally burned; Athens, about 20 miles northwest, was pillaged and plundered.

In Madison County, the only significant conflict took place at Madison Station, then a whistle-stop on the M&C railroad. On 17 May 1864, a large number of Confederate troops crossed the Tennessee River near Triana and attacked a 300-person garrison holding the station. The Federal troops first withdrew, but then were reinforced by 100 soldiers from Huntsville and drove the Confederates back across the river. It was reported that there were 80 killed or wounded in the Federal soldiers but only 8 in the Confederates.

Possibly the most effective bush-whacker was Milus E. Johnson, a Methodist minister before and after the war. When the Union Army mistakenly burned his barn near the village of Vienna (later called New Hope, a Huntsville suburb), he formed and led a local group in very effectively attacking Union troops in the

Huntsville region. Johnson and his rangers eventually became Company E of the 25th Alabama Cavalry Battalion (commonly called Lemuel Green Mead's Partisan Rangers). Although General Robert E. Lee surrendered to General Ulysses S. Grant on 9 April 1865, effectively ending the Civil War, it was some time before all conflicts ended. Lt. Col. Johnson surrendered his rangers at Monte Sano Mountain on 11 May 1865.

When the war started, most southerners in the U.S. military resigned and joined the CSA; however, some officers remained in the U.S. Army and Navy. The Medal of Honor – the highest award given in the U.S. military and initiated during the Civil War – was received by three men originally from Alabama.

During the war, some 125,000 men from Alabama served in the CSA Army; 35,000 of these died and about the same number were wounded – many disabled – as a result of this conflict. Madison County suffered 147 men killed in battle, and 214 others died from diseases and weather exposure – often while prisoners of war.

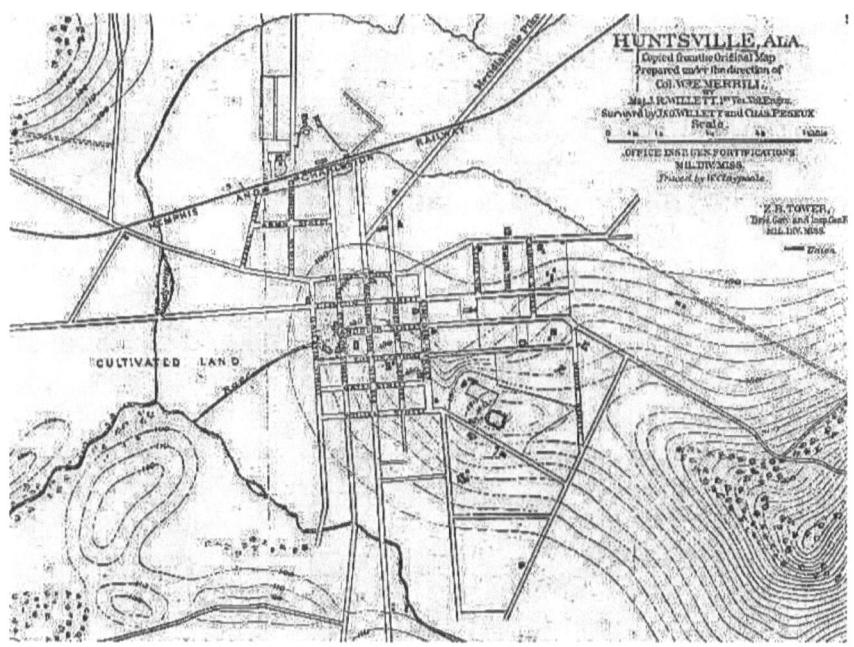

1865 Huntsville Map – Drawn by Federal Occupying Troops
Note: Due North is along Meridianville Pike

RECONSTRUCTION

Following the Civil War, Huntsville and Madison County had their share of scalawags and carpetbaggers – both were derisive labels applied by the native Whites to persons who they believed to be profiting from the South's misfortunes. Scalawags were White southerners who, for various reasons, saw an advantage in backing the policies of Reconstruction. Carpetbaggers were opportunists from the North looking to exploit and profit from the region's depressed condition. There was great resentment and open hostility against both types from those seeking the restoration of the South's economy. Many carpetbaggers acquired land through payment of unpaid taxes and outright purchases at bankrupt prices.

After the Emancipation Proclamation in 1863 and the 13th Constitutional Amendment in the following year, many slaves deserted the farms and sought protection under the occupying Union Troops. In March 1865, Congress established the Freemen's Bureau, intended to assist former slaves in finding employment, providing medical services, and opening schools. A Freemen's Bureau center was set up in Huntsville and was primarily operated by scalawags and carpetbaggers; there is little evidence that this Bureau was very helpful.

Acting as an employment agency, the Bureau set wages at rates that the struggling farmers could not afford, actually reducing the possibilities of ex-slaves returning to the work that they knew best. In the last decades of the 19th century, a large part of Huntsville's Black population was crowded into ghettos called Georgia and Jonesville, where they received little assistance and hope for improvement was low.

During the 1880s and 1890s, a number of leaders did emerge in the Black population. Elected as aldermen in 1880, Daniel S. Brandon and Thomas W. Townsend were the first of nine Blacks who served on that board (the 1901 State Constitution disenfranchised most Black voters from such elections). Born a slave, Burgess E. Scruggs received his basic education in Huntsville, attended medical school in Nashville, and then returned to practice in Huntsville where he was respected by both races. Charles Hendley, Jr., was a teacher, editor of the Huntsville Gazette newspaper, and outstanding chess player. Henry C. Binford was another respected teacher and writer for the Gazette. William H. Councill, although largely self-educated, was a lawyer and Methodist minister, and was the founder and first principal of the Colored Normal School at Huntsville, forerunner of Alabama A&M. Samuel R. Lowery was a prize-winning silkworm culturist and the first Black attorney to practice before the Supreme Court.

New Businesses

Not all Northerners who came to Huntsville following the war were carpetbaggers; a number were wealthy businessmen and industrialists who recognized further opportunities in the recovered South. The O'Shaugnnessy brothers, Michael J. and James F., from Cincinnati used their father's wealth to start in Nashville one of the South's first cottonseed oil factories. In 1881, they came to Huntsville to open another of these factories.

The O'Shaugnessys led in forming the North Alabama Improvement Company, a corporation to develop the real estate, mineral, and transportation resources of the region. This organization's transactions were so extensive that the courthouse had to hire extra workers to keep up with the records. James later returned to the North, but Michael stayed and contributed greatly to Huntsville's economic growth

As noted earlier, Huntsville physician Thomas Fearn – discoverer of the nature of quinine – led in forming a health colony on Monte Sano Mountain, located just east of the city, in 1827. (The name Monte Sano comes from a Latin or Spanish phrase meaning Mountain of Health.) A trail up the north side of the mountain was made into a road, and the town of Viduta was officially formed in 1833.

One of the North Alabama Improvement Company's notable achievements was developing the Monte Sano Hotel at Viduta. Built on a bluff overlooking the city, the five-story, 330-room, Queen Anne-style Monte Sano Hotel represented the New South. Opened in June 1887, it attracted wealthy investors and vacationers from across the nation. It also touted itself as a health resort. The hotel had the most modern of service facilities; included was its own manufactured-gas plant, and it also had service from the local power company – with lines up the mountain – for electric lighting. The hotel's overall operating cost, however, was huge; it closed operations in 1900.

Railroads

There was a good railroad system throughout the southern states prior to the Civil War, but major sections of the tracks and bridges were destroyed by both armies. As a result, the period immediately following the war was mainly spent in rebuilding. Being one of the most important lines in the South, the Memphis and Charleston Railroad (M&CR) that served Huntsville was an early restoration.

William H. Echols made major contributions to the M&CR restoration. A native of Huntsville and a graduate of West Point in 1854, he served as an engineer in the U.S. Army, then entered the Engineering Corps of the CSA where he was promoted to Major. After the Civil War, Echols was employed by the M&CR to lead their local engineering activity. Later, he was co-owner / official with the Bell Factory Cotton Mill, and also Director of the Huntsville National Bank. Echols Hill was named after his residency there.

Although small, a local railroad must be noted. In 1887, to transport guests between the downtown Huntsville and the Monte Sano Hotel, a special railroad was built up the mountain; this was 7 miles in length and had an elevation change of near 1000 feet. The line had three passenger cars pulled by a 2-ton Baldwin engine. To make the turns, the special-designed coaches were 42-feet long, shorter than the normal 50- to 60-foot coach length.

In 1873, the Nashville and Chattanooga (N&C) Railway had reorganized as the Nashville, Chattanooga and St. Louis (NC&StL) Railway. In 1887, the NC&StL opened a line between Nashville and Huntsville, with a spur extending to Whitesburg on the Tennessee River. Starting in 1893, freight cars could be rolled onto a barge and pushed by the steamboat *Guntersville* between White's Landing and Gunter's Landing some 25 miles upstream. This opened another rail route between Huntsville and cities further southeast in the State. The NC&StL later became the Louisville and Nashville (L&N) Railway.

Train Barge

Another railroad operating in Madison County should be noted. In 1887, an attempt was made to build a railroad running from Decatur through Madison County to Fayetteville and on to Gallatin, Tennessee. Only the first 17 miles through Madison County were completed before the project was abandoned. In 1893, the line was restarted under the name Middle Tennessee & Alabama (MT&A) Railroad. In 1896, the 27-mile stretch between Madison Crossroads (present day Toney) and Fayetteville opened for passenger and freight traffic. In 1897, the MT&A was sold to the NC&StL, and the line was completed 10 miles further south to Capshaw. This railroad operated with some success for three decades, but was closed in 1929. The Madison County Highway Department eventually purchased the property in this county and turned it into the Old Railroad Bed Road.

POST-WAR AGRICULTURE

Many factors had caused a reduction in agricultural production throughout the South during the Civil War. These factors included the absence of the land owners and loss of workers.

Cotton Farming

In 1850-1860, the average cotton production throughout the South was 990,000 bales per year; in 1866, the production was only 297.000 bales. The blockade of ports led to a loss of sales and the intentional reduction of planting; however, the price then increased to many times that of before the war, compensating for the loss of volume.

With the surrender of the CSA in 1865, cotton farming in the South essentially started anew. The economy was broken, the labor force scattered, and there was a dearth of horses and other farm animals. Most of the plantation owners were forced to break up their lands into smaller farms; these were worked by families, share-croppers, and tenant farmers, mainly growing crops that required minimal labor. In share-cropping, the owner usually provide a house and about ten acres of cotton land, seed, and use of a mule; the owner would then receive half of the crop. Although cotton was still in demand, particularly in England, cotton production was very low; the new landowners did not have the necessary know-how and the banks – now often controlled by Northern interests – were not receptive to making risky operational loans.

In 1869, a technological event took place in Egypt that also had a considerable impact on cotton farming in the South – the Suez Canal opened. The canal gave a new waterway connecting the Red Sea and the Mediterranean Sea, providing an alternative to the long, dangerous shipping path from Asia around southern Africa. This almost immediately led to large-scale, low-cost cotton shipping from India to Great Brittan, and significantly lowered the demand from manufacturers in England for the southern cotton that for years had been shipped from Mobile and New Orleans.

There were also beneficial changes in this period. Many of the former slaves – finding that work was not available elsewhere – returned to employment as hired hands and supervisors on farms. (The total Black population in Madison County increased from near 14,800 in1860 to over 15,700 in 1870.) The cotton gin was greatly improved in design; new machines that were faster and required less operator labor came on the market. These gins were mainly from manufacturing firms in Southern states, and were thus quickly accepted in the cotton-growing region. The three principle manufacturers were the Gullet Gin Company of Amite City, Louisiana; F. H. Lummus of Juniper, Georgia; and Daniel Pratt in Prattville, Alabama; the Pratt plant eventually became Continental Gin Company, the largest in the World.

By the 1880s, cotton farming in Alabama had largely recovered; Gins were no longer privately operated on plantations but were centrally placed for common use by the many smaller farms. In 1889, there were about 2,761,000 acres producing 426,555,000 pounds of lint, an average productivity of 158 pounds per acre. This compares with 440,530,000 pounds in 1859, the third largest in the United States. (In 1849, Alabama had been the most productive State in the Nation, with 225,770,000 pounds of lint.)

In Madison County, however, there was not a comparable recovery. In 1889, with near 75,200 acres in cotton producing about 6,273,000 pounds of lint, the average was only 83 pounds per acre. This productivity was barely half the State average of 158 pounds, and only about a quarter of the estimated 330 pounds per acre when cotton was first grown in the county. As previously noted, the productivity had been steadily decreasing as cotton was continuously planted, depleting the soil.

Agricultural Advancement

In general, farming methods in Alabama during the last of the 19th century was done essentially the same as had been done for essentially two centuries – by hand labor and using horses and mules for power. Most basic equine-powered mechanized equipment had been invented, but there had been little that could replace the time-consuming handwork in cotton farming, particularly harvesting. Although there was little change in the methods, there were other agricultural advancements.

Until the late 1800s, there was essentially no assistance available in Alabama for farmers in improving their agricultural efforts. The Alabama Department of Agriculture was formed in 1883, but there was little extension service; the farmers primarily learned from each other and the cotton buyers. Fertilizing the soil to improve the productivity of food plants had been known and practiced for centuries, and it was finally tried in cotton farming. Manure from farm animals was some help, but the importing of guano (bat manure) from Chile – although requiring scarce money – provided the greatest relief. Nitrogen-based artificial fertilizer later became available.

Agricultural Education

In 1862, the Morrill Land-Grant Act of Congress had provided eligible States with 20,000 acres of Federal land that could be sold and the proceeds used to establish and operate what were called land-grant colleges. CSA states were not initially eligible, but, after the Civil War, they were included. If a state could not guarantee admission of students regardless of race, a Morrill Act of 1890 allowed such states to set up a separate school for persons of color. In 1891, the Alabama Polytechnic Institute (API, now Auburn University) and the State

A&M College for Negroes (now Alabama A&M University) became Alabama's land-grant colleges.

Tuskegee Institute, a private school but with partial State funding, later became an affiliate; this school had the great heritage of its founder, Booker T. Washington, and one of America's best known-agriculturalists, George Washington Carver.

The land-grant colleges were required to teach agriculture, engineering, and military programs. There was, however, a lack of agricultural research on which to base this teaching. In 1887, Congress passed the Hatch Experiment Station Act, funding experiment stations in each state. Also established was an extension system, placing agents in every county to convey agriculture improvements directly to the farmers. Old Rotation, conducted by an experimental station at API, was the first experiment in the United States to demonstrate the effectiveness of rotating cotton with other crops to improve yields.

Yield Improvement

In 1899, the cotton crop of the United States was the largest ever reported. There were slightly more than 24,275,000 harvested cotton acres, producing about 4,717,000,000 pounds of lint, giving a productivity of some 194 pounds per acre. In Alabama, there were about 3,202,000 acres producing 546,849,000 pounds, giving an average productivity of 171 pounds per acre, possibly indicating that less fertilizer was being used than in other states. Alabama ranked behind Texas and Georgia in cotton acreage.

In Madison County, the 1899 figures were 70,000 acres in cotton and 10,421,000 pounds of lint, giving 149 pounds per acre; this was a significant improvement over the 83 pounds per acre of 10 years earlier, perhaps largely due to artificial fertilizer, but still only some 45 percent of that from the original virgin soil.

Agricultural Diversity

In this period, there was also some diversity in Madison County agriculture: With an abundance of fresh water, watercress cultivation prospered; although not a large part of the local

Watercress Farm economy, for a time Huntsville called itself the Watercress Capital of the World. There was also attention given to raising highly productive

Lily Flagg

Jersey cattle; a Madison County farm was the home of Lily Flagg, the 1892 world-champion butter-producing cow. Northern investors established commercial nurseries northeast of the city; the Huntsville Wholesale Nurseries was, at that time, the largest wholesale nursery in the United States.

INDUSTRY EMERGENCE

For many years, there had been blacksmith shops in Huntsville that did mechanical work, from shoeing horses to making and repairing early farm equipment, but these provided services, not products. There were also early enterprises for making brick, lumber, and other building materials; for making furniture, stoves, and other items for the household; or even making small quantities of clothing, guns, and water-supply items; but these were workshops – outgrowth of cottage industries that were operated by an owner and perhaps a few other persons. For use here, a factory is a building with machinery and a number of employees manufacturing products to be sold.

Iron Foundry

Madison Foundry

Opened in the 1850s, the Madison Iron Foundry was the first Huntsville firm that might be called an industry. At different times, this used other names, such as Madison Iron & Brass Foundry and Madison Foundry & Machine Works. Their foundry capability included a blast furnace, one of the first in the South. Business matters were often handled by a broker, J. R. Young & Company.

It was previously noted that just before the Civil War started, Madison Iron Foundry received a contract to build howitzers for the CSA. The war began before manufacturing got underway, and the equipment was sent to a safer city. After the war, the equipment was eventually returned to Huntsville, and foundry and machine shop work continued for a number of years.

Cotton-Seed Oil Mill

Roughly 60 percent of the weight of harvested cotton is in the seed. Originally, cotton seed was considered essentially worthless except for planting new crops. By the 1870s, a means for extracting oil from the seed was perfected, and it was found to be suitable for cooking. Since it came from a byproduct, it cost much less than olive oil or oil from other plants such as sunflowers; thus, it quickly became popular in the U.S. and Europe.

Michael and James O'Shaughnessy started in Nashville one of the South's first cottonseed oil factories, and in 1881, they came to North Alabama to open another large factory, the Huntsville Cotton Oil Mill on north Church Street . This was the county's first major post-war industry and soon had 80 employees. Within a few years, it evolved into a cottonseed oil empire with nine plants throughout Alabama. Also built at the Huntsville facility was a cotton-gathering machine that was invented and patented by James O'Shaughnessy.

Horse-Drawn Vehicle Works

In this time period, there were three firms in Huntsville building coaches,

Skinner Carriage Store Downtown

buggies, wagons, and similar horse-drawn vehicles. These were the J. W. Skinner Carriage Works, the Columbus Buggy Works, and the Alabama Wagon Works. Skinner Carriage had a two-story factory near the railroad depot employing a number of wheelwrights and blacksmiths and displayed their products on the square in downtown; a number of patents were issued on their products. Another local manufacturer, the Coyle Saddlery and Harness Works, supplied these vehicle firms with auxiliary items.

Farm Machinery Factory

Burwell J. Curry, owner and operator of a large farm in Madison County, recognized that mechanical agricultural equipment would be vital for economical farming. A graduate from the University of Virginia and a practical mechanic (mechanical engineer), Curry used his personal wealth to start a firm in Huntsville to develop and manufacture such equipment. Located near the Huntsville Depot, the plant had a range of iron works and machine shops.

The first of Curry's inventions, the Cotton Cultivator and Chopper (Patent US238028), was shown in 1881. Drawn by a horse or mule team, in a single pass this machine formed the bed, opened the drill, sowed and covered the seed, and an attachment distributed fertilizer. After the plants came up, the same machine could be used for chopping. It was advertised as doing the work of 10 or more men in performing these tasks.

In 1885, Curry released his Cotton Compress (Patent US327435). Powered by a coal-burning steam engine, this could take up to 500 pounds of ginned cotton and compress it to a density of 28 pounds to a cubic foot, giving a bale volume of about 17.9 cubic feet, measures that are still considered standard. Curry also developed a Portable Hay Press and a number of other press machines. In addition to his plant in Huntsville, there were plants in Florence, Alabama, and Holly Springs, Mississippi. All of Curry's equipment met with good sales, and rights were eventually sold to newly evolving national firms.

UTILITIES ADVANCEMENT

The technologies for utility services are always continuously evolving. For a small Southern city, Huntsville had excellent utilities from its early days. Following the Civil War, these continued to develop.

Gas Works

In 1872, the gas plant was moved from the Big Spring area to a three-acre plot on the corner of Dallas and Holmes Streets; with the move, it was completely rebuilt, converting to coal as the fuel and greatly increasing the gas production.

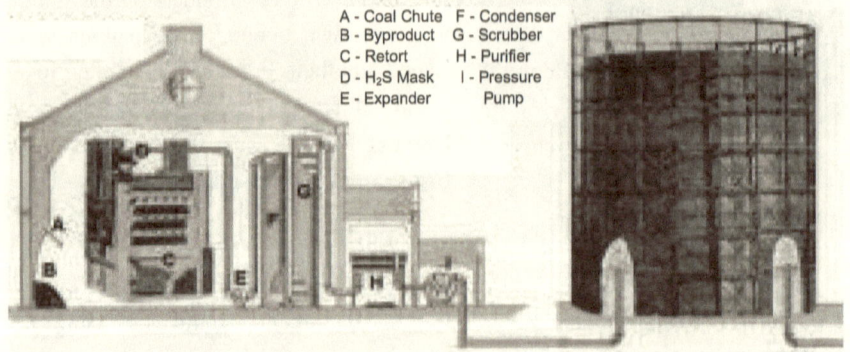

A - Coal Chute   F - Condenser
B - Byproduct   G - Scrubber
C - Retort   H - Purifier
D - H$_2$S Mask   I - Pressure
E - Expander   Pump

Typical Small Manufactured Gas Plant

The basic manufacturing process was described in a previous section.

In the late 1880s, electric power became available for lighting, and the company then promoted its manufactured gas for cooking and water heating. In 1902, the assets and franchise of the Huntsville Gas Light Company were acquired by another private firm, the Huntsville Gas Company; Cyrus S. Sugg was the principal owner.

Waterworks

In 1867, a new fire engine with a steam pumper and 1,000 feet of rubber hose were added at the fire station on the square. The city issued bonds for major waterworks improvements; more iron pipes were laid and a 600,000-gallon reservoir was installed on Echols Hill (formerly Pope's Hill) in 1887. The top was 120 feet above the square, allowing service throughout the city. A 1889 report showed that connected to the system were 591 hydrants, 162 water closets (toilets), 24 urinals, 63 baths, 89 sprinklers, and 7 soda fountains. There was no metering at the consumer – the water was sold at a flat rate.

By 1894, there were ten miles of cast-iron water mains in the system. In 1898, a new building at Big Spring with a steam-driven pump was installed, handling up to 3-million gallons of water daily (about a fifth of the total flow). Since there was no longer a need for a turbine, the old dam was removed and the area around Big Spring greatly improved in appearance.

When the Monte Sano Hotel was being built in 1887, a special force-pump was installed at Big Spring to raise water up 1,000 feet for supplying the hotel atop the mountain. In 1889, there was an outbreak of typhoid fever in the city, and the Medical Society attributed it to the water aquifer being contaminated by horses hitched on streets around the courthouse square. These streets were paved to protect the underground water.

Electric Power

In 1878, Thomas Edison developed an electric power system, and in 1882, Edison Electric Company opened a power plant in New York City as a replacement for gas lighting. This was a system involving direct current (DC) and transformers could not be used to change the voltage level; consequently, power generators were needed at about one-mile intervals in the transmission lines to compensate for voltage losses in the lines. Alternating current (AC) was a competing technology that was primarily developed by Nikola Tesla and promoted by Westinghouse; this could use transformers and allowed use of a single, large, centrally located power generator. From the beginning, the frequency for AC systems was 60 cycles/second (cps, later hertz). Throughout the remainder of the 19th century, there was great debate concerning DC versus AC power systems.

At the same time that Edison was developing the incandescent lamp (a light bulb containing a filament that produced light when heated by an electric current), James Jenny of Fort Wayne, Indiana, perfected a glass-enclosed electric-arc lamp. The arc lamp contained a carbon rod with a gap; voltage applied across the gap caused an arc, producing an intense light. Jenny Electric Company was formed in Indianapolis to produce arc lamps, and also designed an electric dynamo to supply DC power. (Some lamp manufacturers used an alternator generating AC power, but this produced a humming sound in the arc-lights that was objectionable to most customers.) Arc-lamps gave off a brilliant blue-white light, similar to that from an arc-welding machine. In 1883, arc lamps were used to illuminate the first night-time professional baseball game, and Jenny Electric gained great publicity.

Over the years, dissatisfaction grew with the old gas street-lighting in Huntsville, and a group was appointed by the mayor to examine existing electric lighting systems. In early 1887, the Jenny Electric Company invited the group to visit Indianapolis and see the system that they had installed in that city. Pleased with what they found, the Huntsville Electric Company was incorporated as a private enterprise on 26 April 1887; Charles H. Halsey was president and Robert E. Spragins the secretary and treasurer.

In May 1887, a contract was awarded to Jenny Electric for "32 street arc-lamps and 300 incandescent lights with sufficient power to double the capacity when needed." There was no further specification for the power, but since Jenny Electric built dynamos, the generator must have produced direct current (DC); the output voltage is not known, but was likely 100 volts (an early standard). For driving their dynamos, Jenny Electric used steam engines rated at about 10 horsepower built by Mansfield Machine Works.

The equipment arrived in early June, and a coal-fired boiler, steam engine, and a generator were set in place on Miller Street by Jenny Electric workmen. In July, the arc-lights between towers around the square were turned on; other street lights slowly followed.

Street Arc Lamp

In 1892, a franchise was awarded William S. Wells to provide lighting in Huntsville homes; his rate was $1.00 per month per 16-candlepower incandescent lamp. (Candlepower is an obsolete unit of radiated light, and cannot be directly converted to wattage of a modern incandescent lamp). Incandescent lamps from Jenney were rated from 16 to 150 candlepower; Jenny advertised that their generators could provide an average of 200 candlepower per generator horsepower.

The first electric lighting was installed at the home of Samuel B. Moore, just in time for a great party celebrating his World champion cow, Lily Flagg. In 1898, lines were run to atop the Monte Sano Mountain, providing power to the hotel being built there. As the 1890s passed,

Cyrus F. Sugg gradually acquired Huntsville Electric Company stock, ultimately becoming the sole owner.

Electrical Communications

Communications using electrical devices connected by wires originated shortly after it was discovered that an electrical charge could be conveyed from one point to another using a conducting medium. From his kite experiment in 1752, perhaps American Benjamin Franklin might be recognized for this basic discovery, but a number of European scientists in the early 1800s more directly deserve this credit.

Telegraph Improvement

As earlier described, a business-man, Samuel Morse, is generally credited in America with inventing the telegraph in 1837. Morse's assistant, Alfred Vail, developed what is called the Morse Code, making the telegraph broadly usable.

Stock Ticker

Railroads became the first large-scale users of the telegraph; the Memphis & Charlestown Railroad installed a terminal in Huntsville in 1856.

A machine for printing telegraphic information on paper tape was invented in 1856. This was used in devices, called stock tickers, to provide stock and commodity prices ("quotes") by telegraphic means. The most successful of these, the Universal Stock Ticker, was developed by Thomas Edison in 1869. Western Union acquired the rights to the Universal Stock Ticker, and by the mid-1880s, thousands of these were installed, allowing very rapid trading; this had a major influence on the financial market. Western Union provided telegraph services for Huntsville, and brokers in the city were early users when the stock ticker became available.

In 1874, Thomas Edison made the most revolutionary invention in telegraphy – the quadruplex telegraph. With this, four messages could be sent simultaneously over a single line. Acquired by Western Union, it enabled the company to greatly increase its messaging capacity at a minimum of cost; by 1878, there were 13,000 miles of quadruplex lines. The quadruplex telegraph continued to be used well into the twentieth century.

Time Signal

The U.S. Naval Observatory, the keeper of standard time in America, started sending out a time signal on a dedicated wire in 1865; this was a telegraph "click" each second that started a minute before the hour, paused for a few seconds, and then gave a final click marking the new hour. Western Union picked up this service, renting to subscribers large clocks that were set every hour by using the hour pulse to slap the second and minute hands together. Besides the train depot, banks and other firms wishing to impress their clients with the precisely correct time were early subscribers in Huntsville.

Telephone

Credit for the invention of the telephone in America is usually given to Alexander Graham Bell. Like Edison, Bell was working on a telegraph apparatus for simultaneous communications when what became the telephone instrument originated. On 10 March 1876, his assistant in a separate room heard over the apparatus Bell saying, "Mr. Watson, come here! I want to see you!"

The Bell Telephone Company was organized the in 1877, initially producing instruments for single point-to-point operations. The usefulness of the telephone was greatly expanded by the introduction of the telephone switchboard. Bell Telephone quickly incorporated this technology and began a wide expansion by organizing local and regional operating firms. One of these was the Southern Bell Telephone & Telegraph Company, started in 1879. Western Electric had been a strong competitor of Bell, and, in 1881, this firm was acquired by Bell as their manufacturing subsidiary.

In 1883, the Southern Bell Telephone & Telegraph Company brought the telephone to Huntsville; R. A. Moore was the local manager. Initially, there were only 32 individual and business subscribers. This was one of the first locations for incorporating a new telephone manufactured by Western Electric; called a magneto wall-set; this had internal batteries for power and a crank on the side used to "ring" the operator. Long distance service became available in 1886, and by the end of the century, subscribers in Huntsville increased to over 250.

Wall-Mounted Local-Battery Magneto Telephone

SPANISH-AMERICAN WAR

The Spanish-American War lasted only from 25 April to 10 December 1898. When it began, Joseph Wheeler, former CSA Major General and hero in the Civil War, was serving in Congress representing the 8th District of Alabama; this included Madison County. He applied for a commission and entered the U.S. Army as a Major General (the only person to ever hold this rank in both the CSA and U.S. Armies). Sent to Cuba, he led the cavalry troops, including those of Theodore Roosevelt at battle of San Juan Hill. In one battle, Wheeler is said to have shouted, "Charge on, men; we have these damn Yankees on a run!"

As the war closed, an armistice was not immediately signed. Consequently, the U.S. Army decided to return the troops to the States, but keep them at ready and stationed at sites where they could be quickly returned to Cuba. Huntsville was selected as one of the sites. A training camp initially called Camp Joseph Wheeler was opened around Huntsville in August

Spanish-American War Troops Parade for General Wheeler in Downtown Huntsville

1898. Some 14,000 troops from Alabama, Florida, Georgia, Maryland, Michigan, New York, Ohio, and Pennsylvania camped at sites across the city. When General Wheeler returned to the States in late 1898, he was appointed camp commander and immediately renamed it Camp Albert G. Forse, honoring a cavalry Major killed in action at San Juan Hill in Cuba.

Camp Forse was closed in 1899. Wheeler served briefly in the Philippines, and then retired to his Lawrence County farm in 1900. He wrote a number of books on military history and strategy, and died in 1909. Overall, the Spanish-American War had little lasting effect on Greater Huntsville.

Selected Bibliography / Additional Information for Chapter I

Archives Section, the Huntsville-Madison County Public Library (HMCPL); this was an invaluable source of information, particularly for Chapter I.

Adams, Bob (President) and Jacquelyn Proctor Reeves (Editor), *The Huntsville Historical Review*, 2008; Special Issue: "A History of Early Settlement: Madison County Before Statehood 1808-1819."

http://huntsvillehistorycollection.org/hh/hhpics/pdf/hhr/HHR-Madison_County_Before_Statehood.pdf

Adams, Bob (President) and Jacquelyn Proctor Reeves (Editor), "Special Issue: Remembering Sarah Huff Fisk," pioneering historian on early Huntsville; *The Huntsville Historical Review*, Vol. 33, No. 1 (Winter-Spring 2008). http://huntsvillehistorycollection.org/hh/hhpics/hhr/pdf/Volume_33_1_Winter-Spring-08.pdf

Bayer, Linda (editor), "Huntsville: 1866-1927," *The Historic Huntsville Quarterly*. Vol. V. No. 4 (Summer 1979); http://historyconnections.info/hh/hhpics/pdf/hhq1/HHQ-Vol-V-4-Sum79.pdf

Berne, Safford, Handbook of Alabama: *A Complete Index to the State*, Mobile Registry Print, 1878

Betts, Edward Chambers, *Early History of Huntsville, Alabama, 1804-1870*; Brown Printing Company, 1909, revised 1916

"Big Spring Park," *Wikipedia*; http://en.wikipedia.org/wiki/Big_Spring_Park_(Huntsville,_Alabama)

Blake, Tom, "Madison County: Large Slaveholders of 1860 and African American Surname Matches from 1870"; http://freepages.genealogy.rootsweb.ancestry.com/~ajac/

Carney, Tom, "The Mystery of John Hunt," http://huntsville.about.com/bljhunt.htm

"Cherokee," *Wikipedia*; http://en.wikipedia.org/wiki/Cherokee

"Chickasaw," *Wikipedia*; http://en.wikipedia.org/wiki/Chickasaw#Treaty_of_1818

"Cotton," *Wikipedia*; http://en.wikipedia.org/wiki/Cotton

"Cotton Gin," Wikipedia; https://en.wikipedia.org/wiki/Cotton_gin

Cox, Dwayne, "A Short History of Alabama Agriculture, 1820-1945"; http://www.lib.auburn.edu/archive/aghy/alag.htm

Daniel, Larry J., and Riley W. Gunter, *Confederate Cannon Foundries*; Pioneer Press, 1977. ISBN-13: 978-0913150382

De Land, T. A., and A. Davis Smith, *Northern Alabama Historical and Biographical*; Donohue and Henneberry Printers, 1888. ISBN 587-1-68563

Fisk, Sarah Huff, *Civilization Comes to Big Spring: Huntsville, Alabama 1823*; Pinhook Publishing Company, 1997. ISBN 978-0-9655917-3-7

"Heritage of Madison County, Alabama", by Madison County Heritage Book Committee, submitted by Ralph B. Garrison, Heritage Publishing Consultants, 1998. ISBN 1891647180

"History of Huntsville Water Works 1823-1973," Annual Report of Huntsville Utilities, 1973

"History of Manufactured Gas," *Wikipedia*; http://en.wikipedia.org/wiki/Main_Page

"Huntsville, Alabama," *Wikipedia*; http://en.wikipedia.org/wiki/Huntsville,_Alabama

Hurt, Douglas R., *American Agriculture: A Brief History*, Purdue Press, 2002. ISBN 1-55753-281-8

"James (John) Ditto," http://huntsvillehistorycollection.org/hh/index.php?title=Person:James_Ditto

Kvach, John F., Charity Ethridge, Michelle Hopkins, and Susanna Leberman, *Huntsville*; Arcadia Publishing, 2013. ISBN 978-0738-598918

Lakwete, Angela, *Inventing the Cotton Gin: Machine and Myth in Antebellum America*; Johns Hopkins University Press, 2003. ISBN 978-0801-873942

"LeRoy Pope," *Wikipedia*; http://en.wikipedia.org/wiki/LeRoy_Pope

"Madison Country, Alabama," *Wikipedia;* http://en.wikipedia.org/wiki/Madison_County,_Alabama

McNeil, Ian, *An Encyclopedia of the History of Technology*; Taylor & Francis, 1990. ISBN 0-203-33017

Mitchell, Charles C., "The Development of Cotton from the Old World to Alabama," Alabama Agricultural Experiment Station, No. 286, Mar. 2008

Morison, Elting E., *From Know-How to Nowhere: The Development of American Technology*, Basic Books, 1974. ISBN 0-465-02580-3

"Navigation on the Tennessee River," Tennessee Valley Authority; http://www.tva.com/river/navigation/

Pruitt, Ranee G., editor, *Eden of the South: A Chronology of Huntsville, Alabama 1805-2005*; Huntsville-Madison County Public Library, 2005. ISBN 978-0970736826

"Public Land Survey System," Wikipedia; http://en.wikipedia.org/wiki/Public_Land_Survey_System

Record, James R., editor, "Commemorative Album Celebrating our City's Sesquicentennial of Progress: Huntsville, Alabama, 1805-1955"; Committee for Huntsville's 200th Anniversary, 1955

Record, James, *A Dream Come True - The Story of Madison County and Incidentally of Alabama and the United States*; Vol. I: "Madison County 1492-1865"; Hicklin Printing Company, 1970; Vol. II: "1866-1967 History"; "1811-1977 Our City Government"; "1798-1818 Our Territorial Government"; and "1804-1977 Other Data"; Hicklin Printing Company, 1970

Reeves, Jacquelyn Proctor, *King Cotton to Space Capital: The Huntsville-Madison County Story*; Huntsville-Madison County Historical Society, 2013. ISBN 201-394-8503

Richter-Haaser, Elfriede, "Madison County," ALGenWeb; http://www.algw.org/madison/

Roberts, Frances Cabaniss, "Background and Formative Period in the Great Bend and Madison County"; Dissertation, University of Alabama, 1956

Stephens, Elise Hopkins, Pictorial Research by Patricia H. Ryan, *Historic Huntsville: A City of New Beginnings*; American Historical Press, 2002. ISBN 1-892724-31-6

"Surveying," *Wikipedia*; http://en.wikipedia.org/wiki/Surveying

Taylor, Thomas Jones (Judge of Madison County, 1866-1894, and the county's first historian), *History of Madison County and Incidentally of North Alabama, 1732-1840*, dated 1886, Manuscript typed by WPA in 1940. Edited by W. Stanley Hoole and Adie S. Hoole; Confederate Publishing Company, 1976

"Tennessee River," *Wikipedia*; http://en.wikipedia.org/wiki/Tennessee_River#Name

"Tennessee Valley Authority," *Wikipedia*, http://en.wikipedia.org/wiki/Tennessee_Valley_Authority

U.S. Census of Agriculture, published every five years since 1840 by the U.S. Department of Agriculture's National Agricultural Statistics Service; http://www.agcensus.usda.gov/

Chapter II

MANUFACTURING ERA

Beginning near the end of the 1800s, Greater Huntsville evolved from being agriculture-centered to being manufacturing-centered. Chapter Two of this book concerns what is herein called the Manufacturing Era; this covers the period from the last decade of the 19th century until the end World War II and the wartime industry closedown.

In this era two very different manufacturing activities dominated Greater Huntsville: textile mills and wartime munitions plants. Therefore, this chapter has two major sections called Textile Period and Munitions Period. While the emphasis is on these two types of manufacturing, other important activities that influenced the overall technological evolution are included.

As the 19th century drew to a close, Huntsville and Madison County had largely recovered from the Civil War. Agriculture had changed from being dominated by large plantations with slaves to smaller farms worked by families, tenant farmers, and share-croppers. While no longer "King," cotton was still the primary crop. Some of the former slaves that had left the farms seeking a new life in the city had returned to the land as hired hands, disappointed in what the Freemen's Bureau – noted in the previous chapter – had offered. In Huntsville and Madison County, as in many other Southern areas, there were still a large number of unemployed workers, both Black and White.

In addition to an available work force, Huntsville had good railroad and river transportation, up-to-date utilities, and an excellent banking and commerce structure. In addition to the North Alabama Improvement Company, a Board of Trade had been formed in 1892, and two years later the Chamber of Commerce was established.

In the 1890 census, Huntsville had a population of 7,995, and the total in Madison County was 38,119. Through its development years, Huntsville maintained only three square miles within its city limits. The opening of the cotton textile mills and their associated mill villages primarily added to the surrounding population, not to the City itself. In the census of 1920, Huntsville's population was officially reported as 8,018, essentially the same as 30 years earlier, and the total Madison County was 51,268. At the same time, the population within what was called Greater Huntsville – then encompassing eight square miles – was estimated to be 23,349. In 1925, the Chamber of Commerce reported 65 firms as engaged in manufacturing.

TEXTILE PERIOD

From medieval times, homes had spinning wheels where wool, cotton, flax, and silk fibers were spun into spools of yarn; a loom was then used in weaving the yarn into fabric. In 1673, the flying shuttle was invented in Great Britain, leading to automated weaving looms. The next year, a multi-spool spinning frame was invented, making it possible to simultaneously produce many spools of yarn. These inventions were quickly incorporated into large cotton mills concentrated in Manchester, England. The technologies were closely held in Great Britain.

A New Englander, Francis Cabot Lowell, visited the cotton mills in Manchester and, in spite of restrictions, returned to America with details of the equipment and established a similar factory. Opened in 1814, the Boston Manufacturing Company at Waltham, Massachusetts, was the first in the Nation in which all operations for converting cotton lint into finished cloth could be performed in one building. The firm also manufactured and sold spinning and weaving equipment for use in other mills. The Massachusetts areas of Waltham and Lowell became the center of fabric manufacturing in the northeast.

HUNTSVILLE and TEXTILES

Prior to 1890, little attention was given to the manufacturing of cotton fabric in the Southern states. Although the first cotton mill in Alabama – the Bell Factory – had operated near Huntsville in 1832, it initially used slave labor and was unable to adjust to post-war labor conditions, finally closing in 1885. At about this time, investors began to recognize the potential cost benefit of having cotton mills near the cotton farms. The comparative costs of delivering a bale of

cotton from the gin to the mill were as follows: $0.50 to a local mill, $3.00 to a northern mill, and $5.00 to $7.50 to a foreign mill. The advantage to the producer of having local mills was obvious, particularly to wealthy investors.

As shown in future sections, Tracy W. Pratt (1861-1928) had a recognized influence on the early industrialization of Greater Huntsville. He built and operated the relatively small West Huntsville Cotton Mills, but envisioned the area as a major textile center

Tracy Pratt

and had a significant role in making this happen.

The Economy

In the 40-year period from the start of the 1890s through the 1930s, Huntsville was a city with its economy almost completely dependent upon cotton manufacturing – mills that turned cotton lint into cotton goods. Following a pattern found over much of the South, this started with outside investors establishing a number of large cotton mills. After this, the city's economy was greatly dependent upon the price of cotton, and so was the economy of Madison County and the cotton farmers.

With a world-wide increase in cotton production, the price per pound was 4 cents in 1894, and the highest was 13 cents in 1910. This fell to 7 cents in 1914, but the approaching World War took it to 27 cents in 1916, and then 35 cents during the war in 1919. In the decade following the war, it fell to 15 cents in 1920, then for a decade it rocked up and down between this and 28 cents. As the Great Depression started, the price shrank to a low of 5 cents in 1931. President Franklin Roosevelt's New Deal stabilized the price at around 12 cents for a few years, but it was back to 8 cents by 1937.

Although always in the top growers of cotton, Alabama was never better than about fifth in the nation for cotton fabric manufacturers. The best year was 1916, with about 70 cotton mills employing around 16,000 operators.

Textile Mills

In the latter part of the 1800s and early 1900s, a number of cotton mills opened in Huntsville. Although some had different names over the years, the largest were best known as Dallas, Lincoln, Merrimack, and Lowe. Smaller cotton mills included Huntsville Cotton Mill, Huntsville Spinning Company, West Huntsville Cotton Mills, Admiral Braid Mill, Huntsville Knitting, and Erwin Manufacturing. The larger mills were all on property outside the city limits.

Inside a Cotton Mill

The advent of cotton mills made a radical change in Greater Huntsville's work organization. For the first time, large numbers of workers were used in industrial tasks, necessitating a hierarchy of supervisors and managers. Although the equipment involved was built elsewhere, it required engineers and mechanics for plant layout and hardware maintenance. Unfortunately, the

Lowe Manufacturing Steam Boiler

census records do not show these as occupations, but the cotton mills initiated the local need for technical and managerial specialists.

All of these mills initially used steam power, but converted to electrical machinery in the early 1920s.

Work operating the spindles and looms was reserved for Whites; these were often impoverished sharecroppers and tenant farmers who had abandoned worn-out farms for the hope of steady employment in the mills. Entire families worked in the mills, and children were expected to work; employees, who were often called "lint heads," sometimes included children as young as 8 years and working up to 12 hours a day. The State Law required children to attend school eight weeks a year; mills often interpreted this to be satisfied by two hours of school a day.

Starting in 1908, Madison County required the mills to obtain an affidavit permitting employment of children between 12 and 17 years. Because they had small hands, young girls were employed in the spinning room. Doffers changed out full spindles, and were often young boys who were able to climb onto the huge machines to free up the threads; this was a very dangerous job, sometimes

Children Spinners and Doffers

resulting in an arm lost to the machinery. The youngest children mainly worked as sweepers.

Huntsville Cotton Mill

Huntsville Cotton Mill, established in 1881 by D. L. Love of Greenville, Mississippi, was the first such mill built in Huntsville. Recognizing the potential for such mills, Love first toured the East to raise capital, and found

Huntsville Cotton Mill

Joshua Coons who also agreed to be the Superintendent. The factory was in a 45- by 110-foot, two-story building on Jefferson Street. The Huntsville Board of Aldermen passed a 10-year tax exemption that set the founding of future mills. In 1918, Huntsville

Cotton Mill was renamed Margaret Spinning Mill, then Fletcher Mill just before it closed in 1933. At its peak, it employed 300 workers.

Dallas Mills

Dallas Manufacturing Company was chartered in 1890 by Trevanion B. Dallas of Nashville and several other investors from New York. These investors were solicited by the North Alabama Improvement Company and the O'Shaughhessy brothers. As part of the enticement, the City agreed to provide 500,000 gallons of water per day without charge for 10 years. Dallas Mills began

Dallas Manufacturing Mills

operation in 1892, and was for several years Alabama's largest cotton mill. The mill had 25,000 spindles and 750 broad looms; it mainly produced cotton sheeting, but also had other cotton and woolen goods. One of the building features was a spiral fire escape The mill was located between Meridian Street and Andrew Jackson Way on the east side of the Southern Railroad track.

Lincoln Mill

Madison Spinning Company opened in 1900, but folded in 1906. Reopened as Abingdon Mill, it failed again in 1918, but was bought out of bankruptcy by William Lincoln Barrell of Lowell, Massachusetts. Renamed Lincoln Mill, it

became Alabama's largest textile mill with about 800,000 square feet of production space in several connected buildings. It was located on the southeast corner of Meridian Street and Oakwood Avenue, across the Southern Railroad tracks from its rival,

Lincoln Mill

Dallas Mill. During WWII, it was a major supplier of "duck" canvas for the military. During the early 1950s, it had a resurgence of military orders for the Korean War

Merrimack Mill

Although Tracy Pratt and Joshua Coons operated smaller mills themselves, they convinced the Merrimack Manufacturing Company of Lowell,

Massachusetts, one of the largest textile manufacturers in the nation, to open a plant in Huntsville.

Merrimack Mill Company built a plant two miles southwest of downtown on Triana Road, opening it in February 1899. The land purchased by Merrimack included the area of the present-day Brahan Spring Park. Initially employing 750 workers, the

Merrimack Mill

company began plans for a major expansion; in early 1905, a second building was opened to 1,100 new workers. The two buildings of Merrimack Mill housed 90,000 spindles and 2,000 looms.

Lowe Manufacturing

Lowe Manufacturing

In 1901, another Massachusetts investor, Arthur H. Lowe, opened Lowe Manufacturing Company on 8th Avenue in West Huntsville; it had 25,000 spindles producing cotton yarn. The next year, Eastern Manufacturing Company built a large weaving mill adjacent to Lowe. The two mills worked together to produce high-grade cloth for clothing manufactured nationwide. The two firms legally merged in 1907, forming Lowe Manufacturing, Inc., with a total of 171,000 square feet of space. In 1932, during the Great Depression, the firm declared bankruptcy and, as Lowe Mills, Inc., the stock majority was acquired by Avondale Mills of Birmingham. It changed hands again in 1936, becoming Lowe Corporation, but was finally dissolved in 1937.

Mill Villages

All of the major mills in Greater Huntsville also owned adjacent mill villages. Built at the same time as the affiliated mill, their primary purpose was to provide low-cost housing for the employees, partially compensating for the low wages that were common. They were located within easy walking

Merrimack Mill Village

distance to the mill buildings. Most of the houses in each village followed the same architecture: two-family dwellings, initially with outside toilets and water from several common wells (most were later connected to the Huntsville Water Works). In their beginning, they rented for as little as $0.50 per month. Merrimack Village was the largest, with about 280 two-story houses.

An attempt was made to have the villages as complete as practical, making trips to downtown unnecessary. All had one or more general stores, a facility for meetings, and recreational areas; most had libraries, churches, and medical clinics; Merrimack had a hospital with a physician and two nurses.

Merrimack Hospital

In 1900, Merrimack opened a large cemetery on the northeast corner of its property; many former employees and family members are buried there. The recreational area of Dallas was the best in the city and was open to teams of all races; semi-pro teams from mill villages drew crowds numbering in the thousands.

Although young children were often employed in the mills, it was required

Rison School

that they also attend school, even if just a few hours a week. To accommodate this, there was a school in every village. Lowe, two smaller mills, and several citizens gave money to start the West Huntsville Grammar School near the Lowe Village in 1916. Merrimack established the Joseph J. Bradley School in 1919, eventually giving all grades, and Dallas built the Rison School, initially just at the elementary level, but later including high school; these two schools were named for the respective mill's long-time superintendents, The city, with financial assistance from Lincoln Mill, built the Lincoln School in 1929. In addition to education, these schools were at the center of social and community activities in their respective villages.

McCormick YMCA in West Huntsville worked closely with the southwest area schools, particularly in athletics. This YMCA, built by very wealthy Mary Virginia McCormick, also operated a kindergarten and a community nurse station. (Miss McCormick was an heir to the fortune from the McCormick Reaper – the first successful mechanical crop harvester; she also funded the building of a

McCormick YMCA

Colored Hospital, across the street from the first Huntsville Hospital.)

Streetcars

Few of the mill workers could afford a horse and buggy. To provide local transportation, an electric streetcar operation was proposed. In 1899, the Huntsville Railway, Light and Power Company was formed by purchasing Cyrus Sugg's stock in the Huntsville Electric Company. This took over the lighting system, but it was mainly to establish the streetcar operation, and was primarily financed by Tracy Pratt as a part of his textile industry expansion endeavor. The development of a local streetcar system started in 1899, and it began operation the next year.

The first practical electric streetcar was developed by Frank Sprague at Richmond, Virginia, in 1888. Although many other people had built a variety of electric-powered rail vehicles, Sprague's was recognized for safety and reliability, and set the standard of such systems for many years. Sprague's inventions had included a regenerative or dynamic breaking through which the kinetic energy of the moving vehicle is converted to electrical energy and then dissipated by passing it through banks of resistors under the car.

The Loraine Steel Company of Loraine, Ohio, producer of rail for railroads, saw major business in rails for streetcars, and entered this field, offering not only rails but full streetcar systems. The new power company contracted with Lorain for providing all of the equipment as well as the full installation.

Building Tracks Downtown

The streetcars operated on 600 volts direct current (DC), supplied from a single overhead wire and connected to the vehicle using a trolley pole; the return connection was through the wheels contacting the rail. The cars were built for Loraine by the Johnson Steel Rail Company. For the power, Loraine supplied a plant with a 200-horsepower Payne steam engine directly connected to a 200-kilowatt Bullock dynamo. At the streetcar, the incoming current passed through a rheostat that the motorman (operator) adjusted to vary the speed of the motor.

The Huntsville streetcar line connected the Lowe and Merrimack mills in the southwest to the Dallas and Lincoln Mills in the north, passing through the downtown, and covering a distance of about five miles. When fully operational, the system had ten cars – five closed

Huntsville Streetcar

cars for the winter and five open cars for the summer; each car could seat 28 persons, and there were holding straps for standees. There was a single track, of 4-foot, 8.5-inch gage, accommodating two cars at a time with a passing point near downtown. At the line end-points, the car would reverse direction by the motorman switching the direction of the trolley pole and the polarity of the voltage into the motor.

Streetcars ran every 15 minutes between early morning and midnight; the fare was initially 5 cents. They not only provided all-week services for residents of the mill villages, but also allowed workers to have homes at some distance away from work, leading to the development of sub-divisions around the city.

In 1915, the Alabama Power Company acquired the power system in Huntsville, including the streetcar system. They continued operating the streetcar lines until February 1931; at that time Alabama Power converted the operation to a bus transit service. In 1940, the local service was sold to Crescent Motors, a national firm with bus operations in a number of Alabama cities.

Strikes and Mill Demise

Wages paid by the Southern mills were low – averaging 38 percent less than that paid in the Northern and Eastern plants; this was one reason for the textile manufacturing to transfer to the South. In 1930, with the start of the Great Depression, mills began a reduction of workers, and remaining employees had to work longer hours to keep production up. In 1933, Congress passed President Franklin Roosevelt's National Industrial Recovery Act, calling for voluntary acceptance of a 40-hour workweek, minimum weekly wage of $12 ($13 in the Northeast), and the elimination of employing persons under age 16.

All of this gave rise to the labor unions in the mills. In a year of intensive organizing, the United Textile Workers of America increased nationwide membership by 270 percent and called for a general strike. On 14 July 1934, a wildcat strike started in Guntersville. This quickly spread to Huntsville where on 16-17 July, an estimated 4,300 workers in six mills went on strike; Lowe had already settled with the union. The movement rolled across the state, and in a few days 20,000 textile workers had walked out. By 15 September, an estimated 400,000 workers nation-wide were involved, making this the largest labor conflict in American history.

National Millworkers Strike

The strike brought violence to Huntsville – assaults, shootings, and bombings; carloads of strikers roamed the streets, intimidating anyone who appeared to be going to work; girls who crossed the picket lines had their hair cut off. On 22 September, national union leaders reached a settlement and the great strike was over. No charges were ever filed concerning the hundreds of acts of lawlessness. The strike, however, was the beginning of the end of Huntsville as Alabama's textile center.

The remainder of the 1930s was filled with strikes and confrontations; the good relationships between the mill owners and the employees were not recovered. As the war approached, the government made major purchases of fabrics for uniforms and bedding. In December 1941, and the start of World War II, the Huntsville mills were fully loaded with work, but as the end of the war approached, this work slowed, then essentially stopped after the defeat of Germany and Japan.

For a few years, the local mills upgraded to compete with those in other parts of the country, but the advantage of cheap labor was largely gone; then the Asian fabric manufacturers took over the market. Dallas Mill closed in 1949. At Lincoln Mill, workers sued for the right to unionize, and in a Landmark case, the U.S. Supreme Court upheld the worker's rights. As threatened by its owner, Lincoln Mill was closed in 1955.

Merrimack was bought by M. Lowenstein and Company of New York, and reformed as the Huntsville Manufacturing Company in 1946. Huntsville Manufacturing was modernized, increased to 3,437 looms, and had 1,600 employees, making it the largest textile mill in the Southeast, but it could not compete with Asian mills. Sold to Springs Industries in 1988, it was closed the next year.

At various times, the houses in the mill villages were sold to employees and the areas were brought into the city limits. With the houses refurbished through the years, most remain in use today.

WORLD WAR I

World War I (WWI) started as a local conflict, with the Austro-Hungarians invading Serbia on 28 July 1914. Russia mobilized and Germany invaded Belgium and Luxembourg; they then moved on toward France, leading Britain to declare war on Germany. It shortly became two primary combatants: the Allies, with the United Kingdom, France, and the Russian Empire, and the Central Powers of Germany and Austria-Hungary; many other nations eventually joined both sides.

The United States initially pursued a policy of non-intervention, avoiding conflict while attempting to broker a peace. From the start, the German submarines (U-boats) were a major threat to the shipping of supplies to England; then, after the U-boats sank seven U.S. merchant ships, Congress declared war on Germany on 6 April 1917. In July, the drafting of American civilians started, the National Guards were activated, and the first U.S. troops were sent to France. The United States was a major participant until an armistice was signed on 11 November 1918. During the war, the United States built to over 4,700,000 troops, including about 2,800,000 draftees; of the total American military personnel, 116,708 died from all causes.

Almost 95,000 persons from Alabama saw service during WWI, of whom 6,262 were killed; 19 men from Alabama were awarded the Distinguished Service Medal. The records do not show the portions of these numbers for servicemen from Madison County. Aside from personnel in the armed services, Greater Huntsville had little direct involvement in the war.

There were many advances in military technology during WWI. Wireless communications, armored cars and tanks, automatic weapons and submachine guns, and precision artillery were improved for land combat. The widespread use of chemical warfare – chlorine, mustard gas, and phosgene – was a distinguishing feature of close combat. Germany developed 420-mm (17-in) howitzers – nicknamed Big Berthas – able to bombard Paris from distances of over 100 km (62 mi). Aircraft were first used for air-combat, reconnaissance, and bombing. New types of submarines, particularly the German U-boats, made their appearance, and these led to the development of depth charges, hydrophones, and blimps.

OTHER ACTIVITIES

In the 60-year period between 1890 and 1950, while textile mills and, later, munitions plants were at the center of local activities, overlaying these were many other important activities influencing the technological evolution of Greater Huntsville.

Utilities

Essentially from its formation, Huntsville had utilities earlier and superior to most comparably sized cities. For most of the 18th century, the utilities were primarily to serve homes and small businesses. As Greater Huntsville grew, the utility services were expanded accordingly. With the emergence of large factories, the utilities were further expanded to meet their needs.

Electrical Power

In 1887, electrical power was introduced into the city by a private firm, the Huntsville Electric Company. The electrical power came from a local steam-powered generating plant. In the late 1890s, Tracy W. Pratt led the planning for an electric streetcar service, primarily to serve the developing cotton mills and their associated mill villages.

In 1899, the Huntsville Railway, Light and Power Company (HRLPC) was formed; it acquired the stock and assets of the Huntsville Power Company, which at that time was wholly owned by Cyrus Sugg. A new facility was constructed near the MC&StL depot. The original 100-volt DC system from Jenny Electric was retained for the street lights. This had a capacity of up to 100 arc lights and was driven by a 10-horsepower Mansfield steam engine.

The streetcar system was purchased by HRLPC from Loraine Steel Company in 1900. This included the cars, about five miles of track, and a central power system consisting of a 200-horsepower steam engine from B.W. Payne & Sons, directly connected to a 200-kilowatt DC dynamo from Bullock Company. Although not recorded, the output was likely 600 volts, a level then commonly used by streetcars. The streetcar system was put into operation in early 1901.

The HRLPC existed from 1899 to 1909. As the power system enlarged and had longer lines, DC presented a problem: Longer lines had more resistance, causing a reduction in voltage, and DC could not be passed through transformers to change the voltage level.

In 1909, the Light & Power Company was formed; still privately held and located in Huntsville, it purchased the stock and assets of the HRPLC and converted the central generators and distribution lines to AC. The new system consisted of a 50-horsepower steam engine from American Ball, driving a 75-kilowatt, 2,300-volt, 3-phase generator from General Electric. For providing the steam, there were two boilers, one in operation at a time and the other in reserve. The DC power for the streetcars continued to be separately provided.

Alabama Power Company – AC power distribution has another advantage: It can be connected to a network to receive power from a distant source. AC power networks also have an advantage in the ability to couple generating plants at different locations, increasing the available power as well as the system reliability. This requires exact synchronization of the alternating frequency – this was adopted as 60 cycles per second (hertz) in America near the end of the 19th century.

The Alabama Power Company was founded at Gadsden, Alabama, in 1906, with a primary intent of developing hydro-electric power. In a short time, it began to acquire small power companies in North Alabama, tying them into a network

Alabama Power acquired the Huntsville-based Light & Power Company in June 1915. At the close of World War I in 1919, Huntsville was connected into an Alabama Power transmission network that delivered power generated at plants on the Tennessee and Coosa Rivers and the Little River Canyon. The local steam-driven generation plant then ceased operation.

The fabric mills in Huntsville converted from steam-driven mechanical power to electrical driven machinery in the early 1920s. It is also likely that the power for the streetcars was provided by Alabama Power using one or more motor-generator sets near the tracks to convert AC to DC.

In 1924, Alabama Power constructed the first rural electric power line in Alabama, running it out of Huntsville to Whitesburg. During the years of the Great Depression, small communities throughout Madison County received this utility.

Rural Power

Tennessee Valley Authority – As a major element of President Franklin Roosevelt's New Deal in combating the Great Depression, the Tennessee Valley

Guntersville Dam

Authority (TVA) was established by Congress in 1933. Among many other projects, TVA built dams and hydro-electric generating plants along the Tennessee River, and sold power at low cost to locally owned utilities. As an additional benefit, the building of dams finally opened all of the Tennessee River to boat traffic.

The City of Huntsville purchased the electrical system in Madison County from Alabama Power on 19 July 1940, forming a part of the emerging Huntsville Utilities. An Electric Utilities Board was established to be responsible for the electric power system. A 20-year contract was made with TVA wherein electrical power for all of Madison County would be purchased from that agency. Under this agreement, profits from the sale of electrical power by Huntsville Utilities could be used only for system extensions, improvements, or rate reduction. At that time, there were 5,810 Huntsville and Madison County consumers connected to about 250 miles of electrical distribution lines.

Water System

After the waterworks were acquired by the City in 1858, the pipe and pumps were continually improved and expanded; however, Big Spring remained as the sole source. After a severe outbreak of typhoid fever in 1917, Huntsville's first Public Health Officer, Carl A. Grote, Sr., decided that open toilets on the square above the spring were to blame and directed that a sewer line be installed; this, plus a new, efficient chlorinator at the pump station, resulted in a very clean water system. Even with the additional demand for water during the next three decades, the Huntsville Water System and its supply from Big Spring continued to be satisfactory

By 1950, it was evident that the city could no longer depend upon Big Spring for the sole water source. The City Commission then formed a Water Board to oversee the future development and maintenance of the Huntsville water system. Plans were drawn for accessing the aquifer through wells, and consideration was given to a purification plant for water drawn from the Tennessee River. The wells were accomplished quickly, but the purification plant would wait until the next decade.

Gas Works

The Huntsville Gas Company and its initial gasworks facility near the Big Spring dated from 1856. It was upgraded and a new facility built on Dallas Street in 1872, then remained essentially unchanged to near the mid-1900s. The initial use of gas was for lighting. With the advent of electric power in Huntsville, the gas market changed to residential customers using it for cooking, heating homes, and water heating. By 1913, there were over 12 miles of pipes serving customers in the city.

The local gas company was acquired by Alabama Gas Corporation (Alagasco) in 1946. The old plant, producing manufactured gas through processing coal, was replaced by a facility using liquid propane and air to produce gas for distribution. Propane, a natural petroleum product, is in liquid form when under pressure and is then 270 times more compact. It was shipped to Huntsville in large tank trucks, then, at the plant, allowed to become gas and mixed with air for distribution.

The city of Huntsville bought the gasworks system from Alagasco in 1950. A Utility Gas Board was formed under Huntsville Utilities, and a contract was made with Alabama-Tennessee Natural Gas Company to pipe natural gas into the city.

Aviation

In the late 1800s, William L. (Will) Quick had a woodworking and machine shop on the bank of Flint River a few miles east of Hazel Green. He diverted a portion of the river flow and built a water turbine to provide the shop's mechanical power. A creative and inventive person, Quick became very interested in flight and started the design of a flying machine fashioned after those of nature.

First Airplane in Alabama

At that time, the recognized authority in America on flight was Samuel P. Langley and his 1891 book, *Experiments in Aerodynamics*. With Langley's book providing the theoretical basis, Will Quick and his sons began the construction of a powered aircraft in 1900. This was a monoplane having a 38-foot wing span, with the wing covered only on the top side (for accepting the aerodynamic lift force). It had a square, uncovered fuselage with the pilot's seat inside, an overall length of 18 feet, and a tricycle-type landing gear. An automobile steering wheel and a foot-bar allowed wing and rudder control. Completion was held up by an unresolved problem: the availability of an engine with sufficient power and light enough to be carried in the craft.

In 1903, the Wright brothers flew an airplane that they had designed and built – the first in America. Quick continued his search for a suitable engine, finally settling on one from a 1907 Ford Model R automobile. This was a four-cylinder, L-head engine producing 18 horsepower. Stripped of its transmission, it was mounted in reverse, with the drive shaft extending forward to the propeller. Much effort was devoted to shaping the propeller, with testing of the thrust on the strapped-down aircraft.

Will Quick's 16-year old son, William Massey Quick, was selected to make the first test flight; this was done in April 1908, the first airplane to be flown in

Graphic of Quick's Airplane

Alabama. After becoming about 10-feet airborne and flying near 70 feet, the pilot leaned to the side to see his position and lost control; the craft tipped to the side and came down, destroying the landing gear. The wreckage was returned to the shop and placed in storage; there it remained until 1956. It was eventually reconstructed for museum display by Huntsville members of the Experimental Aircraft Association with assistance from Will's son, Joseph Quick.

Will Quick continued designing aircraft; a 1912 newspaper article attributes him with 16 patents. Six of his children became licensed pilots, including daughter Cady Quick Burns – possibly the first female pilot in Alabama. Another daughter married Terah Maroney who, after helping Will build his plane, built one himself. Flying in Seattle, Washington, in July 1914, Maroney took lumberman and shipbuilder William E. Boeing for his first airplane ride, generating an interest that ultimately led to the Boeing Aircraft Company. In 1924, Will's son Curtiss converted a war-surplus Hisso Standard J-1 biplane with equipment for crop dusting, offering out of Huntsville what was possibly the nation's first cotton-dusting service to kill boll weevils.

Huntsville Airfields

Thomas Quick, another son of Will Quick, was a leader in developing Huntsville's first air field. Located in the area south of Bob Wallace Street between Whitesburg Drive and the L&N railroad track, the 150-acre Mayfair Flying Field was officially dedicated in June 1931. Three years later it was listed as a commercial field with four dirt runways, the longest being a 2,400-foot northeast-southwest sod strip. Airmail deliveries using this field began in May 1938.

Huntsville's second airport, located about a mile west of the original Mayfair site, opened in 1941. It had two paved runways, the north-south one initially 4,000 feet long.   Waterman Airlines was the initial operator in 1942, and Pennsylvania Central Airlines (P.C.A.) inaugurated intrastate commuter service in 1944.   Southern Airways came in 1945, followed by Eastern Airlines in 1946. A full terminal and control tower were added in the 1950s, but all operations ended when a new airport opened 10 miles west of the city in 1967.

First Terminal

Agriculture

As the 20th century got underway, Alabama was still primarily an agricultural state, and cotton was still the number one cash crop. During 1899, Madison County had some 70,000 acres in cotton, and produced about 10,400,000 pounds of lint; the yield was 149 pounds per acre, far less than earlier times, indicating soil depletion caused by continually planting the same crop. To assist in correcting this, the first farm demonstration agents were hired in Alabama in 1906. Included was Thomas M. Campbell, a graduate of Tuskegee Institute and the first Negro county agent in the Nation.

Soil depletion was initially addressed by the use of guano (bat manure), but by the early 1900s, fertilizers such as nitrate of soda, superphosphate, and kainit (a potassium-bearing mineral) came into use. Acidic soils were corrected by the use of pulverized limestone rock and slag from the growing iron and steel industry. Some farmers rotated cotton with corn, soybeans, or peanuts, and rather than leaving the field bare, planted cover crops of rye, clover, or vetch to add organic matter and reduce erosion.

Although steam-driven tractors had been introduced in other parts of the nation before the turn of the century, very few had been used in Alabama. Following WWI and using technologies developed for tanks and other military hardware, much smaller, gasoline- and diesel-powered tractors became available. Tractors began to be used in Madison County in the late 1930s, but the major part of cotton farming remained hand-work. It typically required about 42 labor-hours to obtain 100 pounds of lint cotton from farming each half-acre using two mules, a one-row plow, a one-row cultivator, and the slow process of hand-hoeing and hand-picking.

Agricultural machinery was expensive and few farmers could afford it, but by 1930 there were more than 4,600 tractors on Alabama farms. As of 1940, that number had almost doubled, and near 46,000 tractors were in Alabama fields by 1950. The mechanical cotton picker was another time-saving innovation, perfected by International Harvester in the early 1940s. A mechanical harvester could pick almost 1,000 pounds of cotton per hour compared with the 15 to 20 pounds per hour a human could pick.

The 4-H organization, for rural youths between ages 5 and 21, started about the turn of the century. Researchers and teachers found that adults in farming communities often did not readily accept new agricultural discoveries, but saw that youths would experiment with these new ideas and then share their experiences and successes with the adults. Thus, 4-H became a way of introducing new agricultural technologies. From their start, there have been 4-H Clubs throughout Madison County.

The author of this book was a 4-H club member in Calhoun County, Alabama; at the age of nine, he grew his first small bale of cotton. About this same time (the mid-1930s), the local County Agent distributed cuttings of kudzu, a fast-growing vine recently brought from Japan to counter soil erosion in the South.

Boll Weevil Disaster

In 1909, there were about 3.7 million acres in cotton throughout Alabama, with some 1,130,000 bales produced. In that year, however, boll weevils started invading southeastern Alabama, and ten years later they had infested all cotton-growing regions of the United States, causing the greatest agricultural disaster in American history.

The boll weevil (*Anthonomus grandis*) is an insect about six millimeters (about one-quarter inch) in length that feeds solely on cotton buds and cotton flowers. They crossed from Mexico into South Texas in 1892, and, spreading about 40 to 160 miles per year, first reached Alabama in 1909, and by 1917, had covered the state. In 1919, there were only 718,000 bales produced statewide, a reduction of about 35 percent in one decade.

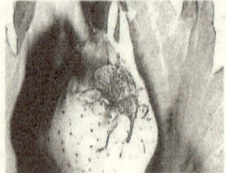
Boll Weevil

As boll weevils started to take their toll, the county agents pushed the farmers to change their practices – planting different types of crops and using crop rotation. George Washington Carver at Tuskegee had earlier shown the value of peanut farming. Coffee County took this up, and by 1919, was the largest producer of peanuts in the Nation. (A monument to the boll weevil was erected in Enterprise, the Coffee County Seat.)

Calcium arsenate was the first insecticide available for killing boll weevils. Hand-cranked or mule-drawn blowers were first used to spread calcium arsenate dust; the process was called dusting, and experimental dusting from Army Air Service aircraft showed this method to be effective. In 1924, Curtis Quick (son of Will Quick) modified for dusting a war surplus Hisso Standard J-1 biplane, and started in Huntsville what was possibly the first crop-dusting service in the Nation.

Crop Dusting (1925)

Agricultural Improvement

Crop dusting using aircraft led to an immediate improvement in cotton production, particularly in large fields. Alabama produced about 984,000 bales in 1924, a 37 percent recovery over the low of 718,000 bales in 1919, but still 13 percent less than the 1,130,000 bales when the disaster started in 1909. By 1930, 1,467,000 bales were produced, with a yield of 197 lb/acre. However, calcium arsenate insecticide never reached its full potential; there was insufficient manpower to carry out the extension services and convince the farmers to change. It was used until DDT became available in the mid-1940s.

In spite of the boll weevil, farmers were hesitant to give up on a crop that had served them well for generations. They continued to plant cotton as their primary cash crop; cotton acreage in Alabama remained more than 3 million acres until the beginning of the Great Depression. A 755-acre experimental station at Belle Mina in adjacent Limestone County was established in 1929, and has been highly beneficial to Madison County farming.

In response to the depression, President Franklin D. Roosevelt enacted the Agricultural Adjustment Act of 1938, mandating price supports for cotton and other crops. The Act also created the cotton allotment program, which required farmers to plant a specified number of acres of cotton and established a quota system to balance supply and demand. As a result of the allotment system, farmers applied more fertilizer and thus produced more cotton on less land

DDT (dichlorodiphenyltrichloroethane) is an organochloride known for its insecticidal properties. Discovered by Swiss chemist Paul Hermann Müller in 1939, DDT was first used in large quantities during the later years of World War II to spray fields before invasions. After the war, DDT quickly became highly popular in killing insects that transmitted diseases (particularly malaria and typhus) and to increase crop yields by reducing insect pests.

Madison County Agriculture

For the early years of the 20th century, only limited agricultural statistics are available for Madison County. As noted earlier, Madison County had some 70,000 acres in cotton and produced about 10,400,000 pounds of lint in 1899, giving a yield of 149 pounds per acre. There was a decline in cotton planting as well as yield during the boll weevil years starting in 1910, but, with crop dusting, this largely recovered in the 1920s. By 1939, the production was close to 42,000 bales and the cotton acreage had about doubled. During much of this time, C. F. Striplin was the County Agricultural Agent; he was followed by J. B. Mitchell between 1928 and 1948.

The official agricultural data for Madison County in this period – dates are when the census was taken – included the following:

	1929	1949
Number of Farms	7,178	5,004
Total Area in Farms, Acres	387,612	415,332
Percent of County	74	80
Average Farm Size, Acres	54	83
Cotton, Acres	129,800	109,400
Cotton, Bales	41,700	50,800
Cotton Yield (pounds/acre)	154	223
Corn, Acres	62,300	56,400
Soy Beans, Acres	---	5,400

This shows a decrease in farm land of about 27,700 acres between 1929 and 1949. Some of this reduction was due to the allotment system, but a large portion resulted from the U.S. Army taking about 19,400 acres of agricultural land when forming Huntsville and Redstone Arsenals in 1942.

Although cotton acreage between 1929 and 1949 decreased, the number of bales produced increased by about 22 percent. Also, the average weight of a bale of cotton lint had gradually increased from the early 350 pounds to 480 pounds. The 45 percent increase in yield was mainly due to increased fertilizing and the application of DDT. It is noted that acreage in corn – primarily grown for cattle feed – decreased, while acreage in soy beans began to be counted.

Civilian Conservation Corps

During the Great Depression, one of the many activities under President Franklin Roosevelt's New Deal was the Civilian Conservation Corps (CCC). Established in 1933 by Congress through the Emergency Conversation Work Act, this put 500,000 men to work on 35 project areas such as forestry, road improvement, and building national parks.

Functioning in a military-like operation, volunteers were assigned to companies of about 200 men each, led by a Captain and two Lieutenants. CCC camps usually had about 20 buildings, but the men often slept in tents. They were paid $30 per month, of which they were required to send about $25 home to a dependent. The first camps opened in late 1933, and all CCC activities ended in mid-1942.

There were 28 CCC camps in Alabama, not all open at the same time; two

Typical CCC Camp

camps were located in the Huntsville area. Camp Clement, opened in June 1935, was located atop Monte Sano Mountain where they primarily worked on improving the Monte Sano State Park under Project TVA-12 and 13. A stone entrance to the camp still exists at Highland Plaza Street. Camp Silver Dollar, opened in August 1935; it was about 1.5 miles southeast of downtown on Tennessee Street in what is today the Blossomwood area. Men from this camp also worked on the State Park and nearby roads under Project TVA-7.

CCC companies 3486 (6/19/1935) and 5402 (8/16/1935) were at Camp Clement, and 3483 (8/21/1935) and 2449-C (Colored - 9/30/1938) were at Camp Silver Dollar. When Huntsville Arsenal was started in 1941, Company 2449-C was assigned to assist in its establishment. Clement closed during 1940, and Silver Dollar during 1942.

Commercial Expansion

It was previously noted that a number of commercial industries were started in Greater Huntsville during the latter part of the 19th century. Except for the cotton mills – which boomed and dominated the local economy – few others were such that they continued well into the next century. Some experiences in commercial expansion in this period will be noted.

Petroleum Exploration

The U.S. petroleum industry is considered to have begun with the drilling of a 69-foot oil well near Titusville, Pennsylvania in 1859. With the petroleum mainly used for oil lamps and refining to kerosene, for the remainder of the 19th century this industry had only modest growth, with wells centered in Pennsylvania. "Oil booms" began early in the 20th century with the introduction of automobiles and the internal-combustion engine.

Little is known about it, but some serious petroleum exploration took place in the Huntsville area during this time. There was speculation that since petroleum was under the Appalachian chain of mountains in Pennsylvania, it might be expected beneath the end of this chain in Madison County. In the 1910s, with possible encouragement from the State geologist, industrialist / entrepreneur Tracy W. Pratt drilled about 20 shallow wells around Madison County, including one in West Huntsville. In most of these, traces of natural gas and very small pools of oil were found, but the venture was abandoned.

In 1930, geologists revisited the area, and concluded that deep wells could bring in great oil finds. The Tennessee Valley Oil & Gas Company was formed. With great publicity in *The Huntsville Daily Times*, it raised over $2 million in New York venture capital, and began a drilling operation near Gurley in early 1931. An

Oil Rig in West Huntsville

earlier well there had been tapped to supply natural gas to light a street corner in Gurley. At about 500 feet, oil in sufficient quantity was found to provide a sample. Visiting consultants from Houston reported the indications were that high-gravity oil would be struck in paying quantities at a depth between 1,500 and 2,000 feet. Plans were initiated to drill up to 100 wells in the region between Scottsboro and Huntsville.

Locally, the Greater Huntsville Oil & Gas Company was formed, with the purpose of extending to a greater depth the well started earlier in West Huntsville. Initial drilling gave positive indications, and a deep-well casing was ordered. Unfortunately, there is no further record of either the Greater Huntsville or the Tennessee Valley oil firms. It is therefore obvious that all of their drilling was unsuccessful.

Commercial Industries

The Huntsville Chamber of Commerce noted having 65 industries in 1925, most being textiles mills or related firms. By the start of the second half of the 20th century, there were only three textile mills still operating, and all were gone in a few years. The city did gain a few new industries; two will be described.

John Blue Company – In 1886, the John Blue Company was formed on a farm near Laurinburg, North Carolina, to repair cotton gins and farm equipment. The founder, himself a farmer, was creative and soon had developed a variety of farming implements. The first successful product was a stalk cutter that was patented in 1891.

In the early 1940s, the foundry in Laurinburg burned, and John Blue, Jr., then the owner, examined sites for moving the full operations. In Huntsville, he found potential workers, available buildings, and saw the potential for a locally operated farm equipment manufacturer. John Blue Company opened in Huntsville during 1945, with facilities on a large lot at the intersection of Bob Wallace and First Avenue. From the start, the grounds had a herd of cattle, indicating farming as the company's target market.

John Blue Plant

The rural-like setting belied the robustness of the company; within a few years it was a powerhouse in the city's economic leaders with several hundred

John Blue G-1000

employees. Among other products, John Blue made fertilizer spreaders, cotton wagons, and agricultural and industrial pumps. For a while, they had a production line manufacturing a tractor – Model G-1000 painted bright blue – developed by Ervin West and Wesley Cagle, their engineering VP. Eventually, John Blue had branches in five other states; a fleet of trucks and a company plane were indicative of their success.

As Huntsville grew, the original plant with its cows moved to the outskirts of nearby Madison in 1986. Stiff competition resulted in downsizing, and the firm eventually became a division of Virginia-based Advanced Systems Technology, but the John Blue operations remained in Madison.

Martin Stamping and Stove Company – In 1905, two Martin brothers opened a cast iron foundry in Sheffield, Alabama. They expanded their business in 1918, acquiring a stove factory in nearby Florence, Alabama. In 1939, the Martins purchased a bankrupt manufacturing plant in Huntsville on West Clinton Street (later Governors Drive); they reopened this as Martin Stamping

and Stove Company, initially producing a line of unvented gas heaters. A spur railroad track came directly to the Martin building.

During World War II, all of the Martin companies manufactured radiant heaters for the Army, and Martin Stamping also made bomb crates and related materials for the Army's Huntsville Munitions Plants. Following the war, the companies returned to manufacturing wood, coal, and gas heaters, and Martin Stamping added electrical heaters to the Huntsville line. The electric heaters were highly successful, leading to opening a facility in Athens, Alabama, as well as leasing 200,000 ft^2 of additional space in Huntsville.

Martin Stove Company

As America grew and more modern homes were built, the market for space heaters declined. In 1974, the various Martin holdings were consolidated into Martin Industries, Inc., with administration, engineering, and marketing centralized in Florence. Huntsville operations continued, producing gas grills, gas fireplaces, and free-standing vent-free gas heaters. The plant in Huntsville was eventually closed in 2000.

Commercial Engineering and Land Surveying

From the beginning of Madison County and Huntsville, services in land surveying and elementary engineering had been offered by individuals. In 1809, the position as County Surveyor was one of the first appointments made by the Territorial Governor, and Hunter Peel worked as both a surveyor and water-works designer starting in 1816. Other well-known technical activities, such as the building of railroads and the design of the Indian Creek canal, required individuals performing what would today be called civil engineering. Available historical records, however, do not show any local firms offering engineering and/or surveying services in the better part of the 19th century.

GW Jones & Sons Engineers - In 1886, George Walter Jones (1866-1946) started the firm G.W. Jones, offering services in surveying and, later, broadened to civil engineering. In a family that dated in this area from the early 1800s, he had not attended college but learned from an uncle who did private surveying work. As the firm matured and his four sons graduated from college (three in civil engineering and one in geology), they joined their father and the firm was eventually renamed GW Jones & Sons Engineers, Inc. They initially concentrated on land surveying and general engineering projects such as bridges, paving, and water and wastewater systems.

G.W. Jones

As the first civil engineering firm in Huntsville / Madison County, GW Jones & Sons was a major force in the region's development and growth from an agricultural-based, rural environment to an urban, highly technical economy. The firm served as Engineer for the City of Huntsville for 36 years and also as County Engineer for Madison County for many years. In 2007, it was inducted into the State of Alabama Engineering Hall of Fame.

Professional Licensing in Alabama - In 1935, the Alabama Board of Registration (later Board of Licensure) for Professional Engineers and Professional Land Surveyors was created. The ensuing Code of Alabama does not provide for the licensing of firms – only for licensing of individuals; firms offering engineering or land surveying must be under appropriately licensed technical owners / leaders.

To obtain a license, an individual must have an approved college degree in the specified field (engineering or land surveying) plus at least four years of intern experience under a licensed mentor. An in-depth examination approved by the Board is also required – this is now a national standard examination. Initially, an alternate to the degree was a greater amount of experience; later, the degree requirement became firm. (This book's author has been a Registered Professional Engineer in Alabama since 1958.)

MUNITIONS PERIOD

The second portion of Greater Huntsville's Manufacturing Era, called herein the Munitions Period, primarily concerns the 1940s decade. The Second World War (WWII) dominated the first years of this decade, and brought about some of the most significant political and technological changes in history. Therefore, the activities in the Munitions Period start with a brief description of this event.

WORLD WAR II

The Second World War started 1 September 1939, with Adolph Hitler's *Wehrmacht* invading Poland. Great Britain immediately declared war on

Remember Pearl Harbor

Germany, but the United States hesitated until after Japan attacked Pearl Harbor on 7 December 1941. Following declaration of war against Japan on 8 December and against Germany and Italy three days later, America became the leader of the Allies (United States, United Kingdom, Soviet Union, and 19 other nations) in a conflict against the Axis powers (Germany, Japan, Italy, and three other nations) that eventually covered much of the world. An estimated 60 to 85 million people, the majority of them civilians, were killed, making it, in size, the deadliest conflict in human history. During the war years, 16.1 million Americans served in active military duty; of these 291,557 were killed or missing (1.81 percent) and over twice this number were wounded. The European and Atlantic war ended 8 May 1945 (V-E Day), but the war in the Far East and Pacific continued four more months, ending 2 September 1945 (V-J Day).

Huntsville Involvement

Even before the United States' official involvement in the war, Greater Huntsville was already engaged in the development of two huge facilities for the production of chemical munitions: the Huntsville Arsenal and the Redstone Ordnance Plant. Throughout the war years, at times employing as many as 11,000 regular workers, these two facilities were at the heart of Huntsville's wartime involvement. In 1943, the estimated total employment in Greater Huntsville was 30,000; aside from those directly working at the Government munitions plants, essentially no others were involved in this activity – the city had no supporting infrastructure.

The cotton mills were heavily engaged in making cotton material for military uniforms and bedding. Textile workers and management came together to fill orders, and unionization was at a minimum. As of the end of 1944, Dallas had 725 employees, Merrimack had 850, and Lincoln had 1,200. Madison County farmers were tasked with producing as much cotton as possible, as well as grain and meat for supplying the military, all under the limitations of 12 rationing programs. Although there were no local business activities requiring special security, Big Spring, as the primary local source of water, was under a wire enclosure and guarded around the clock.

The first war casualty from Madison County was Luther James Isom of West Huntsville; he was killed aboard the battleship *U.S.S. Arizona* in the Japanese attack on 7 December 1941. Throughout the war, some 6,000 persons from Madison County served in the military; of these an estimated 110 were killed or missing (using the 1.81 percent national average). Carl M. Crabtree of Huntsville was the first person accepted by the Draft Board when it started in October 1940 (the nation's first peacetime draft); he was killed on Luzon in the Philippine Islands in 1945.

Many from the County served with distinction; Cecil H. Bolton and Paul L. Bolden were each awarded the Medal of Honor. Bolden was Madison County's most decorated WWII veteran; an 18-mile stretch of Alabama State Route 53, from Research Park Boulevard in Huntsville to just south of Ardmore, Tennessee, is designated as the "Paul Luther Bolden Memorial Highway" in his honor.

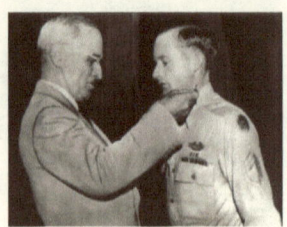

Truman Awarding Bolden

Before the war, the Axis had superior military technologies, but the Allies, particularly America, quickly responded, pulling ahead with innovation and production. Two of the most important technologies were radar and the atomic bomb – it is often said that radar won the war and the atomic bomb won the peace. Huntsville and Madison County, however, had no role in these technological advancements.

Missile Precursor

The long-range missile was another technology that came into being during the war; this was essentially developed solely in Germany. Only after the V-1s and V-2s began raining down on England were the Allies aware of the existence of these developments, and the U.S. Army initiated a desperate attempt to replicate the V-1. It was in this that Greater Huntsville had its first and only wartime involvement with missile technology.

When the German V-1 "Buzz Bomb" – forerunner of the cruise missile – became known to the Allies, the U.S. Army Air Forces (AAF) and contractor Republic Aircraft began the development of a similar weapon designated the JB-2 (JB for jet bomb), and commonly called the "Loon." Earlier, there was a JB-1 "Bat" built by Northrop and using a General Electric J31 engine – the first jet engine produced in the U.S. – but the JB-1 was never successfully flown. The engine for the JB-2 was reverse-engineered from a 900-lb thrust Argus-Schmidt pulse-jet found in a downed V-1. The AAF asked the Army's Chemical Warfare Service (CWS) to be responsible for the propellant. In support of this, Huntsville had its first, and only, involvement with missile technology during WWII.

The Huntsville Arsenal (described later) operated throughout the war as an organization under the CWS. To meet this need of the AAF, CWS turned to Huntsville Arsenal. Between January and September 1945, Huntsville Arsenal conducted investigations of three systems of liquid propellants: hydrogen peroxide-permanganate, fuming nitric acid-aniline, and mononitromethane-catalyst. Called the FRED Project, Major Frederick Bellinger headed the team performing the study. Of special interest was the potential of Huntsville Arsenal manufacturing these propellants. The only reference to this project states that it was successful; otherwise, there is no known documentation concerning details of the activity. The project concluded in September when JB-2 jet-powered missiles were successfully tested at Eglin Field, Florida.

In 1945, no locals would have ever imagined that within five years some of the most capable German scientists and engineers who had developed the V-2 would come to Huntsville and make major contributions to this area's becoming a world center for defense and space technologies.

MUNITIONS PRODUCTION

Following the First World War and the devastation caused by both sides using chemical weapons (primarily mustard gas), the Geneva Protocol prohibited the first-use of such weapons, but not their manufacture or in-kind retaliation. Recognizing that potential adversaries were continuing with research and production in this field, the United States also continued; the Army's Chemical Warfare Service (CWS) was the responsible unit. Edgewood Arsenal in New Jersey was the only existing source, and, as another major war loomed closer, a second source was needed. The CWS dated from 1918, as did the Edgewood Arsenal.

Search for a Site

In early 1941, a national search was made for an inland location for a second arsenal. James Center, industrial agent for the Nashville, Chattanooga and Saint Louis (NC&StL) railroad in Nashville, was contacted about available land with existing rail and water transportation, and on 8 June he brought Lt. Col. Charles E. Loucks and a civilian civil engineer to Huntsville. They first visited an area just south of the Tennessee River, but found it too hilly; they then visited the flat farmland immediately southwest of the city. Following their visit, the search team said that the site adjacent to Huntsville was a perfect location for the new arsenal. James Center must be credited with initiating the activity that so drastically changed the future of Greater Huntsville.

Upon returning to Washington, Loucks filed a report to Maj. Gen. William N. Porter, Chief of CWS, recommending this location. In a few days, Porter and Col. Paul X. English from Edgewood Arsenal personally reviewed the Huntsville location. Other locations that had been surveyed and were being considered were Florence and Tuscaloosa, Alabama; Kansas City and St. Louis, Missouri; Memphis, Tennessee; Toledo, Ohio; El Dorado, Arkansas; and Charleston, West Virginia.

On 3 July 1941, the selection of the site was announced with headlines in *The Huntsville Times*. The selection was based on four major factors: immediate availability of suitable, low-cost land; availability of a lower-paid, production work force; availability of good rail and river transportation; and availability of a plentiful supply of electric power (from the TVA).

The influence of John J. Sparkman, then a U.S. Representative from Huntsville, must be noted. In 1938, Sparkman had introduced a bill directing the Secretary of War to survey this area for a possible arsenal. Following that, Sparkman and Alabama Senators J. Thomas (Tom) Heflin and J. Lister Hill had worked for months with federal officials to convince them that the land and labor market of this area were ideal for new federal facilities.

The Site and Construction

The Huntsville site included 32,244 acres of land just southwest of the city, buffered on the south by 6,400 acres owned by the TVA along the Tennessee River. Existing railroads bordered the north and east edges. The primary land was acquired through condemnation by the Federal Government, and land-use agreements were initially made with the Tennessee Valley Authority for an additional 1,200 acres along the Tennessee River. Most of the land was relatively flat, with an average elevation of near 680 feet. Unusable land within the primary area included about 10,000 acres of swamps and some 2,800 acres around Ward and Madkin Mountains (both located near the top of the site, with elevations of 905 and 1,250 feet, respectively).

Displaced were about 6,000 men, women, and children; between 70 and 75 percent Black; comprising up to 1,000 families; and occupying about 550 dwellings. Some of the families were tenant farmers, but many, both Black and White, were landowners who had worked the fertile soil of the region for decades. Altogether, 81 tracts of farmland were bought. Farmers were allowed to continue using their land until all of the crops were harvested. The Alabama Relocation Corporation was formed to assist families with finding new homes. There were employment opportunities in the construction of arsenal facilities.

Although there were no towns or villages in the 64 square miles of acquired land, there were about 40 miles of unpaved roads connecting many named communities (Pond Beat, Mullins Flat, Union Hill, Elko, Cave Hill, Hickory Grove, Horton's Ford, Bettle Slash, Cedar Grove, Silver Hill, and Center Grove). Patton and Rideout were main north-south roads, Martin Road ran east-west across the middle, and Shield Road connected to Buxton Road (earlier called the Triana-Whitesburg Road) running across the bottom near the river. Near the north end, Goss Road ran east-west between Ward and Madkin Mountains from Patton Road to Rideout Road. Dodd Road ran north-south in the center part from Martin Road to the junction of Shield and Buxton Roads. There were 12 churches, 5 schools, 2 lodges and several other meeting buildings, and at least 46 small cemeteries on the land.

Initially called the Siebert [sic] Arsenal Project, this land was designated for a complex to manufacture and store poison gas, smoke generators, and incendiary materiel. It was found that the name Siebert was intended for a CWS training post near Gadsden, Alabama; thus, the name was changed to Huntsville Arsenal. (Maj. Gen. William L. Sibert, a native of Gadsden, had served as the first chief of the Chemical Warfare Service in 1918.)

The Army Ordnance Department had begun plans for extending its Ammunition Division. When the Chemical Warfare Service decided on the Huntsville area for its new arsenal, it was recognized that an ordnance plant in the same general area would be very beneficial. The Chief of Ordnance, Maj. Gen Charles M. Wesson, sent Major Myron Leedy to examine potential sites. Leedy was accompanied by Major Carroll Hudson, who would later command the new plant. Leedy recommended an area about 10 miles south of Huntsville and adjacent to the southeast corner of the Huntsville Arsenal land. This was a rolling, rural terrain, wholly agricultural in nature. There were no interior paved roads, but the NC&StL Railroad had a spur track along part of the eastern border and a main Southern Railroad track was near the top of the area.

On 8 July 1941, the War Department announced that an ordnance facility, designated Redstone Ordnance Plant, would be built on a 4,000-acre tract southeast of, and adjacent to, the chemical munitions arsenal. Huntsville Arsenal would function as a "works" facility, producing basic materials, while the Redstone Ordnance Plant would operate as a "plant" (commonly called LAP – largely assembly-and-pack) making finished munitions. Later, there would be another related operation: Gulf Chemical Warfare Depot.

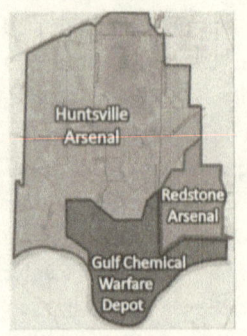

Redstone Land Areas

There has long been a question as to the origin of the name "Redstone." In 1955, the Army conducted a study concerning the origin. It turned up a Major H. Sachs who said that he was in the Ordnance Department when the Arsenal was being planned, and was asked to give the facility a name. To find more about the location, he consulted Lt. Col. Jack A. Goodwin, who earlier, as a Captain, had led a Huntsville-based CCC Camp that had worked on roads in the area. Goodwin told him, "It was beautiful country with red rocks predominating and was sometimes called the Redstone Area." Sachs then submitted the name Redstone Ordnance Works.

Before the end of July, the War Department awarded a cost-plus-fixed-fee contract to Whitman, Requardt, and Smith of Baltimore, Maryland, for architectural and engineering services for designing the Huntsville Arsenal and Redstone Ordnance Plant facilities. In September, as designs were completed, cost-plus-fixed-fee contracts were awarded to C. G. Kershaw Contracting Company of Birmingham, Alabama; Engineers Limited of San Francisco, California; and the Walter Butler Company of St. Paul, Minnesota, for the construction of buildings.

As of the previous January, it had been directed that, wherever possible, Ordnance Department facilities be of temporary, rather than permanent, construction. The importance of the new facilities in Huntsville, as well as the priority granted to the CWS, is shown by

Arsenal's First Building

the extensive use of steel – a critical material – in designing the facilities for Huntsville Arsenal. While many of the buildings were of temporary, wood-frame construction, the main production buildings were designed as steel-frame structures supporting steel roof-decking and truss-work. Economical materials, however, were in corrugated asbestos metal cladding for exterior walls and roofs, and in the substitution of hollow tile masonry for brick.

By early October, about 3,500 construction workers were involved at the facilities; by the end of 1941, this had swelled to near 12,000. A total of 1,016 buildings and structures were eventually built for the Huntsville Arsenal, Redstone Ordnance Plant, and Gulf Chemical Warfare Depot. These were connected by about 66 miles of paved roads, 25 miles of gravel roads, and 75 miles of railroad tracks.

Two railcar classification yards were built, one for the Southern Railroad on the west side and one for the NC&StL Railroad on the east side; tracks connected the two yards, about seven miles apart. Eventually the arsenal-owned railroad system would have two diesel locomotives, 10 hopper cars, a locomotive crane, and a maintenance car; at its peak, about 2,000 cars per month would move along the tracks.

While workers came from all over the nation to build the plants, most were from Madison and other North Alabama counties. All available housing within travelling distance was quickly taken. To accommodate traffic from nearby towns, Madison County built its first four-lane road, Jordan Lane connecting Madison Pike and Athens Pike. The Government brought in trailers and set up trailer parks; large ones were on West Holmes Street and on Triana Road near the Merrimack Mill.

After the munitions plants opened, there continued to be a major shortage of available housing for the regular employees. The Government funded several housing projects, both in and outside the operating area. One of the largest was

Redstone Park Housing

the 300-unit Redstone Park; located in Farley, Alabama, near the Redstone Ordnance Plant and opened in late 1942. A spur track was built out to Farley and in January 1943, three old passenger railroad coaches were added for transporting Redstone Park employees. In June, a 300-unit housing project named Binford Court was opened close to the city; this was for Black employees of the arsenals. For employees from throughout the area, there was good bus service to the arsenals from Decatur, Athens, Guntersville, and Arab, Alabama.

The Army Corps of Engineers had the overall responsibility for construction of the two arsenals and the depot. When the Corps of Engineers left in mid-1943, it turned over the largest chemical warfare manufacturing operation in the world. By the end of World War II, the cost of all construction, including land, totaled $63,431,925 – about $857 million in today's dollars.

HUNTSVILLE ARSENAL

Colonel Rollo C. Ditto arrived as the first commanding officer of Huntsville Arsenal on 4 August 1941; in early October, he was promoted to Brigadier General. Ditto had served in the Army since 1907 – first in enlisted status and then as an officer in the CWS starting in 1922. His prior assignments had included Assistant Commander of the Chemical Warfare School at Edgewood Arsenal. Initially, the administrative activities were conducted at the Huntsville National Guard Armory, the gymnasium of

Rollo Ditto

the Huntsville High School, and the Virginia McCormick YMCA in West Huntsville, which had immediately been offered for use.

The recruitment and hiring of operating personnel involved a major initial effort. A cadre of specialists was brought from Edgewood Arsenal, but many professionals and production workers were needed. Recent college chemistry graduate John L. McDaniel was one of those hired in February 1942, with a daily salary of $6.24; when McDaniel retired from the Government almost three decades later, he was the highest ranking civil service employee in the Huntsville area.

The first construction centered on roads and railroad tracks; buildings were started in September. Huntsville Arsenal's first production facility was activated in March 1942, just 7 months after Ditto's arrival. While most of the buildings

Squirrel Hill Buildings

other than the plants were of temporary construction, two brick-and-mortar facilities – a hospital (for use by both arsenals) and a headquarters building – were built on what was commonly called Squirrel Hill south of Goss Road. Also on Goss Road were military quarters and barracks. Elsewhere were ancillary facilities for sewage treatment, clothing renovation, and other service functions.

Facilities and Products

The production plants were in three distinct areas: Plants Area 1, 2, and 3. Areas 1 and 2 were essentially duplicates, based on the concept of one surviving after a major attack. Plants Area 1 was located at the northeast corner of Rideout and Martin Roads, while Plants Area 2 was about a mile away at the southwest corner of Patton and Martin Roads. These were primarily plants for chemical gas munitions. For community safety, the plants were located near the center of Huntsville Arsenal. Plant Area 3 was on the west side of Patton Road about a mile north of Area 2. Smoke munitions filling, incendiaries, and non-lethal tear-gas (adansite and chloroacetophenone) munitions, but no actual chemicals, were made at Area 3; the buildings were spread apart because of the explosive nature of their products. Each of the three areas had its own administrative units for engineering, personnel, property, storage, and transportation.

The first facilities were six plants for producing mustard gas; these facilities were mainly duplications of those at Edgewood. Engineers E. C. Thompson and L. W. Green from Edgewood were responsible for the technical design.

Commonly called "H" or "HM", mustard gas was an organic compound of chlorine, carbon, hydrogen, and sulphur. Mustard plants were complex

Chlorine Plant

facilities; each had a sulphur-monochloride building, an ethylene-generator building, a mustard-reactor building, two gas holders, a tail-gas scrubber building, and a disposal reactor, all functioning with sequential processing. The end product was heavy, viscous, liquid slurry. Two chlorine plants were constructed to support the mustard-gas

plants. Altogether, the plants were designed to produce between 140 and 240 tons of HM slurry daily, all going into storage. Full production began in April 1942 and continued until March 1944, when both plants were placed on standby. Two mustard-gas filling plants were completed in early 1942, mainly loading 105-mm and 155-mm artillery shells.

Mustard Gas Shells

Plants for making the poison gas lewisite were also built at Huntsville Arsenal. Four of these operated between May and October, 1943. Two other lewisite plants were essentially completed but never put into operation. Two support facilities, a thionyl chloride plant and an arsenic trichloride plant, produced products necessary for the manufacture of lewisite. For a year starting in early 1944, Huntsville Arsenal operated a plant making the poison gas phosgene. An adjacent phosgene-filling facility produced 500-pound and 1,000-pound bombs.

Between May 1942 and August 1945, a white phosphorus-filling plant operated on the west-central area of Huntsville Arsenal, producing artillery and mortar shells, grenades, and similar munitions for incineration and smoke generation. Huntsville Arsenal also produced two types of tear-gas grenades, four types of incendiary-oil munitions, and colored smoke for grenades and

Incendiary Production

canisters. A facility for producing iron carbonyl was transferred from Edgewood Arsenal; as a vapor, iron carbonyl is a highly toxic agent.

In 1943, a 5,000-ft airstrip and several small supporting buildings were built on the northern portion of the Arsenal, primarily to assist the Army Air Forces in testing incendiary devices in preparation for firebombing Japanese cities. Six Air Forces personnel and two planes – a B-26 and an L-20 – were stationed at the Huntsville Arsenal Airstrip (later named Redstone Airfield).

More than eight million pounds of munitions were dropped on Huntsville Arsenal test areas during the war. One test area, on the west side of the Arsenal, was called "Little Tokyo" and had three streets, about 50 small wooden houses and buildings, and a 200-foot structure for proof-testing large bombs; these were totally obliterated by late 1944. There was also a thick, 500-foot-square concrete mat for testing penetration capabilities of dropped bombs.

Organization and Administration

In August 1941, the initial organization was simple. It consisted of an Engineering Division as its major element with Lt. Col. W. J. Ungetheum as Chief, and a number of specialized units such as Civilian Personnel, Adjutant, Procurement, and Signal, all reporting directly to the Commanding Officer. In January 1942, transition to the production phase began; this eventually led to an Operations Division under Col. L. W. Greene; an Engineering Service Division under Lt. Col. R. A. Phelps; a Transportation Division under Maj. Roy A. Burt; and a Fiscal Division under Capt. I. M. Breller. Later in 1942, a reorganization under the "task force" principal gave a level of self-sufficiency to major operating elements, with each having an

REDSTONE NEEDS --
LABORERS *at* _once!_

Patriotic Citizens of Huntsville, Madison County and Alabama. Not Presently Employed In War Work— Are Called Upon To Apply For Work At REDSTONE ARSENAL At Once! In Order To Avoid Delay In Final Victory . . . And Needless Loss Of American Lives . . .

Advertisement for Personnel

engineering, personnel, property, transportation, and storage section.

By May 1944, Huntsville Arsenal's need for production, maintenance, and administrative personnel had accelerated greatly. That month civilian employment at the arsenal reached a WWII peak of about 6,700, divided 63 percent male (52 percent White and 11 percent Black) and 37 percent female (26 percent White and 11 percent Black).

During the war years, 27 million items of chemical munitions were produced at Huntsville Arsenal. The coveted Army-Navy "E" Award was received four times.

Through the years, the following were Huntsville Arsenal's commanding officers:

Col. Rollo C. Ditto (Brig. Gen. as of October 1941) – Aug. 1941 - May 1943
Col. Geoffrey Marshall – May 1943 - Aug. 1945
Col. E. C. Wallington – Aug. 1945 - July 1946
Col. Sterling E. Whitesides, Jr. – July 1946 - Dec. 1947
Col. James M. McMillin – Dec. 1947 - Feb. 1949
Lt. Col. Allen H. Williams – Feb. 1949 - June 1949

GULF CHEMICAL WARFARE DEPOT

Approximately 7,700 acres in the southern portion of the arsenal along the Tennessee River were initially intended as a depot site. Initially this was the Storage Division of Huntsville Arsenal. 1st Lt. William C. Behrenberg, who had early worked for Colonel Ditto, was brought in as the Division Chief. In March 1942, the depot was activated as a separate installation named the Huntsville Chemical Warfare Depot. General Ditto (and successive Huntsville Arsenal Commanders) also served as the Depot Commander, with now Capt. Behrenberg as Executive Officer. The Depot had no office buildings; the headquarters were in the James Cooper House, an old mansion originally built in 1818, and one of

Munitions Warehouse

the few houses left standing when the arsenal area was cleared.

In August 1942, the name was changed to the Gulf Chemical Warfare Depot. The depot received, stored, and shipped chemical warfare materiel, including bulk chemicals, decontaminating apparatus, and protective materials. It covered nearly twelve square miles and was divided into three principal areas: the toxic gas yard, the munitions branch, and the warehouse area. All were in operation by October 1942.

By early 1943, the Gulf Chemical Warfare Depot consisted of 7 warehouses, 370 igloos, 55 above-ground magazines, several outdoor storage areas, 12 miles of railroad track, and dock facilities on the Tennessee River.

REDSTONE ORDNANCE PLANT

On 8 July 1941, the War Department had announced that an ordnance facility, designated Redstone Ordnance Plant, would be built nearby to the

Huntsville Arsenal. Major Carroll D. Hudson was named Commanding Officer of the Redstone Ordnance Plant on 25 September 1941. Hudson had served as a Navy enlisted man during World War I, then graduated in Mechanical Engineering from Stanford University, and was commissioned into the Army in 1926. For three years, he was an officer in the Civilian Conservation Corps, and then was recalled into the Army in 1940. Before being assigned to Huntsville, he had served in the Ammunition Division, Office of the Chief of Ordnance, and had made several visits to Huntsville while the site

Carroll Hudson

was being selected. His service at Redstone Arsenal spans from 1941 to 1952. Hudson was promoted to Lieutenant Colonel in 1942, and then to Colonel in 1944.

On 6 October, Hudson arrived in Huntsville and established a temporary office at Huntsville Arsenal. The Plant requirements, as established by the Chief of Ordnance, were for loading and assembly of 75-mm chemical shells and burster charges for this ammunition. Working with the Area Engineer of the Corps of Engineers and representatives of the architects (Whitman, Requardt and Smith) and contractors (Kershaw, Butler, Engineers, Ltd) – both of whom were building Huntsville Arsenal – Hudson established construction plans for the new plant.

Major Hudson officially initiated construction of the Redstone Ordnance Plant on 25 October 1941. A capable engineer, Hudson was personally involved in designing and building the new plant; then throughout the years of operation, made many valuable engineering contributions. Hudson had asked the Ordnance Department for three officers and these would be augmented with about seven key civilian assistants. Only one officer was made available; 1st Lt. William H. Stevens reported on

Hudson Groundbreaking

30 October, and was quickly assigned as the Adjutant and Property Officer (somewhat later he was made Executive Officer).

Four civilians who had trained at Rock Island Arsenal were hired and took over the administrative work. A few key civilians were sent to Picatinny

Arsenal and Charleston Ordnance Depot for intensive operational instruction. It was not until the summer of 1942, however, that a full complement of officers and key civilians was obtained. In addition to 14 Ordnance officers, the initial staff included a Signal Officer, an Intelligence Officer, a Post Surgeon, and an Assistant Post Surgeon; 1st Lt. Leon P. Woodall was Chief of the Production Division.

The first building, which initially served as temporary headquarters, was completed in early November. When Pearl Harbor occurred (7 Dec. 1941), the production lines were still being built, and the total Ordnance Plant's strength was three officers and nine civilians. As the United States entered the war, construction went on a 24-hour, 7-day schedule, and the personnel recruitment greatly accelerated. Essentially all of the buildings, except storage igloos, were of temporary construction.

Facilities and Products

The Plant initially had four assembly lines; there were also supporting storage and administrative buildings, all completed before the end of 1941. Each assembly line had about 15 buildings distributed on some 25 acres. Lines No. 1 and 2, which were completed first, were similar and both used for the loading of burster tubes. The tubes were from private manufacturers. Other Ordnance plants supplied the explosive ingredients – tetryl and TNT; these were mixed to form tetrytol, a very-high powered explosive.

The mixing to form tetrytol had used what was called a pelleting and reconsolidating procedure. In early December 1941, researchers at Picatinny had developed a radically new and safer process called the melt method. Adopting this at Redstone required significant changes in parts of Lines 1 and 2;

Mortar Shell Production

consequently, although the lines were basically completed, the changes delayed the start of production. Line 1 was changed first and began production in March 1942, and Line 2 followed somewhat later.

Lines No. 3 and 4 loaded and assembled chemical ammunitions, eventually centering on 81-mm and 105-mm mortar shells. Plans for these lines were suggested by Picatinny Arsenal, but were redesigned by Major Hudson to meet local problems and conditions. Line 3 was built using Hudson's design, and Line 4 used the Picatinny design. Early efficiency testing using the same number of workers showed that Line 3 was 25 percent more productive;

thus, Line 4 was modified to the Hudson design. Line 1 started full operations in April, and Line 4 in August.

In the fall of 1942, the plant expanded to have an additional large assembly line (No. 5) for 155-mm chemical shells, soon producing up to 190,000 projectiles per month. Igloo, warehouse, and magazine areas for the finished products were also greatly expanded.

Between March 1942 and September 1945, over 42 million units of ammunition were loaded and assembled for shipment at Redstone Arsenal. Demolition blocks for the Corps of Engineers and Airborne Troops were a particular specialty; about 12 million blocks were produced and shipped.

Initial production included the 105-mm M60 white phosphorus (WP) or mustard gas (HS) shells, the 155-mm WP or HS-filled shells, M5 and M6 burster charges, the 100-pound A1 WP bomb, and the 100-pound A47A2 HS bomb. In 1943, production added M4, M8, and M10 burster charges, 115-pound M70 HS bombs, 75-mm WP M64 shells, and 105-mm M84 HC base ejection shells.

Organization and Administration

Major Hudson was promoted to the rank of Lieutenant Colonel in February 1942. The Redstone Ordnance Plant was re-designated Redstone Arsenal on 26 February 1943. At that time there were six divisions: Administration, Production, Inspection, Engineering, Service, and Safety and Security. The Redstone Ordnance Plant / Redstone Arsenal won the Army-Navy "E" Award five times during the war.

Weekly Newspaper

In June 1942, the first edition of the Redstone Ordnance Plant weekly newspaper, the *Redstone Eagle*, came off the press. Hudson had urged the publication of a plant newspaper as part of the program to maintain high morale among the civilian employees. The publication continued until September 1946.

On 7 October 1943, the Office of the Chief of Ordnance (OCO) informed Redstone Arsenal that a decision had been made to turn the operation of the Arsenal over to a subsidiary of the Continental Can Company. The Arsenal immediately began to prepare for the changeover. Hudson received orders to report for duty at the Office of the Field Director of Ammunition Plants in St. Louis. Major Leslie S. Solar was sent in and assumed command of Redstone. Preparations continued until 17 November when a letter from the OCO revealed that negotiations with Continental Can had been broken off. Major Solar was transferred, and on 7 December, Lt. Col. Hudson returned as Commanding Officer. In February 1944, Hudson was promoted to the rank of full Colonel.

From the opening of Redstone Ordnance Plant, many women were employed  for the production work. By the close of December 1942, about 40 percent of the people working on the production lines were women. The percentage of female employees at Redstone Arsenal during 1944 averaged about 54 percent; it peaked at 62 percent by September 1945. Throughout the war years, women employees, excluding assembly line workers, were provided special uniforms. In March 1944, 2nd Lt. Eleanor B. Wilson became the first Woman's Army Corps (WAC) person assigned to the Arsenal.

Women's Uniforms

During the second quarter of 1944, Redstone Arsenal was designated the official center for the screening and renovation of chemical ammunition returned from overseas and from depots in the United States. In January 1945, Redstone Arsenal started a major expansion, including extensive renovation to earlier temporarily constructed buildings. Line 3 was increased to process ammunition from overseas. Also included was a new production line (Line No. 6) to handle 81-mm chemical mortar and 105-mm chemical shells.

The road gates to Redstone Arsenal were secured and entrance required a badge or visitor permit, but large areas were not secured and were unfenced. The overall boundaries of the arsenal were only partially fenced, and security was handled by guards who often patrolled the areas on horseback. At its peak, there were about 40 horses; for feeding, corn was planted in unused farmland and hay was cut across many areas.

Civilian personnel at the beginning of each year was as follows: 1942-24; 1943-1,906; 1944-3,422; 1945-4,252, the highest ever. In addition, there was an average of about 25 military personnel each year. As the war in Europe drew down, personnel reduced to 1,049 in July 1945. When Japan surrendered on 14 August, production at Redstone Arsenal ceased three days later. Although there were no items produced or shipped during the last quarter of 1945, Redstone Arsenal expected to continue as a permanent manufacturing arsenal under the Ordnance Department. Pending announcement of future work, all unnecessary workers were put on annual leave. By the end of 1945, there were only about 600 civilian employees and military personnel remaining.

Col. Hudson departed on 15 March 1946. Col. Duncan G. McGregor served as commander from April 1946, until July 1947; during his tenure, in February, Redstone Arsenal went on a standby basis with about 225 personnel. Lt. Col. James O. Branch served from July 1947, until September 1948. Plans were underway for Redstone Arsenal to become the Army's center for rocket development, and Col. Hudson returned to guide this reactivation on 30 November 1948.

SPECIAL ACTIVITIES

During the 1941-1945 years, there were a number of special activities related to the munitions manufacturing plants that should be noted.

Arsenal Casualties

Considering the hazardous products involved at the two arsenals and the depot, it is a testimony to the safety training and practices that there were only eight fatalities in the operations during the war years: seven civilian workers and one Army officer. Easter Posey of Hazel Green was killed on 21 April 1942, in an accidental explosion of an incendiary bomb at Huntsville Arsenal; she is recognized as the first American woman killed in the line of duty during WWII.

In addition to those killed in the munitions plants, three Army Air Forces personnel were killed in a crash of a Martin Marauder B-26 bomber while flight-testing incendiary bombs. Because of wartime secrecy, no details of these casualties were released to the public at that time.

On 27 June 1944, the Martin Marauder bomber, stationed at the Huntsville Arsenal Airstrip, took off with bombs that were to be drop-tested on Little Tokyo, the small test-target village on the western side of the Arsenal. Climbing north, at about 3,000-feet altitude an engine problem developed. The pilot radioed that he would attempt an emergency landing on the Huntsville-Athens Highway (present U.S. 72), but was unable to properly maneuver because of the load weight and engine loss.

The pilot turned back and dropped one of the 500-pound bombs into a vacant field near the present HudsonAlpha Institute. He again attempted a landing, but crashed in a cotton field just north of the highway near the present Memorial Gardens Cemetery. The remaining incendiary bombs then exploded. Killed were 1st Lt. Emmett J. Hale, the pilot; 2nd Lt. Jerome Loeffler, the bombardier; and Tech. Sgt. Antone Valim, the onboard engineer.

It is noted that the B-26 aircraft was often called a "Widowmaker," due to the high rate of accidents that occurred during takeoff and landing. Documents show that the Marauder had to be flown at specific airspeeds, particularly when one engine was out.

Personnel and Security

Since its beginning, the United States Government has had equipment and documents that were "classified"; that is, closely held or secret. However, until just before WWII, there were essentially no protocols for determining such classification. On 22 March 1941, President Franklin D. Roosevelt issued Executive Order (EO) 8381 for this. It provided for three classification levels: Secret, Confidential, and Restricted (no Top Secret), but was silent on determining access to such materials. Access determination was up to the guardians of the material and the personnel supervisors.

The advent of munitions production first brought security in employment to Huntsville. Very few of the potential employees had ever been exposed to secrecy, and no formal process existed for vetting them for their eligibility. Essentially all of the top leaders were military officers, and only a few civilian employees had a "need to know" concerning details of the equipment, manufacturing processes, or testing results – information that might be classified. However, there could be spies or saboteurs among the potential employees.

Most of the civilian employees were natives of the region – mainly "good old boys" and hard-working women who had always been the backbone of Southern citizenry. Thus, the ordinary personnel managers could easily make a basic determination of their "security" qualification. Also, all three of the local munitions organizations had an intelligence officer, qualified and dedicated to identifying inappropriate employees, as well as handling inside police matters. Consequently, no records of wartime security problems have been found for Huntsville Arsenal, Redstone Arsenal, or the Gulf Chemical Warfare Depot.

Special Training

Recognizing that there were insufficient personnel with advanced education to fill America's wartime needs, Congress funded a huge program called Engineering, Science, and Management War Training (ESMWT). Conducted by the U.S. Office of Education, colleges and universities throughout the nation gave off-campus courses for preparing persons to fill this urgent need. From October 1940 through June 1945, 227 colleges and universities provided about 68,000 courses for close to 1,800,000 students. Under the assumption that almost anyone might benefit, students were admitted to courses regardless of their prior education.

Through ESMWT, both Alabama Polytechnic Institute (now Auburn University) and the University of Alabama (Tuscaloosa) offered college-level courses in Huntsville. Taught by local professionals and a few visiting faculty members, and with books and materials supplied by the schools, the tuition-free courses were given nights at West Huntsville High School (predecessor of Butler High School).

Course areas included chemistry, laboratory practices, engineering and architectural drafting, mathematics, quality control, accounting, and other subjects. The first classes began in September 1941. Publicity particularly urged women to take advantage of this training to replace men in the work place who were needed for military service.

Prisoner-of-War Camp

After the Allies retook North Africa until V-E day in May 1945, about 240,000 German soldiers were sent to America for internment at some 500 prisoner-of-war (POW) camps. In Alabama, there were 4 primary camps, including one at Fort McClellan near Anniston, and 16 satellite camps; together, there was a total of about 16,000 POWs. Operations within the camps conformed to the 1929 Geneva Conventions, with adequate food, recreation, and medical treatment.

During early 1944, the Army Corps of Engineers (CoE) built a camp on  Huntsville Arsenal as a satellite to Fort McClellan. The camp was initially designed to accommodate 250 POWs, then increased to 655 by mid-year. The CoE constructed the original camp for 250 prisoners, but the remainder of the camp was completed by POW labor. The camp was located east of Dodd Road near the center of the Arsenal; it was a rectangular-fenced area with three main buildings and about 100 six-man, tent-like structures arranged in rows.

Typical POW Camp

There is essentially no official documentation concerning the camp operations; most of what is known came from scant references in reports and from interviewing one of the former prisoners, Karl Spitzenpfeil, who later visited Huntsville.

The POWs were required to work, but received a pay of 80 cents per day, with which they could buy items from the post commissary. Assignments included the post motor pool (across Dodd Road from the camp); the carpenter and machine shops; the sawmill, limestone quarry, and rock crusher; and in mosquito control (a constant problem throughout summers). A few worked at the officers club and the hospital.

The camp had a "college," through which the prisoners were given courses aimed at "re-education," emphasizing America's history, society, and political structure. They could also take correspondence courses from several universities. On their birthday, each prisoner was given two cases of beer. In general, life in the camp was such that – although there was minimal guarding – prisoners almost never attempted to escape.

POST-WAR DRAWDOWN

After V-E (Victory in Europe) Day (8 May 1945) shutdown of production at Huntsville and Redstone Arsenals began. Following V-J (Victory over Japan) Day (2 September 1945), most of the operating buildings and production lines were placed in standby condition, and large quantities of finished products were placed in long-term storage.

In August 1946, the word "Warfare" was dropped from the Depot's title, and the following January the Gulf Chemical Warfare Depot was abolished as a separate entity and its functions transferred to the Huntsville Arsenal. Although not planned, building modification also began in 1946; there was an extremely severe hail storm that damaged 4,000 windows and the roofs of 700 buildings. Following these repairs, most other buildings and structures at the Huntsville Arsenal were prepared for extended standby status.

At Redstone Arsenal, the reduction in force and readjustment to a standby activity was completed by early 1946; Colonel Carroll Hudson remained the commander until 15 March. The standby organization involved several officers and about 250 civilian employees. Redstone Arsenal was still an official unit under the Ordnance Department at the Pentagon, and major political and commercial efforts were made in searching for government or business tenants for space at both Redstone and Huntsville Arsenals.

In 1973, the Army made the decision to liquidate its railroad assets on Redstone Arsenal. The railway equipment was turned over to other government agencies. By 1976, approximately 58 miles of railroad tracks had been removed; efforts to sell the remaining track continued through 2000,

There were several attempts to commercialize plants originally associated with chemical munitions manufacturing. These included the Solvay Process Division of Allied Chemical and Dye Corporation, and Stauffer Chemical Company, both leased chlorine manufacturing plants. General Aniline and Film

Corporation (GAF), the largest manufacturer of roofing in America, leased the plant making iron carbonyl and continues with this operation on Redstone Arsenal today. Two other commercialization efforts are described:

Keller Automobile Plant

Keller Motors was incorporated on 25 November 1947. The objective was to develop and produce small, inexpensive automobiles and make Huntsville the "Detroit of the South." This firm came into existence after an attempt to form a similar company in San Diego to produce a very small vehicle, called Bobbi-Kar, collapsed for legal reasons.

Central offices of Keller Motors were in downtown Huntsville, and Buildings 471 and 481 (later numbered 4471 and 4481) on Huntsville Arsenal were placed

Keller Assembly Line on Arsenal

under a 15-year lease for automobile development and production. George D. Keller, formerly vice president of sales for Studebaker, was the president, and Hubert Mitchell, a successful entrepreneur from Hartselle, Alabama, provided initial financing. John Liefeld, an experienced automotive engineer who had designed the Bobbi-Kar, led the technical efforts, and Mitchell began the sale of dealer franchises. Some $450,000 was quickly raised and about 65 designers, engineers, and other production personnel were hired by Liefeld. Huntsville native Henry L. Hilson was a lead production engineer.

By early 1949, prototypes for convertible roadsters and 'woodie' station wagons were ready. A $5 million stock issue was approved by the SEC, and half of the stock was quickly sold. On 4 October 1949, a celebration was held in New York; the next morning, 52-year-old Keller was found dead in his hotel bed. An acceptable leader could not be found, and the Keller Motor firm went into history. A total of only 18 Keller convertibles and station wagons had been built. The three remaining Keller vehicles are now valuable items sought by antique automobile collectors.

Keller Station Wagon

DDT Manufacturing

In 1947, Benton H. Wilcoxon, a California expert in chemical manufacturing, came to Huntsville and formed Calabama Chemical Company. The firm leased land and facilities on Redstone Arsenal and began manufacturing the insecticide DDT (dichlorodiphenyltrichloroethane).

The manufacturing process resulted in significant amounts of DDT-laden wastewater being released into a reservoir that drained into the Huntsville Spring Branch; this flowed into the Indian Creek and eventually into the Tennessee River near Triana;. In 1948, the Calabama operation was acquired by the Olin Mathieson Chemical Company (later known as Olin Corporation), and this firm continued the production of DDT. The production was about 12.500 tons per year, and by 1969, an estimated 4,000 tons of contaminated water had been deposited. When investigated, the water in Huntsville Spring Branch had DDT as high as 0.3 parts per million. This resulted in major fish kills, particularly in Indian Creek and the Tennessee River in the vicinity of Triana.

Olin Mathieson DDT Plant

Production of DDT was stopped and the plant was demolished in 1973. From July 1979 to August 1982, the U.S. Army conducted an extensive DDT abatement program. DDT wastes, including highly-contaminated soil and sediment, were excavated from the DDT manufacturing areas, the DDT drainage ditch, lagoon, and former DDT disposal sites. The residents of Triana, along with the Justice Department, filed lawsuits against Olin Corporation. In 1982, an out-of-court settlement resulted in Olin pursuing clean-up operations; these were basically completed by 1987, and were called fully successful in 1995.

Selected Bibliography / Additional Information for Chapter II

Archives Section, Huntsville-Madison County Public Library (HMCPL); the HMCPL Digital Archives was a major source of text information and photo-graphs on the Greater Huntsville cotton mills.
http://digitalarchives.hmcpl.org/cdm/

"Arsenal Given Name by Major H. Sachs," *Redstone Rocket*, 21 June 1955

Baker, Michael E., and Kaylene Hughes, "Redstone Arsenal Complex Chronology, Part I: The Pre- Missile Era (1941-1949)"; U.S. Army Missile Command, December 1988

Brophy, Leo P., and George J.B. Fisher, *The Chemical Warfare Service: Organizing for War*, Vol. 6, Pt. 7.1, U.S. Army in World War II, Department of the Army, 1959

Burkey, Martin, "39,000 Acre Site Grew to Shape History," *Huntsville Times*, 9 June 1991

Cagle, Mary T., "History of Redstone Arsenal," Public Information Office, Redstone Arsenal, undated, about 1955

"Chemical Warfare," *Wikipedia*; http://en.wikipedia.org/wiki/Chemical_warfare

"CCC Camps in Alabama," Civilian Conservation Corps Legacy; http://w33.ccclegacy.org/CCC_Camps_Alabama.html

Cronenberg, Allen, *Forth to the Mighty Conflict: Alabama and World War II*; U. Ala. Press, 1995. ISBN 97-808-17307370

"Easter Posey," *Wikipedia*; https://en.wikipedia.org/wiki/Easter_Posey

"Dallas Mill," *Wikipedia*, http://en.wikipedia.org/wiki/Dallas_Mill

"Edgewood Arsenal, History," Encyclopedia of Historic Forts; http://www.fortwiki.com/Edgewood_Arsenal

"Engineering, Science, and Management War Training," *Wikipedia*; http://en.wikipedia.org/wiki/Engineering,_Science,_and_Management_War_Training

" 'Fred's Folly' Becomes Important Test Apparatus," *Redstone Rocket*, 5 March 1947

Hughes, Kaylene, "Two Arsenals of Democracy," *Huntsville Historical Quarterly*, 17 (Winter-Spring 1991)

Hughes, Kaylene, "Redstone's WWII Female 'Production Soldiers'," U.S. Army, Historical Information, http://history.redstone.army.mil/women.html

Hughes, Kaylene, "History of Redstone Arsenal Airfield, U.S. Army, Historical Information. http://history.redstone.army.mil/ihist-airfield.html

Hughes, Kaylene, "German POWs housed in camp here in WWII," *Redstone Rocket*, 17 April 2013

"Huntsville Alabama Sesquicentennial Commemorative Album," Benson Printing Company, 1955

Joiner, Helen Brents, "The Redstone Arsenal Complex in the Pre-Missile Era: A History of Huntsville Arsenal, Gulf Chemical Warfare Depot, and Redstone Arsenal, 1941-1949," Historical Division, U.S. Army Missile Command, 1966; http://history.redstone.army.mil/ihist-chem.html#ii

"Keller (Automobile)," *Wikipedia*," http://en.wikipedia.org/wiki/Keller_(automobile)

"Lincoln Mill and Mill Village Historic District," *Wikipadia*, http://en.wikipedia.org/wiki/Lincoln_Mill_and_Mill_Village_Historic_District

"Lowe Mill," *Wikipedia*, http://en.wikipedia.org/wiki/Lowe_Mill

McDaniel, John L., "The Memoirs of John L. McDaniel," *Huntsville Historical Quarterly*, 17 (Winter-Spring 1991)

"Merrimack Mill Village Historic District," *Wikipedia*, http://en.wikipedia.org/wiki/Merrimack_Mill_Village_Historic_District

"Prisoner-of-War Camp," *Wikipedia*, http://en.wikipedia.org/wiki/Prisoner-of-war_camp#Allied_camps

"Redstone Arsenal," *Wikipedia*, https://en.wikipedia.org/wiki/Redstone_Arsenal

"Redstone Arsenal Celebrates 50 Years," (contains articles on Huntsville during WWII), *Huntsville Times*, June 1991

"Redstone Arsenal, Huntsville, Alabama," Environmental Protection Agency, National Priorities List, http://www.epa.gov/superfund/sites/npl/nar1396.htm

Reynolds, Barbara, "Town's Population Poisoned by DDT," *Chicago Tribune* (Sec. 3, P. 4), 3 February 1980, http://archives.chicagotribune.com/1980/02/03/page/38/article/health-livelihood-ruined

Ryan, Patrick, "Northern Dollars for Huntsville Spindles," Huntsville Planning Department, 1983

Smith, R. H.; "History of the Boll Weevil in Alabama, 1910-2007"; Bulletin 670, Alabama Agricultural Experiment Station, 2007; http://www.aaes.auburn.edu/comm/pubs/bulletins/bull670.pdf

Snow, Whitney Adrienne, "Cotton Mill City: The Huntsville Textile Industry, 1880-1989; *Alabama Review*; Vol. 63, Issue 4, (Oct. 2010)

"Textile Workers Strike (1934)," *Wikipedia*,
http://en.wikipedia.org/wiki/Textile_workers_strike_(1934)

"The Great Gurley Gas and Oil Strike of 1930,"
http://www.contactez.net/gurleyalabama/GasandOilStrike1930.html

"The Street Car," Merrimack Mill Village Newsletter; May/June, 2010;
http://www.merrimackmillvillage.com/wp-content/uploads.pdf

U.S. Census of Agriculture, published every five years by the U.S. Department
of Agriculture's National Agricultural Statistics Service; (1)
http://agcensus.mannlib.cornell.edu/AgCensus/homepage.do;jsessionid=E763E7
5D11EAB5599E4159EF707035AC ; (2) http://www.agcensus.usda.gov/

Ward, Michael D., *Shaping History: The University of Alabama Huntsville
Foundation*; AuthorHouse Publisher, 2008. ISBN: 978-1-4389-4467-8

Wells, Charles R., "B-26 Crashes Near Huntsville," *Old Huntsville Magazine*,
Issue 208 (June 2010)

Chapter III

ROCKET CITY ERA

Documentation on the use of rockets for military applications dates back to 13th-century China. America's national anthem, *The Star Spangled Banner* composed in 1814, includes reference to the rocket's red glare. Through the early1900s, rockets used solid propellants, and while size *per se* was not the limiting factor, they had a short range (flight distance). To obtain greater range, especially altitude, the technology of liquid propellants came into being.

In June 1949, Redstone Arsenal was officially designated as the U.S. Army's Ordnance Rocket Center, with a mission including research and development of guided rockets. This had essentially no relation to the previous munitions manufacturing in these government facilities, but was the most significant turning point of Greater Huntsville's technological evolution. In 1949, Huntsville had a population of about 16,000 – 13th in size in Alabama – and Madison County had some 73,000 residents, most engaged in agriculture.

In this book, what is called the Rocket City Era is the period starting in mid-1949 and continuing to 1970. The name Rocket City comes from the recognition, in the free world, of accomplishments in this emerging field of rocketry in Greater Huntsville, its governmental agencies, and its high-technology industries. There is no question that within this Era the firm foundation of Greater Huntsville's position as one of America's leading technical centers was formed; therefore, this chapter is the longest and most important in examining the technological evolution of this era.

During the first decade of this era (the 1950s), activities centered on the Army's missile developments. Imbedded in this, mainly without specific official approval, was America's embryonic space program. Local high-technology industry and higher education evolved. During the second decade (the 1960s), major attention in Greater Huntsville was given to the opening of NASA's George C. Marshall Space Flight Center and preparations for lunar exploration. Also in this decade there were significant Army activities, including the opening of the Army Missile Command and the start of Ballistic Missile Defense.

This chapter is divided into four segments: Rocketry Foundations, Army Activities, NASA Activities, and Industries and Other Activities. These are not necessarily time-sequenced, and overlap in time as they occurred.

ROCKETRY FOUNDATIONS

Modern rocketry by no means started in Greater Huntsville. To understand and appreciate the Rocket City Era requires going back in time to examine the prior rocket developments elsewhere, particularly in Germany and at a few facilities in America.

ROCKETRY IN GERMANY

Wernher von Braun (1912-1977) was born in Wirsitz, Posen, at that time a part of Prussian Germany. His father, a wealthy Baron, was a national government official, and Wernher was raised in an aristocratic, cultured atmosphere. In 1932, he earned a bachelor's degree in mechanical engineering from the Technical University at Charlottenburg in Berlin. Concurrently with his studies, he had participated with several other young men in a *raketenwesen* (rocketry) group, building and launching liquid-fueled rockets, some to altitudes up to 360 meters (1,200 feet).

The rocketry group was advised by Hermann Oberth, a school teacher visiting from Rumania who had written a widely read book – but unrecognized by the German scientific community – titled *Wege zur Raumschiffahrt (Road to Space Travel)*. Oberth and his book greatly influenced von Braun.

Upon learning of their success with rockets, the Ordnance Department of the German *Heer* (Army) put von Braun and the group under contract to do rocket research and development. Working at the Kummersdorf Army Proving Grounds not far from Berlin, they participated in a rocket project called *Aggregate*. Built in 1933, and designated A-1, their first rocket is often referred to as the grandfather of modern rockets. The A-1 used an engine designed by Arthur Rudolph, with a pressure-fed propellant system burning alcohol and liquid oxygen. (Although not a member

Werhner von Braun (R) with A-1

of the original Berlin team, Rudolph became a close friend of von Braun and they worked together throughout their careers.) Rockets developed at Kummersdorf were flight tested from Borkum Island in the Baltic Sea. In 1934, A-2 rockets reached an altitude of almost 2.5 km (1.5 mi).

While working at Kummersdorf, von Braun was admitted to the Friedrich-Wilhelms University of Berlin to pursue a doctorate in physics. His professors there included Max von Laue, Walter Nernst, and Erwin Schroedinger, all Nobel laureates. He was awarded the doctorate in 1934; his dissertation was titled "Constructive, Theoretical, and Experimental Contributions to the Problem of the Liquid-Fueled Rocket." The document was immediately classified as Secret by the Army and not published until years later; it was entered on the university's records with the cryptic title "About Combustion Tests."

As the German rocket development continued, a larger, more isolated space was needed. In 1937, a joint *Heer* and *Luftwaffe* (Army and Air Force) center was established at Peenemünde, a peninsula on the northern end of Usedom Island along the Baltic seacoast. The A-4 is the best known of the Aggregate series; specified in 1937, the missile would have a range of about 175 km (110 mi) and carry a payload of 1,000 kg (2,200 lb).

As finally developed, the A-4 used a 75% ethanol/water mixture for fuel and liquid oxygen for the oxidizer. At launch, the A-4 propelled itself vertically for 65 seconds, reaching an altitude of up to 80 km (50 mi) at shut-off; then a program motor controlled the pitch to a specified angle, and it then continued on a parabolic ballistic trajectory. The first models used an analog computer to adjust the azimuth, and the flight range was controlled by engine cut-off

A-4 (V-2) at Peenemünde

timing and integrating accelerometers; some later versions made use of radio control, particularly for engine cut-off. System testing showed a typical range of 320 km (200 mi).

While the A-4 was being designed and tested, the development team included a large number of scientists and engineers; Ernst Stuhlinger is one who should be noted. Born in 1913, he finished his doctorate in physics at the University of Tübingen in 1936, then worked in nuclear physics at the Berlin Institute of Technology. Stuhlinger was drafted as a private into the German Army in 1941; sent to the Russian Front, he was wounded in the battle for Moscow. Upon recovery, he was sent to Stalingrad as part of the force fighting the Russians trying to retake the city.

One of the few members of his unit to survive at Stalingrad, Stuhlinger received orders to return to Germany for a more appropriate assignment. Without means of transportation, he walked some 2,500 km (1,500 miles) across Ukraine and Poland back to German territory during six months in the winter of 1942-1943. In April 1943, he was finally taken to Peenemünde where he worked on guidance systems for the remainder of the war. Stuhlinger stayed with von Braun all of his career, and was possibly the most revered scientist in the group – and he was almost lost before beginning rocket research.

Near the end of 1943, production of A-4s started at the Mittelwerk (General Works) underground factory in central Germany; eventually some 6,000 of these missiles were built, at the cost of an estimated 12,000 forced laborer's lives. Beginning in September 1944, over 3,000 A-4s were launched as military rockets against Allied targets, resulting in 7,250 deaths. The propagation ministry called this missile the *Vergeltungswaffe 2* (Retaliation Weapon 2); hence, the popular designation V-2.

It is noted that the *Vergeltungswaffe 1* (V-1) flying bomb – forerunner of the cruise missile – was also developed at Peenemünde, but by the German Air Force; the von Braun team was not directly involved in this effort.

ROCKETRY IN AMERICA

Robert H. Goddard, a research professor at Clark University, also developed liquid-fueled rockets, the first patented in 1914. Although Goddard's rocket  work was revolutionary, he received little public recognition or financial support; his theories of spaceflight were often ridiculed in the press. As a result, he became protective of his privacy and his work. In May 1935, Goddard launched a rocket at Roswell, New Mexico, that reached an altitude of 7,500 feet (2.3 km), approximately the same as the German A-2, but the best ever for his rockets. Insofar as is known, Goddard and von Braun were unaware of each other's work. A regenerative cooling technique for the rocket nozzle was key for larger liquid-propelled rockets, and this was developed in Germany long before Goddard came up with a similar method.

Robert Goddard

During 1943, the U.S. Army learned the extent of missile weapon developments in Germany, and in September the Rocket Branch was formed in the Technical Division of the Army's Ordnance Department. Although small rockets had always been used by America's military, they in no way competed with artillery weapons. Formation of the Rocket Branch gave recognition to the importance of this technology in augmenting or even extending the existing capabilities of weaponry. Although some missile analysis was done by the initial staff of the Rocket Branch, awareness of German developments spurred the Ordnance Department to enlarge this activity and seek outside assistance.

An operation called the Ordnance Research and Development Division Sub-office (Rocket) was set up at Fort Bliss, a large Army post just north of El Paso, Texas. The small Hueco Range, on the post, would be used for limited testing. Also, contracts were awarded to the California Institute of Technology (CIT) and the General Electric Company (GE) for initiating missile research and development for the U.S. Army.

Initial Activities

Beginning in 1935, a group at CIT's Guggenheim Aeronautical Laboratory had investigated rocket propulsion as a means of conducting high-altitude research. Their work had reasonable success, so in 1943, the Ordnance Department contracted with CIT for rocket research and development under what was called the ORDCIT Project. The Jet Propulsion Laboratory (JPL) was formed in November to conduct this work; for safety, the experimental facility was moved to a remote, off-campus site. The next year, the U.S. Army negotiated with CIT to take ownership of the JPL and thereafter carried it as a part of the Ordnance Department's Technical Division, but operated by CIT under contract.

The first activity under ORDCIT was the development of a liquid-propellant (aniline-red fuming nitric acid) motor with 20,000-pound-force (90-kN) thrust – lbf, the common abbreviation for pound-force, will be used hereafter. This eventually evolved through the PRIVATE missile, which was relatively unsuccessful, to become the CORPORAL. A tactical missile with a range of 75 mi (140 km), CORPORAL is called the embryo of the Army's missile projects; it was eventually put into production by Firestone Tire and Rubber Company.

In June 1945, a facility for testing ORDCIT products was started at the newly established White Sands Proving Ground (WSPG), New Mexico. Covering an area 100 by 40 miles (160 by 65 km) across the Sacramento Mountain range from Roswell, WSPG was the largest military base in America. This facility was later called Launch Site 33 (LS 33).

The U.S. Army Signal Corps asked the Ordnance Department to develop a rocket for taking measurements of the environment at high-altitudes. This was placed under ORDCIT, and the WAC CORPORAL with an objective altitude of 200 kft (60 km) began development; this was based on the CORPORAL design but smaller in size and with less thrust strength – thus, the name WAC. The rocket required a booster, and a TINY TIM aircraft launcher was modified for this use. In September 1945, a TINY TIM was the first rocket static tested at WSPG; a few days later, a partially-charged WAC CORPORAL was test fired. An extended series of modifications and test flights followed. To assist with tracking radar and the application of

WAC Corporal

telemetry equipment, a detachment of Signal Corps personnel was assigned to the Fort Bliss/WSPG operations. It was not until March 1947 that the WAC CORPORAL was finally ready for production.

Rockets of the type such as WAC CORPORAL would become called sounding rockets – vehicles to carry instrument to take measurements and perform experiments during sub-orbital flight. The term "sounding" is from nautical vocabulary to sound – casting a weighted line from a ship into the water to measure the depth to the bottom. In the rocket context, "sounding" is equivalent to "measurement taking."

In the summer of 1944, the Ordnance Department initiated the Hermes Program, under which weapon-grade missiles would be developed. In September 1944, two months after Germany first used the V-2, the Defense Products Group of GE was given a broad-based contract to research and develop a liquid-propelled ballistic missile for use against ground targets and high-altitude aircraft. Called the Hermes Project A, this included the full range of missile and related technologies, under which several different missiles with a HERMES designation were developed. GE had formed the Guided Missiles Department for this and similar work.

HERMES RV-A-5 on Launch Site 33

Hermes A-1 Project was to develop a liquid-fueled surface-to-air missile – the first for America – similar to the German WASSERFALL (a smaller version of the V-2). The outcome of the first effort was purely a test vehicle, the RV-A-5 HERMES. In the Hermes A-2 Project, GE proposed a low-cost, 75-mi (120-km) surface-to-surface missile powered by a solid-propellant engine. Because such engines were as yet unproven, the work led to another test vehicle, the RV-A-10 HERMES. After several successful test firings, further work on the RV-A-10 was assigned to the JPL; this ultimately became the SSM-A-27/MGM-29 SERGEANT, which replaced the CORPORAL.

Project Paperclip

After May 1945 and the close of the war in Europe, the U.S. and the USSR were in great competition to gain the benefits of Germany's weapon developments. As a part of this, about 1,500 key German and other Axis scientists, engineer, and technicians were brought to the United States through Project Paperclip to work under one-year contracts. Colonel Holger N. Toftoy, then head of the Ordnance Department's Rocket Branch, arranged for 125 specialists who had developed rockets at Peenemüde to be included.

In January 1946, the German team led by Wernher von Braun began arriving at Fort Bliss, Texas, where they became contract employees of the Rocket Sub-office. This was a military activity; thus, there was a resident detachment of Army personnel led by Major James P. Hamill to whom von Braun and the German team reported. At this same time, 341 railway cars were received at WSPG; these were filled with missile assemblies (but no full missiles), components, parts, instruments, and many thousands of documents, mainly collected at the Mittelwerk factory just before it was occupied by the Soviets.

Project Paperclip Personnel at Fort Bliss

The German team had entered the United States as "special employees" and without any immigration status. In 1949, anticipating future naturalization, arrangements were made for von Braun and others to cross the border into Juarez, Mexico, and return to register as legal immigrants; this started the process for citizenship that was finally granted in 1955.

V-2 Descendants and Other Missiles

One of the first activities of the German team was sorting out the contents of the railroad cars and assembling several V-2 missiles. This work was at WSPG, about 30 miles away from the Rocket Sub-Office at Fort Bliss. GE's Hermes activities moved to WSPG in early 1946, and their contract was widened to include assisting the German team. On 16 April 1946, the first V-2 launch took place at WSPG Launch Complex 33.

For several years, the German team, together with an increasing number of civil service employees and military servicemen, worked with GE in assembling and improving the V-2s, testing them at the WSPG. The Army's code name for

V-2 Testing Failure at White Sands

this testing was Project Fire-Ball, likely taken from spectacular failures of some V-2 launches.

The Hermes C Program was started in July 1946. Assigned to GE with participation by the German team, this centered on studies of long-range surface-to-surface guided missiles. Before the end of 1947, conceptual plans were developed for a 250,000-lb (110,000-kg) missile with two liquid-propellant stages and an unpowered supersonic glider stage. The first stage would have six engines with a combined thrust of 600-klbf (2,670-kN), and the second stage a single engine providing 100-klbf (444-kN) thrust. The glider would deliver a 1,000-lb (450-kg) warhead to a range of about 2,000 miles (3,200 km). This propulsion power was far beyond any then under consideration for development, so the design was scaled back to provide a range of about 500 miles (800 km). The missile under design was named MAJOR, but development work was given very low priority.

The initial development assignment given to the German team was a broadly defined missile research effort called Hermes II. The first task was to design and fly a test vehicle composed of a V-2 first stage and a ram-jet-powered second stage called RAM. The much smaller RAM had wedge-shaped wings that housed small, rectangular-section ram-jet units. To increase stability of the overall missile, the fins of the V-2 were enlarged. After burnout of the V-2 engine, the RAM would be separated from the V-2 and propelled by the ram-jets to a speed of about 1,000 m/s (3,200 ft/s). The RAM contained a new gyro-inertial stabilization and guidance system developed by the team.

The first flight of the Hermes II vehicle with only a dummy second stage was in May 1947; control of the V-2 was lost, and the vehicle flew off-course and crashed in a cemetery near Juarez, Mexico. After successful, full-system flights, the test vehicle was assigned the designation RTV-G-3 HERMES and used in support of other projects; later it was re-designated RV-A-3.

In late 1947, the Army established preliminary requirements for a liquid-fueled surface-to-surface missile to carry a 1,000-lb (450-kg) payload 150 mi (240 km) with an accuracy of 200 ft (60 m). This development was assigned to GE as the Hermes A-3 Project. The work progressed very slowly, mainly because of frequent changes in warhead and range requirement, but ultimately resulted in the design of another test vehicle, the SSM-A-16 HERMES that accommodated a type W-5 (47 kton) nuclear warhead.

As the WAC CORPORAL continued in development, the German team began an investigation of the potential of this being improved by using a V-2 missile. In June 1947, the GE Hermes contract had this development. In a joint effort also involving JPL and the German team, a rocket named BUMPER resulted; this was a WAC CORPORAL mated to a V-2 booster. BUMPER would be a test vehicle for studying the separation of stages at high velocity and the associated aerodynamics. Round 1 was successfully tested at WSPG in May 1948; this was the first large, two-stage rocket to be launched in America.

On 25 February 1949, the German team – always looking toward rocket-powered space travel – participated in a step in this direction; BUMPER Round

5 became the first U.S. missile to penetrate outer space. It attained a speed of 5,150 mi/hr (2.30 km/s) and reached an altitude of about 244 miles (388 km) – the greatest velocity and highest altitude reached to that date by a man-made object.

Round 7, the last in the BUMPER tests, was fired at the newly opened Long Range Proving Ground, Cocoa, Florida, on 29 July 1950 – the first successful launch of a missile at this Air Force test facility. This attained a speed of Mach 9, the largest in the Earth's atmosphere at that time.

BUMPER Missile

An altitude of 100 km (62 mi) above sea level is commonly used to define the boundary between the Earth's atmosphere and outer space. Called the Karman line, this boundary was named for Theordore von Kármán, a Hungarian-American engineer who calculated that the atmosphere beyond this altitude would be too thin for aeronautical purposes.

In mid-1950, GE's involvement in the Hermes C Program was cancelled. The previously described GE Hermes A-3 Project at WSPG was terminated at the end of 1954; this was the last activity by GE under the overall Hermes Program. Although the total Hermes Program cost had been in excess of $100 million (over $800 million in present-day dollars) and no operational missiles resulted, the information from research and testing was of great value in the successive missile developments.

ARMY ACTIVITIES

By early 1948, activities at Fort Bliss had progressed to a point where the Chief of Ordnance decided to establish a permanent rocket research and development center at a better location. In addition, many members of the initial German team, particularly those who had sent for their families, found the conditions at Fort Bliss to be highly undesirable and requested a change. During the summer of 1948, a survey was made of available Ordnance installations, and in early December, the Chief of Ordnance announced that Redstone Arsenal in Huntsville had been selected.

From the time of closing munitions manufacturing at Redstone and Huntsville Arsenals, Alabama Senator John J. Sparkman and Congressman Robert E. (Bob) Jones, Jr. (from Scottsboro, Alabama) had pressed for a major military operation to occupy the vacated land. The newly formed U.S. Air Force considered this for establishing a major simulation and testing facility.

Senator Sparkman liked to tell the made-up story that President Harry S. Truman had telephoned him with bad and good news:

> The bad news is that the Air Force has selected a site near Tullahoma, Tennessee, for the testing facility, but the good news is that the Army Ordnance Department's rocket work and German team at Fort Bliss are being transferred to Redstone Arsenal.

A major disappointment at the time, the real winner soon became obvious.

A NEW REDSTONE ARSENAL

On 1 June 1949, Redstone Arsenal was officially reactivated, and the Chief of Ordnance designated this as the Ordnance Rocket Center. Its mission included research and development of guided rockets and related items; maintenance of the chemical ammunition loading plant in standby condition; and operation of facilities for storage and maintenance of ammunition items.

In the previous November, Colonel Carroll D. Hudson had been reassigned to his second tour as Redstone Arsenal commander and guided the installation through a facility construction program. During the Korean War, he directed the resumption of ammunition production and expedited the development of rockets and guided missiles. Hudson remained at Redstone Arsenal until May 1952.

At the beginning of 1949, Redstone Arsenal had only 245 civilian and 10 military personnel. During the year, an organization staff of administrative and procurement personnel was hired; physicists, chemists, mathematicians, and engineers in a variety of disciplines were added to form the technical staff of what was called the Rocket Research and Development Center (RRDC). By the end of 1949, Redstone Arsenal had near 700 civilian employees and some 340 military personnel. The administrative and technical activities were concentrated in facilities at the southeastern portion of the site.

Chemical industries were invited to submit bids to have government-owned / company-operated facilities at Redstone Arsenal. Thiokol Chemical Corporation and Rohm and Haas Company were selected, and opened operations on Redstone Arsenal during the summer of 1949. These operations were located in the original Redstone Ordnance Plant area, near the RRDC.

On 30 June 1949, the Chemical Corps deactivated Huntsville Arsenal and the CoE put it up for sale, but the new operations of Redstone Arsenal needed the

Arsenal for Sale

land and facilities. On 1 April 1950, the consolidation of the two arsenals was made official; at this time, the Redstone reservation was expanded to include an approximate 10-by 12-mile area, comprising about 40,300 acres. In June, the U.S. Army was reorganized, with the Ordnance Department becoming the Ordnance Corps.

Ordnance Guided Missile Center

The Ordnance Guided Missile Center (OGMC), a unit of Redstone Arsenal, was officially activated on 15 April 1950. All guided missile activities at Fort Bliss would be transferred to that Center by November, at which time the Missile Sub-office at Fort Bliss would close. The primary mission of OGMC was to serve as the Ordnance center for guided missiles, including component and systems development in the fields of aerodynamics, guidance and control, propulsion, assembly techniques, transport, test, and launching. GE personnel working on the Hermes C program also transferred to Huntsville, but the program was cancelled in September; at that time, responsibility for continued work on the MAJOR missile was assigned to OGMC.

In addition to OGMC, Redstone Arsenal had staffs for contracting and procurement, field service, and special research and development. In May 1950, the Chief of Ordnance assigned Redstone Arsenal responsibility for the preliminary design study of the special-purpose, large-caliber field artillery rocket later named the HONEST JOHN.

By the end of 1950, the OGMC had about 700 civil service, military, and contractor personnel; this included 130 foreign-born contract employees (many of the original 125 had returned to Germany, but others had been added). The majority of the technical personnel were organized on a functional basis; very few were permanently assigned on any specific project.

OGMC mainly occupied facilities near the center of the original Huntsville Arsenal, and a major construction activity was begun to provide more appropriate research and development facilities. MAJ James P. Hamill, who was among the military transferred from Fort Bliss, served as acting OGMC Commander and von Braun was the Technical Director.

The addition of personnel from Fort Bliss, many with families, was coupled with an expansion of the existing personnel at Redstone Arsenal as well as construction workers for the new facilities. This had a major impact on housing and general business in Huntsville. Many of the Germans bought or built homes on Monte Sano Mountain (often referred to as 'Kraut Hill'), saying that it was most like their homeland.

While Wernher von Braun and his German team were mainly remembered for their rocketry accomplishments, it is noted that one of their early local endeavors was in establishing Huntsville's first Lutheran church (St. Mark's) in September 1950, and later were leaders in founding the Huntsville Civic Orchestra (predecessor of the Huntsville Symphony Orchestra) in early 1955. Von Braun himself was a professing Christian and an excellent musician.

Significant Activities

In January 1951, responsibility for the CORPORAL missile was transferred to the Redstone Arsenal. This was the first tactical missile system under the authority of the Army at Redstone Arsenal. In its first function, the Redstone Arsenal Board of Awards gave JPL a contract for the complete development of the Corporal system. In June 1951, Firestone Tire and Rubber Company was awarded a contract for the production of Corporal Systems.

CORPORAL Missile

Redstone Arsenal established a Technical and Engineering Division in August 1951. Major elements of this division were the Rocket Development Group, the Guided Missile Development Group, and the Engineering Group. A Technical Library was opened for maintaining the related reports.

Thomas Moore with JETVEST

The JETVEST, sometimes called the ROCKETBELT, was the smallest and least-known rocket developed at Redstone Arsenal. Thomas T. Moore came to Huntsville from Fort Bliss with the GE Hermes Project, and then became an original member of the Engineering Group of OGMC. As a side project, Moore developed "rockets on a man," a rocket system using hydrogen peroxide as the fuel. With support from von Braun, the JETVEST was built and flight-tested in 1952. Following that, there was no further development of the JEVEST at Redstone Arsenal, but a similar system was later built, patented, and sold by Bell Aerosystems. In recent years, Thunderbolt Aerosystems supplied the rocket for use in movie and television productions.

The Ordnance Corps assigned Redstone Arsenal the responsibility for the research and development phase of the Nike Program in August 1951. This program had begun in 1945, when the Ordnance Department and the Army Air Forces issued a contract to Bell Telephone Laboratories (BTL) to study the feasibility of an antiaircraft defense system using guided missiles.

During 1952, Redstone Arsenal added the Field Service Division and the National Procurement Division (later called Industrial Division). Also in 1952, the Ordnance Corps opened the Ordnance Guided Missile School (OGMS) on Redstone Arsenal for training military personnel in maintaining guided missile systems. Another change that year was the assignment of the first general officer to command the overall Redstone Arsenal, BG Thomas K. Vincent.

Thomas Vincent

Holger Toftoy

The Ordnance Missile Laboratories (OML) was formed in September 1952; this was a major enlargement of the previous Technical and Engineering Division. COL Holger N. Toftoy returned to Redstone Arsenal as Director of the OML, and was soon promoted to Brigadier General. Toftoy, who had been primarily responsible for bringing the German rocket team to America and had personally led in "selling" missile development, was known throughout the Army as 'Mr. Missile.'

The OML was composed of eight units centering on a variety of technology areas needed for development of missiles, from design to launch. Each unit was headed by an Army officer, with, in most cases, a member of the original German team serving as the technical director; Wernher von Braun was the overall technical director.

A test stand was urgently needed, particularly to test the engine from NAA. Funds for facilities were not available; thus, Kurt H. Debus, Technical Director of the Test Laboratory, led the construction of an Interim Test Stand composed of existing materials. Debus would later set up a launch facility at Cape Canaveral, Florida, Air Force Base.

First Test Stand

During the mid-1950s, Redstone Arsenal had the largest construction boom since the start of WWII. This included new buildings for the Ordnance Guided Missile School, a 200,000-ft^2 research and engineering building, a guidance and control development building, and many other smaller facilities.

While leading the development of Army missiles, Wernher von Braun never let his vision of manned space travel waiver. Between 1952 and 1954, von Braun and several of his "space" associates wrote eight articles for *Collier's Magazine,* and also published two books: *Across the Space Frontier* (1952) and *Conquest of the Moon* (1953). The excellent illustrations in these publications particularly created nationwide interest in this subject. In 1955, Walt Disney produced a series of television programs on space that featured von Braun and Ernst Stuhlinger, the German team's lead scientist. Von Braun also gave public lectures across the nation, often starting with, "I want to apologize for my accent – you know I'm from Alabama."

In September 1954, BG Toftoy assumed command of the overall Redstone Arsenal. National mission agencies then at Redstone Arsenal were the Ordnance Missile Laboratories, the Field Services Division, the Industrial Division, and the Ordnance Guided Missile School (OGMS). There were also several Project Offices, a relatively small Munitions Production Division, and the Redstone Depot (one of several such operations under the Ordnance Corps that stored and shipped ammunition, strategic raw materials, and reserve industrial tools).

At the start of the Korean War, ammunition production was resumed by the Munitions Production Division; about 39 million rounds of artillery munitions were produced from July 1951 through July 1955. Products included mortar shells, 3.5-inch rockets, 105- and 155-mm shells, and 81-mm illuminating shells. During 1954, Redstone Arsenal had an average of over 7,000 civilian and military personnel, excluding contractors.

On 11 November 1954, 39 of the original German team, along with two wives, were sworn in as U.S. citizens. Then on 14 April 1955, another 103 German-born specialists and members of their families became American citizens; Wernher von Braun was among this second group taking the oath of citizenship. The ceremony was held at the Huntsville High School.

Citizenship Ceremony

Early Missiles

When the rocket research and development activities were transferred from Fort Bliss to Redstone Arsenal, the responsibility for several then existing or under development rockets and missiles also came; these included the previously described CORPORAL, SERGEANT, MAJOR, and LOKI-DART.

Brief descriptions follow of early missiles that were either initiated at Redstone Arsenal or managed in their completion.

Honest John – The HONEST JOHN was the first missile fully developed at Redstone Arsenal. Authorization for the preliminary design was given in April 1950, only a month following the Arsenal's official formation. Initially designated Nuclear Artillery Rocket XM31, it was later named the HONEST JOHN (MGR-1) by COL Toftoy. The first testing was made in June 1951.

This was an unguided, fin-stabilized rocket designed to deliver a nuclear warhead (type W7, variable 2, 10, or 30 kton) at a range of 24.8 km (15.5 mi) and a scatter of 230 m (760 ft). An alternate warhead was 680 kg (1,500 lb) of conventional explosives; later, Sarin nerve gas

HONEST JOHN Missile

cluster ammunitions were also available. The solid-fuel motor from Hercules Powder Company had a thrust of 150,000 lbf (670 kN). The rocket was 8.3 m (27.4 ft) in length and was truck transported; it weighed 2,600 kg (5,800 lb). Douglas Airplane Company of Santa Monica, California, was the production contractor, with first deliveries in September 1953. Produced through 1965, with 7,000 built, it remained in service in the United States through 1973, and in other nations as late as the 1990s.

Redstone Missile – In June 1950, with the outbreak of the Korean War, Redstone Arsenal was given the mission of developing a surface-to-surface ballistic missile with an objective range of 500 miles (800 km). The development of this missile was by far the most important activity at Redstone Arsenal during the first half of the 1950s. Starting with an upgrading of the

MAJOR missile – which had carried over from White Sands – the design went through a series of improvements. Development of the rocket engine had the longest lead time. To expedite development of this urgently needed missile, the objective range was greatly reduced, allowing adoption of an existing, smaller engine, the A-7 from North American Aviation (later Rocketdyne).

The A-7 produced 78-klbf (350-kN) thrust, reducing the operational range of the missile to between 58 miles (93 km) and 200 miles (320 km). This ultimately became the PGM-11, a single-stage, liquid-propellant missile with the name

REDSTONE Rocket REDSTONE. This used the same propellants as the V-2: liquid oxygen and a mixture of 75% ethanol and 25% water.

131

In 1952, Chrysler Corporation was asked to assist the OML in production engineering. Then at the start of 1953, Chrysler was awarded a prime contract for fabricating and assembling REDSTONE missiles. Twelve prototypes were completed in Redstone Arsenal facilities by Chrysler in cooperation with the OML; these were mainly from components supplied by private industry. Static testing of the REDSTONE began in the spring of 1953. To accelerate the program, 15 more REDSTONE missiles were eventually built at Redstone Arsenal; one of these had the fuel tank extended eight feet for use in the Mercury-Redstone sub-orbital launch in 1961.

The first REDSTONE flight test was at Cape Canaveral in August 1953; this was a partial success, flying for a minute and 20 seconds before suffering an engine failure. The second test, in January 1954, was fully successful, flying 55 miles (88 km). In the third test the following May, the missile exploded on launching. Subsequent tests, however, were completely or partially successful, and the REDSTONE was declared operational in 1955.

The Redstone Arsenal Industrial Division awarded Chrysler a contract for full-scale production of REDSTONE missiles in June 1955. One hundred and one of these missiles were built by Chrysler at the Navy-owned Michigan Ordnance Missile Plant in Warren, Michigan. The Rocketdyne Division of North American Aviation Company provided the rocket motors, Ford Instrument Company of Sperry Rand Corporation produced the guidance and control systems, and Reynolds Metals Company fabricated fuselage assemblies.

The PGM-11 was the Army's first deployed missile with a thermonuclear warhead capability (type W-39, 3.8 Mton). It became operational in June 1958, and was subsequently deployed in Europe to support NATO's shield force on the front lines of the Cold War. On 31 July 1958, the Atomic Energy Commission had a REDSTONE launched from Johnson Island in the Pacific to test effects of a high-altitude nuclear explosion. The W-39 warhead was detonated at 47.7-miles (76.8-km) altitude; it blacked out short-wave radio over the Pacific for up to nine hours.

Lacrosse – In 1947, the U.S. Navy had initiated studies at Johns Hopkins and Cornell Universities on a short-range tactical missile for the U.S. Marine Corps. The missile was called LACROSSE because – similar to the game of that name – after it was launched a forward observer nearer the target would pick up the control. In late 1950, management responsibility

LACROSSE Missile

was transferred to Redstone Arsenal. Glenn L. Martin Company was awarded a contract to complete the design in 1955, early tests started the next year, and MGM-18 LACROSSE entered service in 1959. The missile weighed 1,000 kg (2,300 lb), was radio command guided, had a range of 19 km (12 miles), a speed of Mach 0.8 (0.27 km/sec), and carried either conventional explosives or a type W-40, 1.5-10 kton nuclear warhead. About 1,200 missiles were produced, and it was withdrawn from service in 1964.

ARMY BALLISTIC MISSILE AGENCY

As activities at Redstone Arsenal continued to expand, the Ordnance Corps upgraded the cognizant organization and its level of command to Agency status.

John Medaris

The Army Ballistic Missile Agency (ABMA) was formed on 1 February 1956, taking over from Redstone Arsenal all the facilities and personnel involved in missile development and production. MG John B. Medaris was assigned as the Commanding Officer, and Wernher von Braun was the Director of the ABMA Development Operations Division. Later commanders of ABMA were BG John A. Barclay (March 1958-April 1960), and BG Richard M. (May 1960-December 1961.

In October 1957, ABMA opened a staff-level Redstone Anti-Missile Missile Systems Office (RAMMSO). Led by COL Mathew R. Collins, Director of the R&D Division in ABMA's Ordnance Research Laboratories (ORL), the office had five military and 19 civilians, all from the ORL. Although formed on an interim basis, RAMMSO was the first anti-missile organization at Redstone Arsenal specifically dedicated to ballistic missile defense.

When ABMA was formed, Redstone Arsenal then became an Army post or support activity; commonly called RSA, it supporting the ABMA and, in the future, other resident organizations. RSA was also under the command of MG Medaris, and BG Toftoy was his RSA deputy. The Ordnance Missile Laboratories, now a part of ABMA, still reported to BG Toftoy.

The original Huntsville Arsenal Airstrip had been inactive since late 1945. With the formation of ABMA, Medaris – then in training as an Army pilot – established an aviation activity in ABMA. The airstrip was increased to 7,300 ft, a 25-ft control tower and several support buildings were added, and the facility eventually renamed the Redstone Airfield.

Redstone Airfield

ARMY ORDNANCE MISSILE COMMAND

In only two years after ABMA was opened, there was another major reorganization. The Army Ordnance Missile Command (AOMC) was formed on 31 March 1958; headquartered at Redstone Arsenal, it was commanded by MG Medaris. Initially, BG Toftoy was deputy commander, but in July he was named the Commanding General of Aberdeen Proving Ground. AOMC had three major subordinate elements: ABMA, JPL, and White Sands Missile Range (WSMR – formerly WSPG). In June, another activity, the Army Rocket and Guided Missile Agency (ARGMA), was formed and added to AOMC.

Although not immediately affecting the Army organizations at Redstone, President Dwight D. Eisenhower established the National Aeronautics and Space Administration (NASA) on 29 July 1958. NASA quickly started Project Mercury, with the goal of putting a human in orbit around the Earth, and doing it before the Soviet Union. This was planned and managed by the Space Task Group from NASA's Langley Research Center (LaRC). An initial team of seven astronauts was selected, and NASA began planning unmanned flights using Army rockets. McDonnell Aircraft was selected to be prime contractor for the Mercury one-man spacecraft.

The latter years of the 1950s was one of the most nationally publicized times in Redstone Arsenal's history. General Medaris could frequently be seen on national television or testifying before Congress urging funding for his programs. The magnitude of the operations is reflected in his command's budget; this approached $2 billion ($16 billion in today's dollars) in Fiscal Years 1959 and 1960, about 25 percent of the Army's budget for those years. Medaris remained the Commander of AOMC until he retired on 31 January 1960; several years later he became an Episcopal priest. Following Medaris, AOMC Commanders were MG August Schomburg (February 1960 - May 1962), then MG Francis J. McMorrow (May 1962 - August 1962).

Missile Programs

With the formation of AOMC, the missile programs had been split, with the more powerful systems and their associated activities remaining under ABMA, and the shorter-range systems under the new ARGMA. The following are brief descriptions of the two largest missiles developed under ABMA.

Jupiter Missile – During a briefing to the U.S. Secretary of Defense in September 1955, Wernher von Braun pointed out that a 1,500-mi (2,400-km) missile would be the logical extension of the REDSTONE. Von Braun had particularly wanted this, seeing it as a step toward rocket boosters for future manned flights.

There was considerable debate in the Department of Defense (DoD) as to which military service should have this responsibility. This led to the Army and Navy being assigned a joint program for developing a land- and sea-based Medium-Range Ballistic Missile (MRBM, 1- to 3-km range). This would be designed by the ABMA with participation by the Naval Research Laboratory (NRL). Designated PGM-19 and named JUPITER, the liquid-propellant missile used a Rocketdyne S-3D engine producing 150-klbf (667-kN) thrust; this engine was already being developed for the Air Force THOR missile. Shipboard use required that the missile be relatively short but with a wide girth, providing the necessary volume for the fuel.

In November 1956, the Navy pulled out of the JUPITER development. Later in November, the Secretary of Defense assigned all land-based long-range missiles to the Air Force, with the Army retaining control of battlefield missiles with a range of 200 miles (320 km) or less. The Air Force already had under development the THOR, an Intermediate-Range Ballistic Missile (IRBM, 3- to 5-km range). As a backup to the THOR, the DoD had ABMA continue with their development of JUPITER, and would then turn the system over to the Air Force for production and ultimate use.

Comparative Sizes

To test JUPITER re-entry nose cones, a rocket called JUPITER-C was developed in 1956. This was a three-stage vehicle composed of a REDSTONE missile with a lengthened fuel tank as the first stage, and second and third stages using clusters of scaled-down SERGEANT engines, 11 and 3, respectively. Often called a sounding rocket, the JUPITER-C was successfully used in the development of nose cones.

The first successful JUPITER test launch was on 31 May 1957, at the Cape Canaveral Air Force Station, Florida; an apogee of about 500 mi (800 km) was achieved. The Army turned the JUPITER program over to the Air Force in August 1958, but there was continued development work by the Army. The Air Force awarded Chrysler a production contract.

At the request of the DoD, the Ordnance Guided Missile School (OGMS) at Redstone Arsenal initiated training on JUPITER operation and maintenance. In 1958, the Air Force activated the 864th Strategic Missile Squadron for JUPITER training at OGMS. Later in the same year, the 865th and 866th Squadrons also began operation and maintenance training at OGMS. For some time, it was common to see around Huntsville many men in Air Force uniforms.

Pershing Missile – Although the REDSTONE was just entering service, the Army was looking toward its replacement. The Secretary of Defense rescinded the limitation of 200 miles for Army missiles, and approved the development of REDSTONE-S – a REDSTONE with solid propellants and a longer range. With Arthur Rudolph – a leading member of the Operation Paperclip team – as Project Manager, ABMA began the development of what was soon renamed the PERSHING missile in late 1958.

PERSHING Missile

PERSHING was the first and only solid-fueled MRBM (Medium Range Ballistic Missile) deployed by the U.S. Army. The first version, later called PERSHING I (MGM-31), had an operational range up to 460 miles (740 km). PERSHING had started at ABMA in 1958, and thus had the benefit of the von Braun team in its beginning. However, as the project matured, there was a change in the developmental process. While the concept development and technical control remained under the government and Arthur Rudolph, a greater reliance was placed on the prime contractor and specialty subcontractors for detailed development and testing.

The Martin Company (later Martin Marietta) was selected by General Medaris and Rudolph as the PERSHING prime contractor. In the subsequent performance, considerable credit is given to Philip B. Crosby, Martin's quality control leader; Crosby developed the concept of Zero Defects that greatly enhanced the production and reliability of the system.

The solid-propellant engines were developed by the Thiokol Rocket Division in Huntsville. The first stage used the TX-174, rated at 25.9-klbf (115-kN) thrust, and the TX-175, with 19.1-klbf (85.0-kN) thrust, was on the second stage. Since solid-rockets cannot be turned off once started, the second-stage engine had a squib that would open the thrust reversal ports in the forward end of the stage and ignite the propellant in the forward end, causing the engine to reverse direction.

The missile was steered by jet vanes in the rocket nozzles and air vanes on the engine case. Guidance was provided by an onboard analog guidance computer and an Eclipse-Pioneer ST-120 (Stable Table-120) inertial navigation system from Bendix. Range selection was made by thrust-reversal and case-venting of the engine. The system accuracy was 1,300-ft (400-m) circular error probable. The warhead could be conventional explosive or a W-50 nuclear weapon with three yield options: 60, 200, or 400 kilotons. The Bulova Watch Company delivered the fuzing and arming system.

The initial test launch of the first stage was at the WSMR in February 1960. Complex 30 for test-launching in the PERSHING program had been constructed at Cape Canaveral. The first two-stage launch using the tactical Transporter Erector Launcher (TEL) was in January 1962; the TEL was designed by Unidynamics and manufactured by FMC Corporation.

A development program for an upgraded PERSHING IA (MGM-31B) was begun in 1965; this was a quick-reaction system, having faster vehicles, launch times, and new digital electronics. Martin Marietta received the PERSHING IA production contract in mid-1967. In production through 1969, 754 PERSHING I and PERSHING IA systems were built and deployed; 180 were in Europe where they were also adopted by the German Air Force.

Biological Test Flights

While the JUPITER was still under development, ABMA, in support of Project Mercury, used the developmental missiles in a number of suborbital biological tests flights. The first was in December 1958, when a squirrel monkey, Gordo, was carried in the nosecone. Telemetry data showed that the monkey survived the launch and reentry accelerations, but the nosecone was lost when the recovery parachute failed to operate.

Another biological flight was made on 28 May 1959, with Able, a rhesus monkey, and Baker, another squirrel monkey, aboard. Both survived in good condition the 16-minute flight of an altitude of 59 miles (95 km) and a distance of 1,500 miles (2,400 km) down range. They withstood accelerations 38 times normal gravity and were weightless for about 9 minutes. Able died four days after the flight while undergoing surgery to remove an infecting medical electrode, but Baker lived to age 27. (Miss Baker is buried on the grounds of the Space and Rocket Center in Huntsville.) Wernher von Braun and other advocates of manned spaceflights were greatly encouraged by these biological flights.

Miss Baker

137

ARMY ROCKET AND GUIDED MISSILE AGENCY

As previously noted, the Army Rocket and Guided Missile Agency (ARGMA) was formed and added to AOMC in June 1958. BG John G. Shinkle was the Commanding Officer, and Nils L. Muench was the Chief Scientist. Shinkle served April 1958 - June 1960, and was followed by BG John G. Zierdt (June 1960 - December 1961).

ARGMA had management and technical responsibilities for shorter-range surface-to-surface systems such as the SERGEANT, CORPORAL, and LACROSSE; HONEST JOHN and LITTLEJOHN rockets; as well as NIKE AJAX, NIKE HERCULES, HAWK, and REDEYE surface-to-air systems. Most of these had been developed earlier and have been previously described. Air defense, however, was a relatively new activity and will be discussed.

Air Defense Missile Emergence

In 1944, the U.S. Army Ground Forces had requested that research be started on an "antiaircraft rocket torpedo" – a ground-to-air missile. In January 1945, the Army gave a contract to Bell Telephone Laboratories (BTL) to research such missile systems – BTL had been a leader in WWII radar development. Called Project Nike, the missile itself was called NIKE. Progress in the early years was slow, but picked up with the start of the Korean War in 1950.

Nike Ajax – Government responsibility for the R&D phase of Project Nike was assigned to the Guided Missile Development Group of Redstone Arsenal in August 1951. Later in the year, the system (missile, radar, and ground controls) was shown to be effective against high-flying jet aircraft, and in March 1954, deployment of antiaircraft systems using a missile called NIKE AJAX (M1M-3) began at 40 defensive areas.

The Nike Ajax system was designed to attack conventional bomber aircraft flying at high subsonic speeds and altitudes above about 50 kft (15 km). The system used separate radars for tracking the target and tracking the missile. An analog computer was used to calculate the impact point, and guidance signals were sent to the missile encoded in the tracking-radar signals; the warhead would be detonated on command. Western Electric Company (WECo, half owned by BTL) was the prime contractor. Douglas Aircraft was responsible for the missile; this had a solid-fuel booster, giving 48-klbf (210-kN) thrust, and a liquid-fueled sustainer with 2.6-klbf (12-kN) thrust.

Nike Hercules – BTL/WECo continued with developments, and soon had a missile called NIKE HERCULES (MIM14) that could carry a blast-fragmentation warhead or a variable-yield atomic warhead (W-31, 2 kton-40 kton). This two-stage, solid-fueled missile had a booster engine producing 220-klbf (980-kN) thrust, and a second-stage sustainer engine with 10-klbf (44-kN) thrust. Like in the Nike Ajax system, the missile was supplied by Douglas aircraft. The NIKE HERCULES had a range of about 90 miles (140 km), a top speed in excess of 3,000 mi/hr (4,800 km/hr),

NIKE HERCULES

and a maximum altitude of about 150 kft (46 km). Starting in June1958, the Nike Hercules system began to be deployed. These were the only nuclear-armed surface-to-air weapons ever operational in the U.S. Army; 393 ground systems were manufactured.

Hawk – In addition to the air-defense missiles systems from BTL/WECo, ARGMA was responsible for the final development of the Hawk (MIM-23) system. Northrop developed the launcher, radars, and fire control system, and

Raytheon the missile. This solid-fuel air defense rocket had a speed of Mach 2.4 (1840 mph), an operational range of 45-50 km (28-31 mi), and ceiling of 14 km (45 kft). It had a 119-lb (54-kg) blast-fragment warhead. The Hawk system was first deployed by the U.S. Army in 1959, and by the U.S. Marine Corps in 1960. The radar was highly complex and used vacuum tubes, leading to a MTBF

Hawk System Radar

of only 43 hours. An improved system, the I-Hawk (MIM-23B), entered service in 1972. Worldwide, about 40,000 HAWK missiles were produced.

Redeye – The traditional anti-aircraft guns were made ineffective with the advent of high-speed jet air craft. As a possible alternative weapon, engineers at the Convair Division of General Dynamics in San Diego experimented with small, solid-fueled, ground-to-air missiles guided by a heat sensor. This was patterned on the existing SIDEWINDER air-to-air missile of the Air Force. By mid-1956, their efforts had led to an elementary man-portable infrared-guided missile; they called it REDEYE because of its infrared sensor. In November 1956, this was demonstrated to the Army and Marine Corps.

In July 1957, ABMA issued an RFP for a self-contained, very light-weight, low altitude, air defense weapon to seek and destroy aircraft travelling at speed up to 600 knots at ranges up to 4,100 meters. Energy radiated by the target must be used for detection. Three proposals were received; Francis W. DuVall, Chief of ABMA's ABM Advanced Concepts Studies Office, led the proposal evaluation team. The REDEYE was eventually selected for an in-depth feasibility study, and Convair was awarded the contract in January 1958.

Redeye System

With the formation of AOMC in March 1958, the Redstone Project went with this agency; then with ARGMA coming into existence in June, the project went there. Charles A. Cockrell was named the Project Director, responsible for management of R&D, procurement, and production. The first test rounds were fired in March 1960, and a year later a shoulder launch was made. There were, however, significant technical problems, and the prototype did not meet the specifications. General Dynamics took over as contractor, and there was redesign and testing during 1965 and 1966. Changes included a new gas-cooled detector cell and an improved warhead. Finally, the system was put into production in May 1967, and a year later the FIM-43C Redeye Missile System entered service.

In this same period, the USSR was developing a very similar system: the 9K32M STRELA (Arrow), known in the West as the SAM-7. A comparison of the two systems is very interesting:

Characteristic	REDEYE	SAM-7
Maximum Range	4,500 m	4,200 m
Missile Speed	580 m/s	500 m/s
Maximum Target Speed	225 m/s	260 m/s
Engagement Altitude	0.05-2.7 km	0.05-2.3 km
System Weight	13.3 kg	15.0 kg
Missile Weight	8.3 kg	9.8 kg
Missile Length	1.40 m	1.44 m
Warhead Weight	1.06 kg	1.17 kg
Detector Type	Gas Cooled PbS	Uncooled PbS

Both missile systems entered service in 1968 -- the REDEYE after 12 years in development. Typical development time for small weapons in the USSR was far less. Thus, it is highly likely that the REDEYE was compromised well before it was fielded. The REDEYE was replaced by the more capable FIM-92 STINGER in 1986.

Project Plato, FABMDS, and SAM-D

While the Nike Project was underway, the DoD Ordnance Department was planning an anti-missile missile (AMM) protection for the field army against short- and medium-range threats. Called Project Plato, contracted studies were conducted by Cornell Aeronautical Laboratory in the 1953-1956 period. The studies showed that the PLATO missile would need speeds of Mach 6 to 8, well beyond anything being considered at that time. Sylvania was awarded a development contract in 1956, but the program was cancelled in early 1959 – for lack of funds, rather than technical deficiencies.

The basic requirement to protect the field army still existed; this was taken up by ARGMA and the Field Army Ballistic Missile Defense Systems (FABMDS) project came into being in September 1959. An RFP brought 17 proposed systems, and General Electric was selected. However, it was eventually assessed by the staff of the Secretary of the Army that the FABMDS would not protect from the total missile threat and also had high risks; the project was cancelled in October 1962. It was immediately replaced by a new program called AADS-70 (Army Air-Defense System - 1970). In 1964, AADS-70 was renamed SAM-D (Surface-to-Air Missile - Development), a program which eventually resulted in the MIM-104 PATRIOT.

ARGMA R&D

In addition to the management of systems with smaller missiles, ARGMA initiated advanced research and development in a number of related technologies. Many of the projects were funded by ARPA, and an ARPA Division under William J. Lindberg handled all of this work; much was done in the area of re-entry physics phenomena for applications in the emerging field of ballistic missile defense.

An Anti-Missile Research Advisory Council (AMRAC) was formed involving leading scientists and engineers from across the nation. AMRAC's first meeting was held by ARGMA at Huntsville in December 1960; well-known nuclear scientist Edward Teller, who had a personal interest in missile defense, was a leading participant. Biannual AMRAC meetings, led by Nils Muench and administered by Wallace E. Kirkpatrick, were held during the following years.

The ARGMA Research Laboratory was established by L. Post Hallows; this performed theoretical and experimental science and engineering projects supporting missile programs. Included was Thomas Albert Barr's 20,000-joule plasma-generator for simulating high-velocity aerothermodynamics – this required a dedicated 8 MW power substation. Thomas G. Roberts developed powerful CO_2 lasers for directed-energy sources, and an analytical group led by Ralph L. Edwards was involved in anti-missile optical and radar target signatures and discrimination.

The Downrange Antiballistic Measurement Program (DAMP) was led by John E. Hagefstration. In this, the USAS *American Mariner* operated throughout the South Atlantic, filled with instrumentation for taking optical and radar measurements of U.S. ballistic missiles as they re-entered the atmosphere. Barnes Engineering was the prime contractor for measurements in the optical spectrum from far-infrared through ultraviolet. RCA was responsible for radar measurements using high-performance systems.

DAMP Ship

In other ARGMA research organizations, Stephen L. Johnson became well known for his development of radars for missile guidance and in radar countermeasures. Everett Gilman and Richard Dillard developed new mathematical techniques for predicting missile accuracy. Robert E. Betts developed IR and RF signature techniques for analyzing propellant efficiencies. Robert Lusser formulated methods for determining the economic consequences of unreliability, and formulated Lusser's Law for predicting reliability. Papers from these studies were published in the *ARGMA Quarterly Research Journal*.

BALLISTIC MISSILE DEFENSE ORIGINS

When ARGMA was formed as a unit of AOMC in April 1958, it had been given responsibility for Project Nike. Various segments of ARGMA started technical relationships with BTL, and improvements to the missile and radar were jointly studied. As the Soviet Union developed the intercontinental ballistic missile (ICBM), a much higher-performing system was needed for ICBM defense.

Initial Systems

The study and development of systems – early called Anti-Ballistic Missile (ABM) – for defending U.S. Army assets and civilian population centers against ballistic missile threats from foreign adversaries has been under the responsibility of a continuum of Army organizations centered in Huntsville. The first such official organization was the Redstone Anti-Missile Missile Systems Office (RAMMSO) established by ABMA in 1957. These initial study activities were later continued in ARGMA with analytical work in the ARPA Division and the experimental studies in the Research Laboratory. The ARGMA-coordinated Anti-Missile Research Advisory Council did much in building the body of information about the threat and possible defense.

Supporting the Government in these and other advanced activities, there was a corresponding growth in local high-technology firms. When started in the early 1960s, the Research Laboratories of Brown Engineering Company (BECO) had this as a major area of work. What is now generally called ballistic missile defense (BMD) has made a significant contribution to the technological evolution of Greater Huntsville.

Nike Zeus – To counter the new ICBM threat, BTL was directed to discontinue work on defense against air-breathing targets and turn full attention on anti-ballistic missile (ABM) systems. In January 1958, BTL brought out the

NIKE-ZEUS (LIM-49), an improved NIKE-HERCULES with a 25-kton W-31 nuclear warhead and later a 400-kton W-50 warhead. With a new 450-klbf (2.00-MN) thrust solid-fuel booster and two sustainers, all from Thiokol, interceptions could take place at over 200-mi (400-km) range and over 170-mi (280-km) altitude. ZEUS was first test-launched during August 1959, and demonstrated a top speed

NIKE ZEUS

of greater than Mach 4 (about 2,800 mi/hr / 4,500 km/hr).

At ARGMA, the Nike Zeus system development under COL Glenn Crane began in July 1960, and it was designated as a separate Nike Zeus Project Office in December 1961. Radars were the key development for Nike Zeus system. The powerful Zeus Acquisition Radar (ZAR) was on a large rotating triangle, with the receiving antennas centered in a rotating Luneburg lens. Targets picked out by the ZAR were then illuminated by the Zeus Discrimination Radar (ZDR); selected target information was passed to the Target Tracking Radar (TTR); the

missile was launched, and tracked by the Missile Tracking Radar (MTR), and guided to an interception using radio commands. Running all of this was the digital Target Intercept Computer (TIC).

In June 1962, a full Zeus System was installed at the Test Site. The following month, the system intercepted an ICBM nose cone flown by an SM-65 ATLAS from Vandenberg AFB, California, and in December it intercepted an ATLAS ICBM

Zeus System on Kwajalein

carrying a target vehicle and two decoys. Of 14 tests carried out over two years, 10 were successful interceptions. It is noted that "interception" was determined by lethal proximity to the target, not the actual detonation of the warhead.

Ivey Drewry

The Army Missile Command (MICOM, described later) succeeded ARGMA/AOMC in August 1962. At that time, the Nike Zeus Project Office was established under COL Ivey O., Jr. COL Crane took the responsibility for the Kwajalein Test Range.

Having demonstrated the feasibility of the Nike Zeus System intercepting and destroying an incoming ICBM warhead in December 1962, the Secretary of Defense placed a new requirement on ABM systems: an anti-satellite capability. In May 1963, at the Kwajalein Test Range under an activity called Project MUDFLAP, an orbiting Agenda D satellite was successfully intercepted and destroyed by a Nike Zeus System with its missile armed with a nuclear warhead.

Nike X – With a rapidly evolving Soviet threat, the Zeus System had cost and operational and limitations (it was incapable of countering a massive ICBM attack), prohibiting its full development and deployment. ARPA examined the ABM requirements against the capabilities and shortcomings of Nike-Zeus System; using extensive analog simulations, ARPA came up with Nike-X. This was a conceptual system incorporating phased-array radars, high-performance computers, and separate high-velocity missiles for low-altitude and high-altitude intercepts.

In January 1963, BTL's Nike Program was redirected to address the ICBM threat of the 1970s, with an orientation toward Nike-X. In February, responsibility for the Nike Program was transferred from the jurisdiction of MICOM and reported directly to Headquarters, Army Materiel Command (AMC). At that time, the Nike Office had 228 civilian and 25 military personnel. The following year, this Office was re-designated the Nike-X Project Office (NXPO); COL Ivey Drewry was the Director.

BTL was given a new contract to serve as the Nike-X prime contractor in October 1963. To oversee BTL – including WECo and several major subcontractors – NXPO assembled a staff of technical and administrative personnel working in MICOM facilities on Redstone Arsenal. LTC Norman C. Buchholz was the Deputy Director, Charles E. Richardson the Chief Engineer, and William O. Turney handled contracting.

The Nike X System would incorporate two types of radars: a very large Multifunction Array Radar (MAR) for long-range operation, and a short-range

Missile Site Radar (MSR) for guiding the interceptor missiles. The missiles were SPARTAN (LIM-49A, an upgrade from ZEUS) for exo-atmospheric intercepts, and SPRINT for high-velocity, low-altitude intercepts. A centralized Data Processing Center would serve all of the computer functions. The PAR and MSR radars and the SPARTAN and SPRINT interceptors could be assembled in many combinations, and their deployments could be tailored to meet various threats.

In 1965, still with BTL as the prime contractor, Nike-X was begun as the first ABM system; the next year, BMD was elevated to the Chief of Staff of the Army (CSA) level. To have more local technical support, the NXPO contracted with BECO (later TBE) for the services of engineers, scientists, and other specialists; Herschel A. Matheny was BECO's BMD Program Manager.

In September 1966, the CSA established the Nike-X System Office; located in Washington, DC, LTG Austin W. Betts was named Nike-X System Manager; Betts also served as the Army Chief of Research and Development.
Sentinel System

In January 1967, President Lyndon B. Johnson announced improvements in the Soviet Union's long-range missile capabilities and their deployment of a limited ABM system (Galosh) near Moscow. China exploded their first thermo-nuclear device in June 1967, and, at the same time, the U.S. and the Soviets were having Strategic Arms Limitation Talks (SALT). These led to an agreement known as the Strategic Arms Limitation Treaty; this froze the number of strategic ballistic missile launchers at existing levels. This, however, ill-defined the threat, since both sides had begun to have an increasing number of missiles with multiple independently targetable reentry vehicle (MIRV) warheads.

Sentinel System – The United States revised its ABM system requirements, and the Sentinel System was born, replacing the Nike-X System in November 1967. Sentinel would be a "thin umbrella" defense, widely distributed throughout the U.S. to protect cities from a limited nuclear attack.

A Sentinel System included SPARTAN and SPRINT missiles and the Missile Site Radar (MSR), all upgraded from Nike-X; a new long-range Perimeter Acquisition Radar (PAR); and a Ballistic Missile Defense Center that included a centralized Data Processing System (DPS). The SPARTAN, with a range of nearly 500 miles

SPARTAN Missile

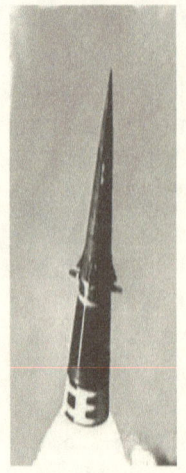

SPRINT Missile

(800 km), was designed to intercept the incoming missiles well outside the Earth's atmosphere and destroy them with a multi-megaton nuclear warhead.

Accelerating at 100 g, the SPRINT would intercept 'leaking' targets at between 5.0- to 100-kft (1.5- to 30-km) altitude in at most 15 seconds, and destroy them with a relatively small nuclear warhead. Martin Company of Orlando, Florida, was the SPRINT contractor. The HIBEX (HIgh Boost EXperiment) missile was a design predecessor and competitor to the SPRINT missile; it was a similar high-acceleration missile in the early 1960s, with a technological transfer from that program to the SPRINT development program.

Located in a pyramid-shaped building, the MSR had over 20,000 elements distributed in arrays on its four faces. Developed by General Electric, PAR had a huge phased-array antenna pointed north to detect nuclear warheads fired by China or the Soviet Union as they passed over the North Pole. With the array

Missile Site Radar

incorporating 6,888 elements, the PAR was capable of identifying and tracking incoming basket-ball-sized targets at ranges up to 2,000 miles (3,200 km).

The primary function of the central DPS was managing system resources and controlling a large radar-tracking and missile-guidance system in real time. The central logic and control could be configured with up to ten processors, each with a throughput of about 1.5 million instructions per second. Every processor had access to each of several read-only instruction memories as well as read-write memories; these had a memory-cycle time of 500 nanoseconds and a double-word size of 64 bits to provide a memory bandwidth in excess of that required for maximum performance of a single processor.

The Army Corps of Engineers (CoE) established a nationwide district based out of Huntsville in October 1967. Designated CoE-Huntsville and under BG (later MG) Robert P. Young, their exclusive mission was to handle the huge BMD site installations expected over the coming years.

SENTINEL / SAFEGUARD SYSTEM COMMAND

In November 1967, A Sentinel System Office was established in the OCS, Washington, D.C., and LTG Alfred D. Starbird was named the Sentinel System Manager. The Sentinel System Command (SENSCOM), located at Redstone

Arsenal, was formed, absorbing the former NXPO; COL Ivey O. Drewry was promoted to Brigadier General and made Commander. SENSCOM was assigned the Kwajalein Test Site as a subordinate element; in 1968, this was renamed Kwajalein Missile Range (KMR). The Sentinel Logistics Command moved from Washington, D.C. to Huntsville.

The Army Ballistic Missile Defense Agency (ABMDA) was formed in 1968, directed by Julian Davidson and reporting to CSA. Although head-quartered in Washington, ABMDA had a large operation in Huntsville and took over Sentinel and some other BMD projects previously under MICOM. ABMDA established the Nike-X Development Office (NXDO), and it and SENSCOM set up operations in a new facility at 110 Wynn Drive in the Research Park, directly

Julian Davidson

adjacent to TBE's campus. Built by a commercial firm for lease to the government, the building complex had 300,000 ft^2 of office space and was also occupied by the CoE-Huntsville and the Sentinel Logistics Command. These were the first of a long series of Army BMD organizations at this location.

In March 1969, newly inaugurated President Richard M. Nixon changed the nation's BMD system from Sentinel to Safeguard, a system composed of the same basic components as Sentinel but intended to protect up to 12 strategic military sites and Washington, D.C. – primarily from low-intensity attacks by China and unintentional launches of Soviet missiles. Organizationally, it was mainly a change in names from Sentinel to Safeguard – SENSCOM became the Safeguard System Command (SAFSCOM)

On 31 July 1969, BG Ivey Drewry, then CG of SAFSCOM, retired; BG Robert C. Marshall was named CG. On 6 August, the Senate authorized (by two votes) the deployment of the Safeguard System. (See next Chapter for Continuation of the Safeguard System.)

MISSILE INTELLIGENCE ORIGINS

In June 1956, General Medaris established a Signals Intelligence Office on the ABMA R&D Staff to obtain missile and space intelligence data and report on foreign activities. With six engineers and analysts, this was the first organizational existence of what would eventually become a major intelligence operation on Redstone Arsenal.

Carl E. Duckett came to Redstone Arsenal among the civil service personnel transferred from White Sands Proving Grounds (WSPG) with Wernher von Braun's team. In July 1956, he joined the ABMA Guidance and Control

Laboratory as a telemetry specialist and also served as an advisor to the commanding officer. It is likely that he also participated in the activities of the Signals Intelligence Office.

During WWII, Duckett had served as a radar technician in the U.S. Army. Later assigned to the WSPG, he participated in the first launches of rebuilt German V-2 rockets and gained knowledge of the telemetry equipment used in this testing. Upon leaving Army active duty, Duckett remained at White Sands as a civil-service employee. During his tenure at WSPG, he learned from the Germans all about the telemetry previously used at Peenemünde.

In early 1957, it became known that the Soviet Union was testing an intercontinental ballistic missile (ICBM) at their Tyuratam range in Ukraine. The CIA opened a listening station in northeast Iran, some 1,000 miles (1,600 km) across the Caspian Sea from the range. Tapes were made of signals obtained by the listening post, and an *ad hoc* activity called Jam Session was started by the CIA for their interpretation; Duckett was asked to be a Jam Session participant. He soon recognized that the Russians were basically using the same telemetry frequencies and formats originally developed by the Germans at Peenemünde.

Telemetry is the radio system used to transmit instrumentation signals generated at one point to another point for analysis. It is vital to rocket launch operations, sending pulsed information from sensors on the rocket – pressure, temperature, acceleration, etc. – back to the ground. Interpretation of these signals gives operational characteristics of the rocket.

With the launch by the USSR of Sputnik I on 4 October 1957, the CIA assembled a Telemetry and Beacon Analysis Committee (TABAC) to analyze the signals recorded from the launch at Tyuratam as well as those heard all over the world from the satellite. The telemetry used on Sputnik I had two transmitters, 20.005 and 40.002 MHz, with pulse-position modulation on one and pulse-duration on the other. The launch-vehicle itself had the Tral telemetry system; this was a much more complex version of pulse-position modulation developed specifically for monitoring by the NII (Research Institute) 885.

Duckett was asked to be a leader in the TABAC effort. The telemetry from the satellites was similar to the earlier German format, but it required about 18 months to calibrate the Tral signals and understand how they related to the launch-vehicle characteristics. The payoff, however, was significant; for the next two decades this provided a major window into the testing and operation of Soviet missiles.

In mid-1958, the ABMA became a sub-unit of the newly formed AOMC. At that time, a small Missile Intelligence Division, led by Duckett, was established as a separate unit of ABMA. A receiving station with a 84-ft parabolic-dish antenna was built by ARGMA atop Madkin Mountain, allowing local reception of signals from Soviet satellites. This antenna could be seen from throughout the city, but its purpose was never given a public disclosure.

When MICOM was formed in August 1962 (described later), the Missile Intelligence Directorate

Madkin Antenna

Carl Duckett

(MID) was established. With about 75 personnel and facilities in Building 4505, an Army Colonel was the Director, and Carl E. Duckett was the Technical Director. The Research Laboratories of Brown Engineering Company was put under contract to provide science and engineering support to the MID.

Duckett joined the CIA in 1963, and eventually, serving in the Number Three position, headed their Directorate of Science and Technology for many years.

In the late 1960s, MID obtained one of the first CDC 170 supercomputers, and many others followed through the years. MID also established capabilities in translating foreign-language publications and hardware documentation.

In 1967, the MID became one of the six major production units of the newly formed Army Intelligence Agency. The MID was assigned primary production responsibility for technical intelligence covering foreign tactical ballistic missile systems, ground systems, and defensive missile systems   This eventually evolved to become the Missile Intelligence Agency in 1970.

Cuban Missile Crisis

In 1959, the CIA initiated a program, called Corona, to build a satellite-based photography system for gathering intelligence imagery; Lockheed served as the prime contractor. The exposed film would be ejected in a capsule, slowed by a parachute in the atmosphere, and snatched by an Air Force C-119 aircraft flying near Hawaii. The first successful operation was in August 1960. Although Corona imagery had excellent resolution, the identification of objects on the ground was highly challenging. Once a suspected target was identified, close-up images could be obtain by U-2 aircraft; these had been overflying selected areas of the world since 1956, but their photographs gave only a small fraction of the coverage of satellite-based photography.

When the MICOM MID got underway, Rankin A. Clinton, Jr. (later the Technical Director) led in establishing a capability in photo interpretation. Having established a reputation for the analysis of missile hardware from

Rankin Clinton

ground-level and U-2 photographs, the MID became a participant in interpreting Corona imagery. The imagery and subsequent analysis reports were sensitive compartmented information, and carried the designation TALENT KEYHOLE. The original TALENT compartment was created in the mid-1950s for the U-2. In 1960, it was broadened to cover all national aerial reconnaissance and the KEYHOLE compartment was created for satellite intelligence.

In September 1962, Corona imagery examined by the CIA indicated possible missile launch sites being built in Cuba; on 14 October, U-2 reconnaissance gave clearer pictures. Rankin Clinton and the MID team met at CIA Headquarters in Langley, Virginia, to analyze the Corona and U-2 pictures. It was soon determined that the sites were for Soviet R-12 DVINA (NATO designation SS-5 SANDAL) and Soviet R-14 CHUSOVAYA (NATO designation SS-5 SKEAN), respectively, medium-range and intermediate-range missiles. Other imagery showed missiles being unloaded in Cuba from cargo vessels.

A Top Secret report was prepared for President John F. Kennedy, giving the missile situation and stating that both types of missiles were capable of delivering nuclear warheads to U.S. targets. This set in motion what became known as the Cuban Missile Crisis, generally regarded as when the Cold War came closest to turning into a nuclear conflict. President

U-2 Photograph of Missile Site

Kennedy imposed a military blockade of Cuba. The confrontation ended on 28 October 1962, when Soviet Premier Nikita Khrushchev agreed to dismantle the offensive missiles and return them to the USSR.

Soviet Hardware Exploration

From its start, the Missile Intelligence organization at Redstone Arsenal has had technical experts and facilities devoted to the exploitation of Soviet hardware, primarily missiles and radars. Little is ever made public concerning this activity – it usually carries a high level of security classification, and is often done in association with IMINT (Imagery intelligence), SIGINT (Signals intelligence), and MASINT (Measurement and signature intelligence). In addition to the civil service engineers and technologists, appropriately cleared contractors were involved, some in the obtaining of foreign-built components and systems. The author of this book participated in this endeavor.

OTHER IMPORTANT ACTIVITIES

Several other activities of the Army during the 1950s/1960s that had major importance in the evolution of Greater Huntsville are described in the following sections.

Explorer – America's First Satellite

As part of its participation in the 1957-58 International Geophysical Year (IGY), the United States planned to place a scientific satellite in Earth orbit. At a meeting of the American Rocket Society in 1954, Wernher von Braun had outlined a plan for launching an artificial satellite using a REDSTONE rocket with a cluster of small solid-fuel rockets atop.

For the IGY effort, Project Orbiter was initiated as a joint effort between the Army's Ordnance Missile Laboratories (OML) at Redstone Arsenal and the Naval Research Laboratory (NRL). The Navy later withdrew participation and began development of the VANGUARD rocket; this was a three-stage system that had also evolved from a V-2 heritage. The DoD Committee on Special Capabilities selected the VANGUARD rocket from the NRL for the IGY launch.

Although not authorized to do so, von Braun and his space team, now at the newly formed ABMA, had continued with what they called the JUNO-1. This was a three-stage JUPITER-C – developed for testing in the JUPITER program – with a small fourth-stage payload carried within the nose cone of the third stage. A SERGEANT rocket motor in this fourth stage could take it to an orbital velocity of 18,000 mph (8.0 km/s); the JPL, directed by William A. Pickering, would supply this fourth stage.

James Van Allen of the University of Iowa had earlier proposed that an experiment in examining cosmic radiation at orbital altitudes be included in the IGY satellite. For the JUNO-1 scientific payload, an omnidirectional Geiger-Müller tube from George Ludwig of Iowa's Cosmic Ray Laboratory would be included. To provide maximum volume for this cosmic-ray detector as well as other sensors, transmitters, and batteries, it was decided to make the entire fourth stage into an orbiting scientific payload called Explorer.

J. Warren Harper and Thomas Alan Barr at ABMA designed a highly unusual antenna for the satellite. Two fiberglass slots/rings were placed around the metallic enclosure, one at the front and the other at the back, and could be excited to give omnidirectional dipole-like radiation around the Explorer axis.

On 4 October 1957, America was surprised and humiliated by the USSR putting up Sputnik I, the first Earth satellite. Until that time, efforts in the U.S. were waiting for success with NRL's VANGUARD rocket.

Testing the Antenna

The NRL schedule was accelerated, but ABMA was not authorized to continue with JUNO-1. The Soviets followed with Sputnik II on 3 November; aboard was a dog, Liaka, the first living creature in Earth orbit.

ABMA was then told to "prepare" for a launching, but there was no full authorization to continue. The first attempt by the NRL to launch a satellite-carrying VANGUARD was made on 6 December 1957. A nationwide television audience saw the rocket lift slightly off the pad, then topple over and explode in a giant fireball.

Toftoy, Medaris, and von Braun immediately pleaded for the opportunity to show what the Army's "space team" could do. Although the NRL had the satellite mission, work on a system with satellite-launching capabilities had continued at ABMA; von Braun said that this launch system could be made ready in two months. The go-ahead was given.

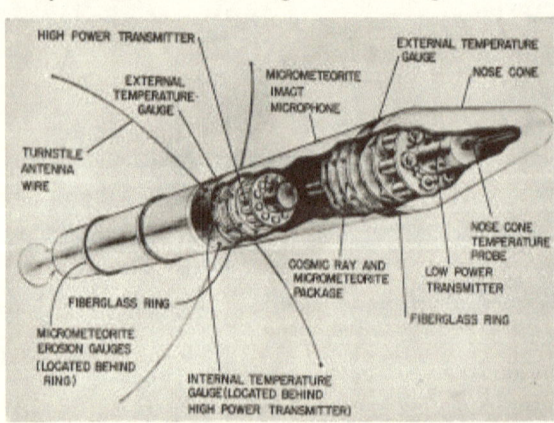

Explorer 1 - America's First Satellite

Completion of the scientific payload was a closely coordinated effort between ABMA, JPL, and the Physics Department of the University of Iowa. Walter Haeussermann and Ernst Stuhlinger led the ABMA research efforts; Robert E. Lindstrom was the ABMA project engineer. In addition to an omnidirectional Geiger-Müller tube to detect cosmic radiation, Explorer-1 had crystal-acoustic and wire-grid sensors to detect impacting micrometeoroids; there were also five temperature sensors.

Explorer-1 had two transistorized transmitters: one 60 mW at 108.00 MHz, and a smaller one 10 mW at 108.03 MHz. In completing the payload, engineers at JPL added a backup to the slot antenna – a flexible turnstile antenna with four wires extending outside the satellite body. Electrical power for the transmitters and sensors was from mercury chemical batteries, taking some 40 percent of the payload weight.

On January 31, 1958, America's first satellite, Explorer-1, was placed into orbit using JUNO-1. Since JUNO-1 looked like and was, in fact, a modified version of the JUPITER C, the media has most often credited JUPITER-C with the launch. Von Braun in particular wanted the JUNO-1 designation, getting away from the name JUPITER that was known to the public as a military missile.

The launch was from Cape Canaveral, Florida, with an orbital inclination of 33.24 degrees. The expected orbital period was about 105 minutes. After being

launched at 10:47 PM EST, it should have gone eastward around most of the Earth and passed over tracking stations in Earthquake Valley near San Diego at about 12:30 AM EST. As von Braun and others waited, no satellite signal was detected by the tracking stations; there was great anxiety – had Explorer 1 failed? By 12:40 and still no signal, everyone was convinced of failure. Thirty seconds later, the signal came in

Worried

loud and clear – Explorer-1 was in orbit!

Explorer Leaders: L-R William Pickering, James Van Allen, and Wernher von Braun

The delay was explained – but never proven – as the four stages having very slightly more thrust than expected, placing Explorer-1 into an orbit with an actual period of 114.8 minutes.

As the Nation watched via live television, Huntsville had an unequaled celebration. This relatively unknown community in Alabama was suddenly at the center of the free-world's space race with the Soviet Union. The government laboratories on Redstone Arsenal, together with Huntsville's emerging industries, immediately acquired recognition as a center for high-technology.

The first successful satellite was followed on 3 March 1958, with the launch of Explorer-2, essentially identical with Explorer-1, but this failed to achieve orbit. Another try was made on 26 March; Explorer-3 orbited and operated for 93 days. It had a highly eccentric orbit with an apoapsis (greatest distance away from the Earth) of 2,800 km (1,700 mi). The cosmic-ray data led to the determination that there was a belt of charged particles starting at about 1,000 km (600 mi) above the Earth and later found to extend to some 60,000 km (36,000 mi); this is the Van Allen Belt, named for its discoverer. The discovery of the Van Allen radiation belt was considered to be one of the outstanding discoveries of the IGY.

Explorer-4 was launched in a secret mission on 26 July 1958; it remained in orbit 71 days, during which time the telemetry sent information on three Operation Argus nuclear weapons tests at high altitude. Explorer-4, with an eccentric orbit that took it as high as 2,200 km (1,400 mi), also mapped the lower boundary of the Van Allen Belt. Explorer-5 was launched 24 August, but failed to achieve orbit.

After NASA was officially formed on 1 October 1958, further missions in the Explorer series were under this new organization in cooperation with JPL; these continued until 6 October 1981, and included 54 successful missions. It is noted that NASA established a policy of not giving a designation name to unsuccessful launches.

Saturn Rocket – A Heavy Lift Vehicle

Following the surprise launch of a satellite by the USSR, the Department of Defense (DoD) studied why the U.S. was falling behind. They concluded that bureaucracy was a significant contributor; the Army, Navy, and Air Force competed with each other in pursuing new technologies and wasted scarce funds in doing so. To partially overcome this problem, a new research and development organization was planned. Focused on space launchers and given wide discretionary powers that cut across traditional Army/Navy/Air Force lines, this would be at the top level of the DoD and report to the Director. This organization, the Advanced Research Projects Agency (ARPA), officially came into being on 7 February 1958.

While ARPA was being formed, the DoD released requirements for a new system that would be one of the first for the new agency. This was for a heavy-lift vehicle to orbit advanced communications and "other" highly classified satellites. Basic requirements were for a vehicle capable of putting 19,800 to 39,600 lb (9,000 to 18,000 kg) into orbit, or accelerating 5,900 to 11,800 lb (2,700 to 5,400 kg) to escape velocity – the speed necessary to leave Earth orbit.

Von Braun already had ABMA planning higher versions of the JUNO rocket, and immediately recognized this heavy-lift vehicle as a way to continue on a path to space projects. He assigned the effort to Heinz-Hermann Koelle, chief of Future Projects Design, and Koelle's preliminary analysis showed that a first-stage booster with a thrust of some 1.5 Mlbf (6.7 MN) would be necessary.

Although a native of Germany, Koelle was not a member of the Project Paperclip group. While still in undergraduate school, his talents had been recognized by von Braun, and, upon graduation, he had been invited to join ABMA in 1955. Although much younger than other members of the German team, Koelle was a "favorite" of von Braun.

In December 1957, ABMA submitted a proposal to the DoD for a SUPER-JUNO first stage, composed of a JUPITER surrounded by eight REDSTONES serving as fuel tanks, and a thrust plate with four Rocketdyne E-1 engines, each producing 380-klbf (1,700-kN) thrust. All but the E-1 engines were off-the-shelf components. This became the first major project considered by ARPA.

In reviewing this proposal, ARPA accepted the clustered-rocket approach, but asked for a lower risk by changing the E-1 engines, which were still under development. ABMA responded by switching to eight Rocketdyne H-1 engines, with 188-klbf (840-kN) thrust each; these were 25-percent improved S3D engines already in use on the JUPITER and THOR. The total from the eight engines met Koelle's thrust requirement and also significantly decreased the cost and delivery time.

Accepting this basic design for the first stage, ARPA issued an order to proceed on 15 August 1958; there was a goal to demonstrate a full-scale, captive firing by the end of 1959. Von Braun called this the JUNO 5; the ABMA team, however, was already referring to the design as SATURN (indicating the succession as the planet after Jupiter). In February 1959, the name SATURN was officially adopted; then within a short time, this basic vehicle was called SATURN I, although there was still a debate as to the upper stages. Plans were started on an even larger first stage and a next generation vehicle, SATURN II, then later III, IV, and V.

In the Foreword to historian Roger E. Bilstein's seminal book *Stages to Saturn*, William R. Lucas, then Director of NASA Marshall Space Flight Center, summarizes the remarkable building of SATURN:

> It was as if the Wright Brothers had gone from building their original Wright Flyer in 1903 to developing a supersonic Concord in 1913. Unimaginable; yet in 10 short years the builders of Saturn progressed from the small, single-engine rockets like Redstone to the giant vehicle with clustered engines that put man on the Moon. Our Earth-to-orbit weight-lifting capability grew in that decade by 10 thousand times.

Project Horizon – Outpost on the Moon

The newly formed ARPA had provided funding for the initial development of a heavy-lift vehicle, SATURN, but there was no funded Army program having a requirement for this vehicle. MG John Medaris, commander of the Army Ordnance Missile Command (AOMC) of which AMBA was a part, asked Wernher von Braun to conduct a study of an appropriate application for SATURN, therefore developing such a requirement.

President Dwight D. Eisenhower had just formed the National Aeronautics and Space Administration (NASA) in which von Braun was expected to become a key member; thus, von Braun was reluctant to be responsible for the study, but offered Heinz-Hermann Koelle in his place. Subsequently, Francis L. (Frank) Williams, a senior design engineer in Koelle's Preliminary Design Section of ABMA, organized the study and proposal development, with Koelle heading the technical aspects.

There were a number of reasons driving the urgency of such a project. The Soviet Union had announced that their 50th anniversary, in 1967, would be celebrated by Soviet citizens on the Moon. The National Space policy intelligence estimate was that the Soviets could land on the Moon by 1968. To be second to the USSR in establishing an outpost on the Moon would be disastrous to our nation's prestige. The USSR in establishing the first permanent base, could claim the Moon or critical areas thereof for its own; then a subsequent attempt to establish an outpost by the United States might be considered and propagandized as a hostile act.

Proponents referred to the lunar outpost as "high ground," meaning a strong advantage to military actions below it on the Earth. The study was authorized in a letter from the Chief of Ordnance dated 20 March 1959; it included the following justification:

> The lunar outpost is required to develop and protect potential United States interests on the Moon; to develop techniques in moon-based surveillance of the Earth and space, in communications relay, and in operations on the surface of the Moon; to serve as a base for exploration of the Moon, for further exploration into space and for military operations on the Moon if required; and to support scientific investigations on the Moon.

In addition to personnel from ABMA, a team of specialists from all 11 of the Army Corps (Corps of Engineers, Ordnance Corp, Medical Corp, etc.) were temporarily assigned to Redstone Arsenal for the effort. The study report, classified Secret and in two volumes, was dated 9 June 1959, and titled "Project

Horizon, A U.S. Army Study for the Establishment of a Lunar Military Outpost."

Horizon Space Dock

Major results of the study were as follows: The permanent outpost was predicted to cost $6 billion ($38 billion in today's dollars), and involve an effort comparable with the Manhattan Project that developed the atomic bomb. It would require 61 SATURN-I and 88 SATURN-II launchings. An important part of the delivery system would be the Horizon Space Dock (a space station), built from spent SATURN vehicles and placed in an equatorial Earth-orbit. Deliveries from the Earth of supplies and fuel by a number of vehicles would be received and held in this Space Dock, and then transferred to a carrier for the trip to the Moon.

Deliveries to the Moon would begin in January 1965, and the first manned landing by two men would be made in April 1965. The build-up and construction phase would be continued without interruption until the outpost was ready for beneficial occupancy and manned by a task force of 12 men in November

Initial Lunar Outpost

1966. During this period, some 490,000 pounds of useful cargo would be transported to the Moon. In the first operational year of the outpost, 64 additional SATURN II launchings would be needed to provide an additional 266,000 pounds of cargo.

The Project Horizon report was endorsed by the Ordnance Corp's Chief of Research and Development, LTG Arthur G. Trudeau, and was passed through and approved by the Chief of Ordnance as well as the Secretary of the Army. Although the estimated cost was extremely high, it was justified by the belief that the Soviet Union would have a similar outpost in less than 10 years.

The plan was ultimately presented to President Eisenhower, who rejected it with the simple comment that any manned mission to the Moon would be under the newly formed NASA.

It was later found that the Air Force Space Systems Division had, during 1958-1960, conducted a similar study. Called Lunar Expedition (code-named Lunex), major milestones were to be a manned circumlunar flight in 1966, a three-man landing and return in 1967, and a Lunar expedition in 1968.

ARMY MISSILE COMMAND

Following the transfer of the major missile research and development capabilities from AOMC / ABMA to NASA in officially forming the George C. Marshall Space Flight Center (MSFC) on 1 July 1960 (described in next segment), the Army's capabilities at Redstone Arsenal for developing new missile systems were fragmented. Some residual elements of ABMA were continuing with larger missiles – such as PERSHING – and ARGMA was responsible for the development of a number of smaller missiles, but there was no central activity to replace the overarching R&D capability of Wernher von Braun's organization.

In 1962, the U.S. Army underwent a sweeping reorganization; one of the results of this was the creation of the Army Missile Command (MICOM). This reported to the Army Materiel Command (AMC), headquartered in Washington, D.C. Activated at Redstone Arsenal on 1 August 1962, MICOM was built from the remaining units of ABMA, particularly ARGMA. MG Francis J. (Frank) McMorrow was named MICOM's first Commanding Officer.

General McMorrow made two significant changes: the

Frank McMorrow   establishment of project managers for missile system programs, and the consolidation of research and development activities in a

R&D Center. For the Center, funds were obtained for a large facility (Building 5400) on Martin Road. General McMorrow died unexpectedly in August 1963. In March 1964, the 205,000 ft^2 (20,300 m^2) building was completed and dedicated as the Francis J. McMorrow Missile Laboratories.

McMorrow Laboratories

With the creation of MICOM, all of the R&D activities of ARGMA, and those residual in AOMC, were consolidated to form the Directorate of Research and Development; John L. McDaniel was the Technical Director. McDaniel, a chemist by education, had come to Huntsville to work in the munitions factory in 1942, and was a technical leader from the start of rocket development.

John McDaniel   Through the following 1960s, MICOM reestablished Redstone Arsenal as the leader for Army missile research, development, and procurement. The commanding generals during the decade and their date of assuming command were as follows: MG John G. Zierdt, September 1963; MG Charles W. Eifler, July 1967; MG Edwin I. Donley, November 1969.

Research and Development Activities

Research and development in a wide variety of disciplines of science and technology was vital to fulfilling MICOM's mission. Shortly after the opening of the Directorate of Research and Development in 1962, groups were established for research in electromagnetics, structures and mechanics, aerodynamics, optics, atmospherics, thermal sciences, propulsion, solid-state physics, plasma dynamics, mathematical modeling, and other areas.

Radar was a major unit in many of the missile systems, particularly those for air defense. This technology was used in detecting and tracking moving targets, as well as in missile guidance. For air defense, the Forward Area Alerting Radar (AN/MPQ-49) was developed and went into production; this was a pulse-Doppler system operating in the 1 to 2 GHz (30 cm and 15 cm wavelength) microwave band, with a range of about 20 km (12 mi).

In general, the higher the frequency of the radar, the more precise its operation; thus, there were research efforts in moving up from microwave to millimeter wavelengths. Also involved was research in signal processing, propagation analysis, counter-and counter-countermeasures, as well as transmitter and receiver components. As previously noted, Stephen L. Johnson made major developments in radars for missile guidance and in radar countermeasures.

Optics, particularly infrared optics, was highly important for passive sensors on missiles. All objects that are above absolute zero in temperature emit optical radiation, with the intensity and peak wavelength indicative of the temperature. Targets, such as a missile or aircraft engine, have elevated temperatures that give peak radiation above red in the visible spectrum and is thus called infrared.

Dating from WWII, infrared optical units were used for detecting heated targets and in generating guidance signals for seeking missiles. The actual detectors were mainly solid-state devices similar in chemical characteristics to transistors. Research in infrared detectors and related technologies – such as atmospheric transmission, jet-plumb effects, and cryogenics – at MICOM was carried over from earlier ARGMA research. The REDEYE (FIM-43) was a well-known missile of this type; it was put into production in the late 1960s.

Immediately after the first papers on lasers were given, these devices were recognized as having major applications in Army weaponry. David J. Salonimer started work at ARGMA on laser guidance concepts in 1961. At MICOM,

Salonimer and Norman Bell examined target designation by a pulsed-laser beam and proposed a guidance system to home in on the reflected laser light.

William A. Davis, Jr., led the development of high-energy lasers, and, in 1963, received the first AIAA Holger N. Toftoy Award for his management of laser programs.

The quest for high-energy lasers led some of the MICOM researchers into a highly diverse activity – the

William Davis

application of lasers in the medical field. Working with the National Institutes of Health, William A. McKnight with Robert C. Hoye, MD, from the National Cancer Institute used pulsed, high-energy lasers for treating cancers in mice, rabbits, and monkeys. This activity became public in 1964. McKnight was awarded a doctorate in physics from Oxford University in 1968, with a dissertation based on his research at MICOM.

Robert Hoye (L) and William McKnight

Missile Systems

During the 1960s, MICOM was responsible for 22 land-combat and air-defense missile systems. The land-combat systems were PERSHING, SERGEANT, REDSTONE, CORPORAL, DRAGON, HONEST JOHN, LITTLEJOHN, LACROSSE, LAW, LANCE, SHILLELAGH, 2.75-Inch Rocket, and, from France, ENTAC, SS-10, and SS-11. The air-defense missile systems included NIKE AJAX, NIKE HERCULES, NIKE ZEUS, CHAPARRAL, HAWK, MAULER, and REDEYE. Some were carried over from earlier developments; several of those developed under MICOM are briefly discussed.

It is noted that performance characteristics of these missiles somewhat vary with the date of reporting; those shown are possibly conservative. Also, some of the warheads are nuclear; insofar as is reported, none of these devices were at Redstone Arsenal.

Several other missile systems were studied in this period. Included were Field Army Ballistic Missile Defense (FABMDS, noted earlier) and its follow-on Army Air Defense System - 1970 (AADS-70). In 1964, AADS-70 became Surface-to-Air Missile - Development (SAM-D), which ultimately evolved to became the PATRIOT (discussed in the next chapter).

LITTLEJOHN – Like its predecessor, HONEST JOHN, the MGR-3

LITTLEJOHN was a free-flight artillery rocket that followed a ballistic trajectory to its target. It was an in-house development, started in 1955, tested in 1956, and manufactured by Douglas Aircraft beginning in 1961. The primary difference in the two JOHN missiles was weight; the "little" version was 350 kg (780 lb) – only13 percent that of the earlier rocket and allowing a light-

LITTLEJOHN Missile   weight, easily transportable launcher. Its reduced range was 19 km (12 mi), and it carried a W-45, 1-10 kton nuclear warhead. Only about 500 were produced.

LAW – Invented in 1942, the M1 Portable Rocket Launcher – commonly called the bazooka – was widely used during WWII as a weapon against tanks

and fortifications. An early activity at Redstone Arsenal was a project to improve on the large, cumbersome bazooka. In one of its first tasks, Rohm and Haas developed a solid propellant for what they called a light anti-tank weapon (LAW). The Hesse-Eastern Division of Norris Thermodore designed a full system – the M72 LAW – and started production in 1963. The fire-and-forget LAW missile has a velocity of 145 m/sec (475 ft/sec) burning

LAW Missile

completely before leaving the muzzle, and a range of 200 m (660 ft). The warhead penetration is 20-cm (7.9-in) of steel plate, 60-cm (24-in of reinforced concrete, or 1.9 m (5.9-ft) of soil. American production ended in 1983.

DRAGON – Another shoulder-launched missile in that time frame was the

M47 DRAGON, an anti-tank weapon. Designed by Raytheon starting in 1966, this used wire-guidance with a night tracker for 24-hour operation. The speed was 300 m/s (660 ft/s), and the effective range was 75 to 1500 m (2,400 to 4,900 ft). The cone-shaped, high-explosive warhead was capable of penetrating armored vehicles, fortified bunkers, concrete gun emplacements, and other hard targets. Manufactured by McDonnell

DRAGON Missile   Douglas starting in 1975, a total of 50,000 missiles were produced for the U.S. Army and U.S. Marine Corps. It was retired in 2001.

SHILLELAGH – Named for a wooden club weapon from Ireland, the MGM-51 SHILLELAGH was a guided, anti-tank missile launched from

161

cannons on Sheridan and Patton tanks. It was developed by Ford Aeronutronics starting in 1958, and was first tested in 1960; production by Martin Marietta began in 1964. About 88,000 rounds were produced through 1971, but very few were used in combat. Guidance was by the gunner using the tank's standard

SHILLELAGH Missile

gunsight with corrective signals generated by an analog computer and sent to the missile through an infrared link. The range was 2,000 m (6,600 ft) with a speed of 320 m/sec (1,060 ft/sec); the warhead weighed 6.8 kg (15 lb) with a shaped charge that could penetrate 15-cm (5.9-in) armor.

CHAPARRAL – When BTL's Nike program was redirected to ballistic missile defense in the late 1950s, there was then a critical need for a high-

performance anti-aircraft missile. Development began on the MAULER (MIM-46) for this purpose, but major problems were encountered and the MIN-72 CHAPARRAL started development as a replacement in 1965. This was an infrared-guided missile based on the Navy's air-to-air AIM-9D SIDEWINDER, modified for

CHAPARRAL Missile

ground launch. It was carried by a tracked vehicle with a launcher holding four missiles and eight stored. The missile had a range of 500 to 9,000 m (1,600 to 30,000 ft), altitude of 25 to 4,000 m (82 to 13,000 ft), speed of Mach 1.5 (1,140 mph), and carried a 12.2 kg (27.8 lb) continuous-rod warhead. Ford developed the launcher vehicle, and Loral Aerospace the missiles. First placed into service in 1969, it continued in use until 1997.

LANCE – The MGM-52 Lance was a mobile, field-artillery, tactical missile with a self-propelled launcher. The engine burned pre-packaged liquid propellant, giving a maximum range of 130 km (81 mi). With an inertial guidance system, it had an accuracy of 100-m CEP. Its warhead was either a W-70, 1-100 kton, radiation-enhanced nuclear bomb (the first on a U.S. weapon) or a variety of conventional munitions. The development initiated in 1962, was first tested in 1965, and deployment began in 1972.

LANCE Missile

Production was by LTV Corporation (later Vought); 2,133 systems were built.

Redstone Scientific Information Center

A technical library has been available on Redstone Arsenal since it became a military installation in 1948. Later, the Defense Documentation Center Field Office was also available with microfilm documents. In August 1964, these two operations merged, resulting in the Redstone Scientific Information Center (RSIC) becoming what was likely the largest technical library operated by the U.S. Government. Included at the time of merger were over 234,000 microfilm documents, 1.2 million special documents such as patents and translations, collections of some 3,000 journals, and about 50,000 cataloged books. Cleo S. Cason headed the library operations at Redstone Arsenal from the start.

Missile and Munitions School

In May 1963, Actress Tallulah Bankhead, a native of Huntsville, dedicated a new building at the Ordnance and Guided Missile School (OGMS), memorializing her father, U.S. Representative William B. Bankhead and her uncle, U.S. Senator John H. Bankhead.

OMMCS Area 1960s

The Ordnance Corps ammunition training, including explosive ordnance disposal training, was relocated to RSA in 1965. OGMS became the directing agency for nuclear weapons instruction, although actual training was conducted at other locations. Following the addition of these new missions the school was renamed the U.S. Army Ordnance Missile and Munitions Center and School (OMMCS).

In 1966, a life-size model of a Viet Cong village was built on Redstone Arsenal as part of a new training program at OMMCS for personnel headed for duty in Vietnam. In addition to the tunnels, huts, shelters, and living facilities found in a real Vietnamese community, booby traps, mines, bamboo spikes, and other war devices unique to the Viet Cong made the training more realistic.

Metrology and Calibration Center

A metrology laboratory was established shortly after MICOM was formed; this was gradually enlarged to become the most complete of its type in the Department of Defense. The John M. Cone Laboratories, named for the commander of White Sands Missile Range at the time of his death, was opened by the Metrology Center in 1966.

In June 1967, AMC formed the Army Metrology and Calibration Center (AMCC) and Standards Laboratory at MICOM. The new center incorporated the former Metrology Center and also absorbed elements from Frankford Arsenal and Tooele Army Depot. The Center was responsible for the program management and technical guidance for the total U.S. Army calibration system.

In January 1971, the AMCC became a direct report to the Army Materiel Command. With close ties to the National Bureau of Standards (the National Institute of Standards and Technology in 1988), the AMCC became responsible for all aspects of measurement standards and associated equipment within the U.S. Army. Among other activities, the AMCC conducted technical measurement audits at 11 Army Area Calibration Laboratories and 23 Army Internal Calibration Laboratories. Over time, this evolved to become the U.S. Army Test Measurement and Diagnostic Equipment Activity (USATA).

NASA ACTIVITIES

Since 1915, the National Advisory Committee for Aeronautics (NACA) had been the federal agency responsible for aeronautical research. After the Soviets launched the world's first artificial satellite (Sputnik 1) on 4 October 1957, the U.S. Congress, alarmed by the perceived threat to national security and technological leadership, urged immediate and swift action; this led President Dwight D. Eisenhower to propose a new federal agency to conduct all non-military activities in space. Advanced Research Projects Agency (ARPA), formed in early 1958, would develop space technology for military application.

The National Aeronautics and Space Administration (NASA) was officially established on 1 October 1958; its stated mission was to "provide for research into the problems of flight within and outside the Earth's atmosphere, and for other purposes." T. Keith Glennan was the first Administrator NASA absorbed NACA intact, including 7,500 employees and three major research installations: Langley Aeronautical Laboratory, Ames Aeronautical Laboratory, and Lewis Flight Propulsion Laboratory. These were renamed Research Centers, and abbreviated, respectively, LaRC, ARC, and LRC. At the same time, the Jet Propulsion Laboratory (JPL) was transferred from AOMC to NASA.

Within the first week of NASA's existence, it was authorized to establish Project Mercury with the primary object of putting a man into Earth orbit. The Space Task Group at NASA LaRC was responsible for Project Mercury, and McDonnell Douglas Aircraft received a contract to build the one-man Mercury space capsule. REDSTONE rockets from ABMA would be used for sub-orbital testing and manned flights, and ATLAS-D rockets from the Air Force would boost the capsule on manned orbital flights.

MARSHALL SPACE FLIGHT CENTER

Redstone Arsenal's role in the space field began to change dramatically when President Eisenhower decided that it would be in the Nation's best interest to make the von Braun team a part of NASA. On 21 October 1959, the President

approved such a transfer. This was accomplished effective 1 July 1960, when 4,670 civilian employees and about $100 million worth of buildings and equipment transferred from AOMC / ABMA to officially form the George C. Marshall Space Flight Center (MSFC). At the same time, a long-term lease was obtained by NASA for 1,841 acres of Redstone Arsenal land. Wernher von Braun was appointed the MSFC director.

MSFC Transfer Ceremony

MSFC was dedicated on 8 September 1960, by President Eisenhower in person. The Center was named in honor of Five-Star General of the Army George C. Marshall – Army Chief of Staff during World War II, United States Secretary of State, and Nobel Prize winner for his world-renowned Marshall Plan.

President Eisenhower and Mrs. Marshall at Dedication

As the Marshall Space Center was being activated, NASA described this as the only self-contained organization in the Nation that was capable of conducting the development of a space vehicle from the conception of the idea, through production of hardware, testing, and launching operations.

Initial Organization

Wernher von Braun had accepted the position as Director with the condition that he would have full authority to organize the technical activities and select all technical leads. All of the initial technical heads appointed by von Braun were individuals who had assisted him in his success at ABMA. Shortly after opening, the technical activities and leaders at MSFC were as follows:

Director – Wernher von Braun
Deputy Director, R&D – Eberhard F. M. Rees
Associate Deputy Director, R&D – Erich W. Neubert
Reliability Office – H. August Schulze
Future Projects Office – Heinz-Hermann Koelle
Light & Medium Vehicles Office – Hans Hueter
Saturn Systems Office – O. Hermann Lange
Technical Program Coordination Office – George N. Constan
Weapons Systems Office – Werner G. Tiller
Launch Operations Directorate – Kurt H. Debus
Aeroballistics Division – Ernst G. Geissler
Computation Division – Helmut Hölzer
Fabrication & Assembly Engineering Division – Hans H. Maus
Guidance & Control Division – Walter Häussemann
Quality Division – Dieter E. Grau
Research Projects Division – Ernst Stuhlinger
Structures & Mechanics Division – William A. Mrazek
Test Division – Karl L. Heimburg

With the exception of Koelle and Constan, all of the technical leaders had come to the United States under Operation Paperclip after working together at Peenemünde. Von Braun knew well the capabilities of these individuals and

L-R: Eberhard Rees, von Braun, and Keith Glennan

had great confidence in them. This confidence was shown to be appropriate; in the following decade of developing hardware and technical operations – there was never a single failure of their designs during manned flight. Arthur Rudolph, a leading member of the German team and future SATURN V leader, did not transfer to MSFC until 1961, remaining with the Army as project manager on the PERSHING missile.

The MSFC administration was conducted by U.S. citizens with extensive experience in governmental operations. The initial activities and leaders were:

Deputy Director, Administration – Delmar M. Morris
Associate Deputy Director, Administration – Harry H. Gorman
Assistant to the Director – Jerry C. McCall
Management Analysis Office – Chauncey W. Huth
General Counsel – William E. Guilian
Public Information Office – Bart J. Slattery
Office of Technical Services – David H. Newby

Office of Management Services – Victor C. Sorenson
Office of Financial Management – Claude E. Stockton
Office of Procurement and Contracts – Wilbur S. Davis

The initial projects were almost exclusively continuations of work initiated earlier at ABMA. In fact, most of the civilian employees were doing essentially the same work in the same facilities as before the organizational separation. Of immediate importance was the final preparation of a REDSTONE rocket that would be used in Project Mercury to carry the first American into space.

The Project Mercury flight was originally scheduled for October 1960, but was delayed until 5 May 1961; on this date, astronaut Alan B. Shepard, Jr., in a Mercury space capsule made a sub-orbital flight. Similar to circumstances of the first satellite launch, the delays allowed the USSR to have the first manned space flight; cosmonaut Yuri Gagarin had orbited the Earth on 12 April 1961.

As the Government transitioned from the Eisenhower to the John F. Kennedy Administration in early 1961, Jerome B. Wiesner, a renowned professor at MIT, was named Science Advisor to the President. Immediately upon taking office, Wiesner released a report making strong recommendations for improving the nation's space program and was critical of NASA's top management. As a result, James E. Webb was appointed the NASA Administrator, and George E. Mueller the Associate Administrator for Manned Space Flight, both highly qualified for their positions.

On 25 May 1961, just 20 days after Shepard's flight, President Kennedy addressed a special joint-session of Congress:

> I believe that this nation should commit itself to achieving the goal, before this decade is out, of landing a man on the Moon and returning him safely to the Earth. No single space project in this period will be more impressive to mankind, or more important in the long-range exploration of space; and none will be so difficult or expensive to accomplish.

Prior to this, MSFC had continued work on the SATURN launch vehicle without a specific NASA application. With President Kennedy's commitment, the lunar program, soon named Apollo, became the application.

Space travel had been the dream of Wernher von Braun since the early 1930s. The activities underway by the government and contractors, and the resulting products and lunar exploration in the Apollo Program, were for him a dream come true.

Organizing for Lunar Exploration

To expand NASA for accomplishing the lunar-landing program, the Space Task Group from LaRC was used to form the Manned Spacecraft Center (MSC). Made official on 19 September 1961, the MSC was initially headed by Robert R. Gilruth and remained at LaRC until the summer of 1962, when facilities became available at a location near Houston, Texas. In 1973, the center was renamed Lyndon B. Johnson Space Center (JSC). Through the years, there were a number of 'turf' battles between MSFC and MSC / JSC concerning mission responsibilities.

Initially, there were two approaches being considered for a mission that would take astronauts to and Moon and then return them to the Earth: a direct mode, and an Earth-orbit rendezvous (EOR) mode. The direct mode was the simplest, involving a minimum of space maneuvers, but a new, more powerful launch vehicle (then called the NOVA) would have to be developed, likely prohibiting the mission from being accomplished "before the decade is out." The EOR mode, favored by von Braun and his team, required two heavy-launch vehicles (SATURN Vs) and complex maneuvers, including joining of the two SATURNs in Earth orbit.

In August 1960, John C. Houbolt of NASA LaRC showed that a lunar-orbit rendezvous (LOR) would be much more energy economical, requiring only a single SATURN V. For almost two years, there was a frantic effort involving NASA Headquarters, MSFC, and the Space Task Group from LaRC, searching for the optimum way to accomplish the lunar-landing mission. On 11 July 1962, NASA finally announced that the mission would be accomplished using the LOR mode.

Following the decision on the LOR mode, the various responsibilities were made known. MSFC would provide the SATURN V, including the inertial guidance system. MSC would be responsible for the transfer stage, the lunar excursion module, and the reentry vehicle. The guidance, control, and navigation for the three spacecraft would be developed at Draper Laboratory of the Massachusetts Institute of Technology (MIT). Both MSFC and MIT would be jointly responsible for the guidance transition from launch rocket to lunar spacecraft.

Between January 1962 and November 1966, Project Gemini was conducted by NASA. Gemini built on Mercury's achievements and extended NASA's human spaceflight program to spacecraft built for two astronauts. Managed by MSC, with George E. Mueller of NASA Headquarters serving as project director, Gemini's 10 flights also provided NASA with more data on weightlessness, perfected reentry and splash-down procedures, and demonstrated rendezvous and docking in space. McDonnell Douglas was the prime contractor for the two-man Gemini spacecraft, with astronaut Virgil I. (Gus) Grissom heavily involved in the design. TITAN II rockets from the Air Force were adapted for the launch vehicle. MSFC had little involvement with Project Gemini.

Central Office Complex

As the Apollo Program got fully underway, MSFC had major growth. Buildings obtained from the Army were enlarged and updated, and a new complex of three high-rise buildings for the headquarters and central administration were built at the northwest corner of the MSFC area. The Huntsville Research Park (described later in detail) was started a short distance north of Redstone Arsenal, and most of the companies associated with the program relocated to the Park, freeing space in the Huntsville Industrial Complex (HIC). MSFC soon had portions of the design activities occupying the HIC facility.

Carried over from ABMA activities, engineers from Huntsville traveled to Florida to initially conduct launch activities at the Cape Canaveral Air Force Station. The first NASA launch facility there (Launch Complex 39) was designed and operated by MSFC. On 1 July 1962, the overall site was given equal status with other NASA centers and was named the Launch Operations Center (LOC) with Kurt H. Debus as Director. In 1963, it was renamed the Kennedy Space Center (KSC).

By 1965, there were some 15,000 people working in MSFC facilities on Redstone Arsenal, about half being government civil-service employees, and the other half employees of the prime contractors Boeing, Chrysler, Douglas Aircraft, IBM, and Rocketdyne, and support contractors including Brown Engineering, Hayes International, Lockheed, Northrop, and Sperry Rand. Their primary mission was the development of heavy-lift rocket vehicles – the SATURN family.

Wernher von Braun remained the MSFC Director, with Eberhard F. M. Rees the Deputy Director for Research and Development. Formal guidance and direction came down from Headquarters to von Braun, as well as other directors of NASA Centers, through NASA's Associate Administrator for Manned Spaceflight, George Mueller

Technical Operations

The technical operations of MSFC were in two parts: Research and Development Operations, headed by Hermann K. Weidner, and Industrial Operations, headed by Edmund F. O'Connor. Weidner was a long-time member of the von Braun team, but O'Connor was a Major General on loan from the U.S. Air Force. (In this period, NASA realized that it had a deficiency in senior personnel with program management experience; to assist, the U.S. Air Force placed on loan to NASA 42 field-grade and General officers to fill various management positions, 12 of whom were at MSFC.) As a part of the R&D Operations, there were eight laboratories:

Aero-Astrodynamics Laboratory – Ernst D. Geissler
Astrionics Laboratory – Walter Häussemann
Computation Laboratory – Helmut Hölzer
Manufacturing Engineering Laboratory – Werner R. Kuers
Propulsion & Vehicle Engineering Laboratory – William R. Lucas
Quality & Reliability Assurance Laboratory – Dieter E. Grau
Research Projects Laboratory – Ernst Stuhlinger
Test Laboratory – Karl L. Heimburg

These laboratories were essentially direct descendants of the older technical divisions. One significant change might be noted: William Lucas, a native of Tennessee and doctoral graduate of Vanderbilt University, had broken into the line of Peenemünde / Project Paperclip technical leaders.

The development activities at MSFC were divided into four overarching programs in the Industrial Operations. These programs and their managers were as follows:

Saturn I / IB Program – Lee B. James
Saturn IB / Centaur Program – Stanley R. Reinartz
Saturn V Program – Arthur L. H. Rudolph
Engine Program – Leland F. Belew

In addition to these laboratories and program offices, there were many more technical and administrative offices that provided indispensable assistance. One that should be mentioned was the Advanced Systems Office, directed by Francis W. (Frank) Williams; von Braun often called upon Williams for special studies and later named him Advisor to the Director. Also, by this time MSFC had expanded beyond Huntsville with the Michoud Assembly Facility, managed by George N. Constan in New Orleans, and the Mississippi Test Facility, spread along the Pearl River in Hancock County and managed by Jackson M. Balch.

Assistance to the MSFC laboratories by support contractors had continued to increase, and, in addition to those previously noted, many other firms providing support services had opened in Huntsville. During 1965, NASA Headquarters directed MSFC to select a single prime-support contractor for each of the eight laboratories; intense competitions resulted.

The Propulsion & Vehicle Engineering (P&VE) effort, centering on SATURN V, would be by far the largest. Although relatively small, the Research Projects (soon to become Space Sciences) effort was the most sought-after, primarily because it involved senior-level research personnel and some of the most forward-looking endeavors in NASA; proposals for this contract were made by 16 university and industrial research organizations from across the Nation.

Prime support contracts for both the P&VE and Research Projects were awarded to BECO, the only firm receiving two awards. In the subsequent activities, William A. Giardini was responsible for the P&VE efforts, and Raymond C. Watson, Jr. (this book's author) led the Research Projects / Space Sciences efforts.

SATURN LAUNCH VEHICLES

As previously described, the development of a heavy-lift rocket, later named SATURN, was initiated as a defense project in August 1958. Funded by ARPA and under ABMA, it was to meet a need to orbit advanced communications and other highly classified satellites. With the formation of NASA and the awaited transfer of capabilities from ABMA, the SATURN development officially became a NASA responsibility on 18 November 1959.

High-Thrust Liquid-Fueled Engines

Liquid-fueled rocket engines delivering high thrust were at the center of MSFC activities. Since some engines were common to more than one rocket, they will be discussed together. Engine development normally preceded the development of the stage in which it was used; consequently, the engine program often became the pacing item. The overall activity leading to the development, selection, and testing of these engines was managed in the MSFC Engine Program Office under Leland F. Belew. As it turned out, all of the SATURN engines were built by Rocketdyne, a division of North American Aviation, in facilities at Canoga Park, California.

H-1 Engine – As previously described, the original heavy-lift vehicle proposed by ABMA and considered by ARPA used the E-1 rocket engine, then under development by Rocketdyne. ARPA suggested that this be changed, and

ABMA switched to the new Rocketdyne H-1 engine that was close to being ready for production. The H-1 was an improvement of the S3D engine already in use on the JUPITER and THOR and was expected to provide about 190-klbf (845-kN) thrust.

With small improvements by Rocketdyne and ABMA / MSFC engineers, the H-1 was ultimately rated at 205-klbf (912-kN) thrust. The H-1 was fueled by RP-1 (refined kerosene) with liquid oxygen (LOX) as the oxidizer. The engine was ignited using an expendable solid-propellant device (essentially a small solid rocket), and thus once started could not be restarted. The H-1 system was about 8.8 ft in length.

H-1 Engine

LR-10 Engine – There was considerable debate as to the propulsion for the SATURN I second stage; NASA appointed a special committee to settle the selection. Led by Abraham "Abe" Silverstein, the committee recommended an engine using liquid hydrogen (LH2) as the propellant. Based on this recommendation, it was initially proposed that the second stage be an Air Force CENTAUR rocket with LH2 and LOX propellants, but this rocket would not be available as soon as needed. Consequently, only the engine (the LR-115) intended for use on the CENTAUR was selected.

The LR-10 (LR for Liquid Rocket) had a long and interesting history. In 1938, the American Rocket Society designed and perfected the regenerative cooling rocket technology, for the first time making liquid-fueled rockets capable of burning for long enough periods to be practical. Spun off as Reaction Motors, Inc. (RMI), they built a 6.0-klbf (27-kN) thrust liquid-rocket engine that powered the Bell X-1 aircraft that first broke the sound barrier in 1947. In the late 1940s, they contracted with the U.S. Navy for a cryogenic liquid rocket, the LR-10, for use on VIKING sounding rockets.

LR-10 Engine

In 1956, RMI merged with Thiokol Chemical, and started the development of the LR-30 engine for the U.S. Navy's experimental aircraft X-15. Experiencing technical problems, the Navy brought in Rocketdyne, and the engine was designated TR-139. Success with this engine resulted in the X-15 having 189 flights and achieving a record mach-6.72 speed at 67 miles above the Earth.

When Rocketdyne received a contract from the Air Force to develop a cryogenic hydrogen-oxygen engine for the CENTAUR, it was designated LR-115. In August 1960, MSFC contracted for a similar engine for use on the SATURN, and it was somewhat upgraded in performance and designated LR-119. In a short while, the Air Force and NASA versions were combined and the designation LR-10 was adopted for both applications. (Other than what is shown herein, there are no known correlations between this Rocketdyne designation and the original engine of Reaction Motors.) After testing and modifications by MSCF, this engine finally became the LR-10A-3. This was rated as providing 15.0-klbf (66.8-kN) thrust, some 30 percent better than similar designs using hydrocarbon fuels.

J-2 Engine – Following the Silverstein recommendation, requirements for a new liquid-fueled cryogenic engine, designated J-2, producing thrusts up to 200 klbf (890 kN) were released by ABMA. Five contractors submitted proposals; these were examined for six weeks by a special NASA board chaired by Hermann K. Weidner, R&D Director of the newly formed MSFC. The design from Rocketdyne was selected, and a contract was awarded to this firm by MSFC in September 1960; this was the first contract to explicitly require the design to "insure maximum safety for manned flight."

J-2 Engine

In close coordination with MSFC, Rocketdyne initiated the development of the J-2 with an analytical computer model that simulated engine operations and assisted in establishing design configurations. Included in the design was a provision for terminating and reigniting the engine during flight. The first prototype J-2 engine had a 250-second test run in October 1962. The J-2 was eventually rated as having thrusts of 109 klbf (185 kN) at sea level and 232 klbf (1,030 kN) in space.

Production of the J-2 began in May 1963, with concurrent testing programs continuing to run at MSFC and Rocketdyne. Delivered in April 1964, the first production engine was static tested on an S-IVB test stage at Douglas Aircraft with a full-duration 410-second run in December. Testing continued until January 1966, with one engine igniting in 30 successive firings.

F-1 Engine – To meet a U.S. Air Force requirement for a very large rocket engine, Rocketdyne began the development of the F-1 in 1955. Although test firings of F-1 components had been performed as early as 1957, the Air Force halted development because it no longer had a requirement for an engine approaching 1.5 Mlbf (6.7 MN) thrust. It was of interest, however, to von Braun's space team at ABMA; in a single-nozzle engine, this was the required thrust calculated by Koelle for the first stage of JUNO 5 that eventually became SATURN. In 1958, ABMA contracted with Rocketdyne to continue the F-1 development, and this carried over when the operation became MSFC.

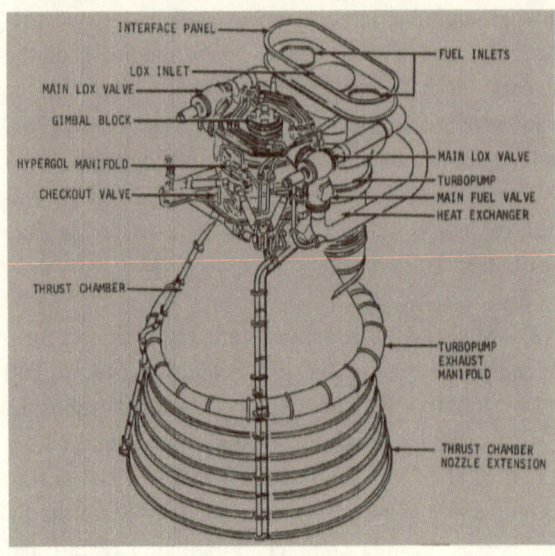

F-1 Engine System

The F-1 engine – the most powerful single-nozzle, liquid-fueled rocket engine ever flown – was 19.1-ft (5.79-m) tall by 12.3-ft (3.76-m) maximum nozzle diameter, and weighed over 18,000 lb (8,300 kg). Fueled by RP-1 (refined kerosene) and LOX (liquid oxygen), it burned 257.9 gallons (976.3 liters) of RP-1 and 413.5 gallons (1,565 liters) of LOX per second during a full 2.5-minute run. The turbo-pumps were the most complex units of the F-1 engine. Designed by Ernest A. Lamont, separate RP-1 and LOX pumps were driven by a gas-powered turbine rated at 55,000 brake-horsepower (41 MW).

In March 1959, the first static firing of a full-stage developmental F-1 was performed at Rocketdyne. This and subsequent firings at the NASA Rocket Engine Test Site at Edwards Air Force Base, California, indicated combustion instabilities, with oscillations of some 4 kHz. By 1961, Rocketdyne and ABMA / MSFC engineers determined how to nullify these oscillations, with combustion eventually so stable that it would self-damp within 0.1 second.

MSFC received the first F-1 in October 1963, and flight-rating tests were completed in December 1964. Testing of the F-1 continued at MSFC through 1965. Small improvements increased the thrust to 1.522 Mlbf (6.77 MN) and run-time to 2.75 minutes.

Instrument Unit

While the rocket engines provided the "muscle" in the various SATURN vehicles, the Instrument Unit (IU) was the "brains." It contained the basic guidance system components – a stable platform, accelerometers, a digital computer, and control electronics – as well as radar, telemetry, and other units.

Basically, the same IU configuration was used on the Saturn I, IB, and V. The IU evolved as a project in MFSC, and, with IBM as the prime contractor, it was the only full Saturn component manufactured in Huntsville.

The SATURN rockets depended on inertial guidance for determining their location along a predetermined path. Central to the inertial guidance package was a precision gyroscope. Like the high-thrust, liquid-fueled engines, this technology primarily came from the A-4 (V-2) development at Peenemünde. There, Fritz K. Mueller devised the SG-66, a three-axis platform with advanced accelerometers and integrators. Brought to America under Operation Paperclip, Mueller ultimately came to Huntsville and ABMA, where he and others developed the ST-80. As the REDSTONE emerged, this became the ST-90 in 1957, and was turned over to Ford Instrument Company for production.

After the SATURN was started, the ST-90 stabilized platform was improved by MSFC to become the ST-124. This was mated with an IBM ASC-15 computer. Originally designed by IBM for the Air Force TITAN II missile, the ASC-15 was a digital serial processor using fixed-point data with 27-bit words. It used a drum memory with 70 tracks, and electronic circuits of discrete resistors, capacitors, and transistors, all contained in welded modules encapsulated in a foam material. IU developments within MSFC were in the Astrionics Laboratory under Walter Häussemann; all of the units for SATURN I and the first four units for the SATURN IB were built there.

IBM served as the IU prime contractor for both SATURN IB and SATURN V. Initially, IBM used facilities at MSFC; starting in February 1964, IBM had its

own facilities in the new Huntsville Research Park. The ST-124 inertial platform was manufactured by the Navigation and Control Division of Bendix Corporation in Teterboro, New Jersey. The 260-in. (6.6-m) diameter, 36-in. (91-cm) high IU structure and the ASC-15 computer were built in

Manufacturing Instrument Units

Huntsville, and the full IU was also assembled and tested there. The launch weight was approximately 4,400 lb (2,000 kg).

The IU featured the first computer application where all electronic circuits were in triple-modular redundancy, providing near ultimate operating reliability. It also had a duplex memory system that operated in harmony during critical mission phases, correcting one another should there be an intermittent failure.

In 1965, it appeared that the IU for the first SATURN IB would not be completed in time for the scheduled launch. Finished IUs had to be carried by barge from Huntsville to the Cape – a voyage of some 2,200 miles (3,500 km) and 10 days. To gain this time for final assembly, Luther E. Powell of the MSFC IU Project Office suggested that the partially completed IU be shipped and completed in route. The barge had assembly-level environmental controls and a temporary clean room was set up. Powell and fellow engineer Sidney J. Sweat, together with several IBM technicians, worked 16-18 hour days as the barge plowed the Tennessee, Ohio, and Mississippi Rivers. Radio calls to MSFC each day ensured that needed parts and assemblies would be flown to ports along the way. The IU was completed before the barge reached New Orleans, and it arrived at the Cape on schedule.

Manufacturing and Testing Facilities

Wernher von Braun believed that personnel designing the space vehicles should have direct, hands-on participation in the building and testing of the hardware. For this, MSFC had facilities comparable to the best to be found in private industries. Included were precision machine shops, huge metal-forming and welding machines, and all types of inspection equipment. For every type of SATURN vehicle, one or more prototypes were fabricated in MSFC shops.

SATURN I in Development

Brown Engineering Company (BECO) received a contract to design and fabricate a gigantic test fixture – called WAAMAC – capable of holding a full Saturn I or IB booster. This was used at MSFC to determine the vehicle's center of gravity and mass moments of inertia about roll, yaw, and pitch axis; and optically align the tanks, engine thrust vectors, and mounting pads with the aerodynamic and geometric centerlines of the vehicle.

At MSFC, automated checkout of individual stages evolved with time. On the SATURN I, the stage checkout was manual; by the end of the SATURN series, automated equipment controlled some 50 percent of the tests. The S-IC stage and the IU used the RCA-110A digital computer, while the CDC-924A from Control Data Corporation was used for the S-II and S-IVB stages. Each stage received a final checkout before delivery to the launch site; the benefit of this procedure was shown in the first three SATURN V vehicles where 40 serious defects were found.

Static test towers had been constructed at ABMA for the REDSTONE and JUPITER rockets. In 1961, the JUPITER stand was modified for the SATURN 1

F-1 Static Test

and 1B stages. A number of other test stands followed, the largest being the SATURN V Dynamic Test Stand completed in 1964. At 475 feet (145 meters) in height, the entire SATURN V could be accommodated. Also completed in 1964, the S1C Static Test Stand was for simultaneous live-firing of the five F-1 engines of the first stage. Producing a total of 7.5-million-pounds thrust, the tests – sometimes as long as 150 seconds – generated earthquake-like rumbles throughout the Huntsville area and could be heard as far as 100 miles (160 km) away. Within about a mile of the test stand, the acoustical intensity was over 125 decibels.

As Saturn activities progressed, facilities other than those at MSFC were needed. In 1961, the Michoud facility, a WWII shipbuilding plant 15 miles (24 km) east of New Orleans, was selected to become the Michoud Assembly Facility (MAF, commonly called Michoud). Used by Chrysler for the Saturn I and Boeing for the Saturn V first stages, it had 43 acres (1,870,000 ft^2, 174,000 m^2) under one roof – one of the largest in the Nation.

MSFC Michoud Facility

Closely associated with Michoud was the Mississippi Test Facility (MTF), a huge area carved out of swampland along the Pearl River in Handcock County, some 47 miles (73 km) by barge north of Michoud. At the heart of MTF were monolithic test stands for the S-IC and S-II stages; this covered about 22 mi^2 (56 km^2), surrounded by an acoustic buffer zone of some 202 mi^2 (517 km^2). In forming the MTF in late 1961, 700 families had to be relocated. Inside the buffer zone, the acoustical intensity of S-IC tests exceeded 110 decibels. In more recent times, the MTF was renamed the John C. Stennis Space Center and is no longer a part of MSFC.

Computer services for both Michoud and MTF were established at Slidell, Louisiana – 20 miles (32 km) northeast of Michoud and 15 miles (24 km) southwest of MTF. The array of digital and analog computers was one of the largest in the Nation at that time.

Transportation

All SATURN vehicles were launched from what was generally called The Cape in Florida, initially at Air Force facilities then later at NASA's Kennedy

Space Center (KSC). The logistics involved in transporting stages and units from their production point to other testing and launching sites was a large part of the overall program. Even at Huntsville, where the distance between manufacturing buildings, test stands, Redstone Airfield, and the Tennessee River port was at most 10 miles, large special-purpose trailers had to be designed and built. For greater distances, water and air transportation systems were developed and used. The size of SATURN stages prohibited delivery by long-distance highways or rail. John C. Goodrum, Chief of MSFC's Logistics Office, had the overall responsibility for this transportation.

Water Transport – When what became the SATURN got underway at ABMA, ARPA agreed to fund a docking facility on the Tennessee River at the south end of Redstone Arsenal, as well as the conversion of a barge for river transport. Both of these were designed to handle the large boosters in a roll-on / roll-off operation – using the ground transporter to carry the stage onto the barge, hold it to its destination, and then wheel it out again.

MSFC's first waterborne carrier was the *Palaemon*, a converted U.S. Navy barge. For propulsion, the *Bob Fuqua* tug operated by Mechling Barge Lines was used. Normally, riverboat tugs pushed barges, but the *Palaemon* had a covered superstructure to protect the cargo, and the tug pilot could not see past this to guide; thus, it was necessary for the *Bob Fuqua* to pull, not push, the barge, and the pilot guided from the barge's pilot house.

Barges at MSFC Dock

Most tugs have a shallow draft to cross shoals and other river obstacles, and, consequently are not suitable for deep water; the *Bob Fuqua*, however, was designed and licensed for coastal work and was seaworthy for Gulf waters. Thus, a SATURN stage could be loaded at the dock in Huntsville; taken down the Tennessee, Ohio, and Mississippi Rivers to New Orleans; thence via the Intracoastal Waterway along the Gulf around Florida; and end up at a dock on the Cape Canaveral Barge Canal – a trip taking 10-12 days.

For transporting the S-IV and S-IVB from the West Coast to Huntsville and then to the Cape, commercial ocean freighters were initially used. As the larger S-II stage emerged, the Military Sea Transport Service provided the *Point Barrow*, a converted Navy LSD; this was used for shipping from California, to the Mississippi Test Facility (MTF), and then to the Cape. Other vessels in the MSFC "nautical fleet" included the *Poseidon*, an oversized barge used to carry huge S-IC first-stage SATURN V boosters between MTF, MSFC, and the Cape;

and smaller barges used to shuttle S-IC stages directly from the factory at Michoud the few miles to the MTF.

Air Transport – The best known air-transport vehicle was the *Pregnant Guppy*, a converted Boeing 377 Stratocruiser. Developed and operated by Aero Spacelines International, the *Guppy* could carry a load of 50,000 lb (22,700 kg), and fly at a cruise speed of 235 mi/h (378 km/h). The wing, engines, tail, nose, and cockpit were unchanged, but the fuselage separated just aft of the wing's trailing edge to accommodate the cargo. The aircraft first flew in September 1962.

Initially, the cargo width of 20 ft (6.1 m) was sufficient for the S-IV stage at 18 feet (5.5 m) diameter. Carrying this stage from California to Florida saved three weeks transit time versus the barge at a cost of $16.00 per mile.

The next transport aircraft, the *Super Guppy*, was built from an Air Force transport version of the commercial Stratocruiser. It could carry a load of 54,000 lb (24,494 kg) and cruise at 253 mi/h, 407 km/h), not significantly more

Super Guppy

than its predecessor. The principal advantage was the increased width in the cargo compartment to 25 ft (7.6 m), sufficient for the 21.7-ft (6.6-m) diameter of the Instrument Units of the SATURN IB and SATURN V. It went into service in August 1965, making use of the Redstone Airfield.

Saturn Launch Complex

As previously noted, rockets developed by ABMA and early MSFC were launched at Cape Canaveral Air Force Station (CCAFS) on the east coast of Florida. After NASA was formed, it began acquisition of land on the adjacent Merritt Island and the design of Launch Complex 39 (LC-39) started. On 1 July 1962, the overall site was named the Launch Complex Center, directed by Kurt H. Debus; then in 1963, it was renamed the Kennedy Space Center (KSC).

Launch Complex 39A & B

The main facilities of LC-39 were two widely separated launch pads (A and B), each with a launch tower and horizontal-access swing arms; the Vehicle Assembly Building (VAB – by volume, one of the largest buildings in the World); the Mobile Launcher Platform and the associated Crawler-Transporter; and a Launch Control Center.

Although LC-39 was no longer a part of MSFC, the Huntsville operations had a major role in the development, with engineering offices on Merritt Island; BECO also had an engineering and manufacturing operation there. Use of LC-39 began with the first SATURN V launch in late 1967.

Saturn Systems

This section provides a brief summary of the SATURN I, SATURN IB, and SATURN V systems, all under the management of MSFC.

Saturn I – The original vehicle, SATURN I, was almost fully developed when taken over from ABMA; it had two propulsion stages and an instrument unit. The first stage (designated S-I) had a cluster of eight Rocketdyne H-1 engines, providing a total of about 1.5-Mlbf (6.7-MN) thrust. To allow vehicle steering, the four outboard H-1 engines were gimbaled.

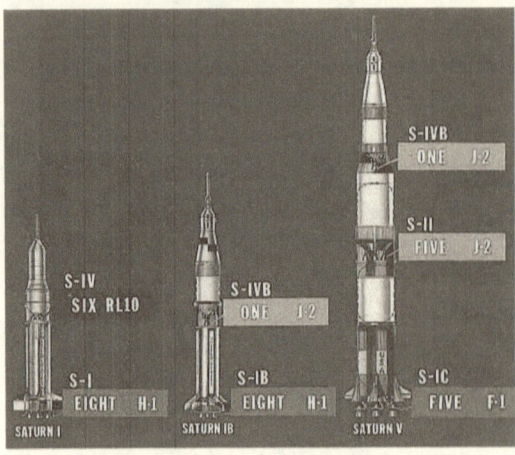

Saturn Family - Stages and Engines

While SATURN was under ABMA/ARPA, there had been debate concerning possible use of an existing rocket for the second stage. It was finally decided to include the second stage with the first-stage design, and to use a modified Rocketdyne cryogenic engine using LH$_2$ (liquid-hydrogen) and LOX (liquid-oxygen) propellant that was being built for the CENTAUR. After final modification design from MSFC, the engine was designated LR-10A-3, and produced 15.0-klbf (66.8-kN) thrust The SATURN I second stage (designated SIV) had six gimbaled LR10A-3 engines, giving a combined thrust of 90 klbf (400 kN).

The first four SATURN I launch vehicles, designated Block I SA-1 through SA-4, consisted only of an S-1 first stage and a dummy second stage. The guidance and control equipment was carried in canisters on top of the first stage and included an older ST-90 stabilized platform built for REDSTONE rockets. SA-1 was launched on 27 October 1961. SA-2 and SA-3 had a water-filled dummy stage that was used in a scientific experiment, Project Highwater (described later).

Block II, SA-5 through SA-10 launch vehicles, included a flight-operational second stage (S-IV) and an Instrument Unit stage. The first of these went up on 29 January 1964, and, three months later, SA-6 was the first to be inserted into orbit. SA-7, launched in September, was the first to use ST-124 guidance. These were mainly for flight-testing of Apollo spacecraft boilerplate (non-functional) units; SA-8, 9, and 10 also carried Pegasus micrometeoroid experiments. SA-8, launched in May 1965, included the first S-1 stage built at the Michoud Assembly Facility by the prime contractor, Chrysler; all of the remaining S-1 flight stages were built at Michoud.

The last two SATURN I Block II flight vehicles were SA-201 and SA-202, starting a new SA series. The second of these flew on 25 August 1966, overlapping with SATURN IB flights. These were to demonstrate Apollo spacecraft reentry adequacy under Earth-orbital conditions. All 12 of the SATURN I flights, all unmanned, were classified as successful with objectives achieved.

Saturn IB – The SATURN IB, sometimes called the Upgraded SATURN I, also had two propulsion stages and an instrument unit. The Space Division of

SATURN 1Bs in Production

Chrysler Corporation was the prime contractor for the two propulsion stages; IBM supplied the IU. As on the SATURN I, the first stage (S-IB) also had eight H-1 engines, with four of them gimbaled. The stage differed from the S-I on SATURN I by having eight fixed fins of equal size fitted to the sides, providing additional aerodynamic stability. Lee B. James was MSFC's developmental program manager for SATURN IB.

The second stage (S-IVB) had a single J-2 engine that gave 232-klbf (1030-kN) in-space thrust, much higher than the combined thrust of the six LR10A-3 engines on SATURN I. The J-2 engine was gimbaled and could also be restarted during flight. As previously noted, the J-2 was developed under NASA's first engine contract that specified "maximum safety for manned flight."

Fourteen SATURN IB vehicles (or partial vehicles) were built by Chrysler at Michoud, then tested at the MTF. The first SATURN IB was flown on 5 July 1966 – before the last SATURN I flight – in an unmanned, four-orbit flight test. Overall, five SATURN IBs were used in unmanned testing of Apollo hardware, and five others served as vehicles in manned missions, primarily during post-Apollo activities. All flight-ready SATURN IB vehicles were assigned numbers in the SA-200 series.

Saturn V – The SATURN V represented the height of rocket developments at MSFC; it was the most vital element in the Apollo Program. The basic configuration for this heavy-lift vehicle was selected in early 1963, based on the requirements for the single-launch vehicle in the LOR Mode for the lunar mission; the name SATURN V was applied at that time. Three configurations that might have led to SATURN II, III, and IV were discarded. Designed under the direction of Arthur Rudolph, the SATURN V holds the record as the largest and most powerful launch vehicle ever brought to operational status from a combined height, weight, and payload standpoint.

Arthur Rudolph

The first stage (S-IC) had five F-1 engines, giving a combined total of 7.5-Mlbf (34-MN) thrust. These engines were arranged in a cross pattern, with the center engine fixed and the outer four gimbaled. Matthew W. Urlaub was the S-IC project manager. The second stage (S-II) had five J-2 engines with the same arrangement as the F-1s and had a total of 1.0-Mlbf (4.5-MN) thrust. The S-II project was managed by Roy E. Godfrey. The third stage (S-IVB), with James C. McCulloch as project manager, had a single gimbaled J-2 engine with 230-klbf (1020-kN) thrust. For making small adjustments to the trajectory, the J-2 engine could be restarted in flight. The instrument unit (IU), essentially the same as on the SATURN IB, was between the S-IVB and the payload. Frederich Duerr was the project manager for the IU.

SATURN V – S-1C in Production

The F-1 engines on the S-IC stage used refined kerosene (RP-1) as fuel, while the engines on the upper two stages used liquid hydrogen (LH2). Although LH2 might have given greater power, it is only half as dense as RP-1, and thus would have required a far larger fuel tank; this would have resulted in unacceptable penalties in tank weight and aerodynamic design.

NASA made the decision to divide the production among multiple firms, believing that this would bring more top people into the program. Boeing was the successful bidder for the S-IC stage, doing the engineering in Huntsville, manufacturing at Michoud near New Orleans, and testing at the Mississippi Test Facility (MTF). North American Aviation (NAA) won the S-II stage, and Douglas Aircraft took the S-IVB stage, both doing the production in California –

NAA at their Seal Beach operation and Douglas at a Huntington Beach facility built for this project. IBM in Huntsville was already the supplier for the IU.

Upon signing the contract in December 1961, Boeing immediately organized a large staff of engineers and support personnel in Huntsville. By the summer of 1962, Boeing had some 500 employees on site at MSFC, and about the same number working at the HIC Building (the former Lincoln cotton mill just north of downtown). MSFC itself was experiencing overcrowding from growth, and also leased engineering drafting space from BECO in the HIC Building.

A total of 15 SATURN Vs were built; 13 functioned essentially without flaws (there was a minor "pogo" flaw in SA-502), and the other two, intended as backup, remained unused. Flight-ready SATURN V vehicles were designated in the SA-500 series, the first being SA-501, launched 9 November 1967, and the last being SA-513, used in the Skylab Program (described later) and launched 14 May 1973. SA-514 and SA-515 were never launched.

Apollo Spacecraft

The primary purpose of building the SATURN V vehicles was to launch Apollo Spacecraft and fulfill President Kennedy's goal of landing a man on the Moon and returning him safely to the Earth before the decade ended. In 1963, Samuel C. Phillips, Major General U.S. Air Force and previously director of the Minuteman Program, was assigned to NASA to head the Apollo Program Office at NASA Headquarters.

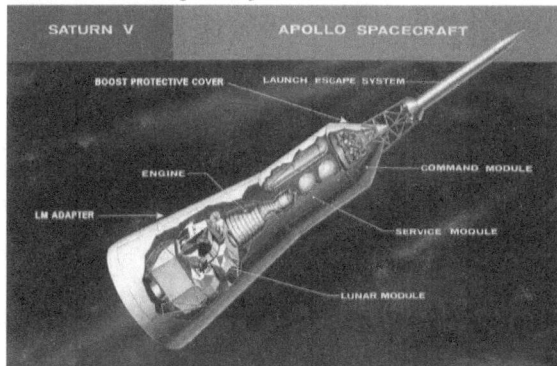

The Apollo Spacecraft was composed of the Lunar Module (LM) and the Command /Service Module (CSM), carried together inside the Spacecraft Lunar

Saturn V Apollo Spacecraft Subsystems

Module Adapter, with the Launch Escape System at the very top. Grumman Aircraft was the prime contractor for the LM, and North American Aviation for the CSM. Thiokol provided the escape solid rocket.

Development of the Apollo Spacecraft and its components was managed by other NASA centers, but MSFC provided the all-important launch vehicles. Thus, the Apollo Missions are a highly important part of MSFC's technology history and the technological evolution of Greater Huntsville.

Special Support Equipment

In addition to the SATURN vehicles, MSFC was responsible for several items of special support equipment that contributed to the success of the Apollo Program. Two of these will be described: the Lunar Roving Vehicle and the Lunar Soil Penetrometer.

Lunar Roving Vehicle – Popularly known as the "Moon Buggy," the Lunar Roving Vehicle (LRV) was developed to provide astronaut transportation for exploring a limited amount of the Moon's surface. This was not in the original lunar exploration planning, but in the early 1960s, NASA assigned this project to MSFC. Initially called the Local Scientific Service Module (LSSM), several different mobility test articles (MTAs) were examined by MSFC and contractors, and MOLAB (Mobile Laboratory) emerged as a leading contender. MOLAB was a large, six-wheeled machine that would have required a separate SATURN V for transporting it to the Moon.

Wernher von Braun, however, thought that a golf-cart-style two-man vehicle would be adequate. With an urgent need to determine the feasibility of such a vehicle, von Braun bypassed normal procurement procedures and had the Advanced Studies Office of MSFC's Propulsion & Vehicle Engineering (P&VE) Laboratory directly task Brown Engineering Company (BECO) to design and build an MTA for an open, light-weight, four-wheeled, two-man vehicle.

The effort began in early 1966. Eduardo C. San Juan, an engineer who had earlier made lunar mobility studies at Hayes International, developed the basic concept. A team of structural, mechanical, electrical, and human factors engineers did the detailed design, making extensive use of analog computers. Specialists in aluminum fabrication were vital participants. Before the end of the year, a working MTA was delivered to MSFC.

Von Braun Driving the BECO MTA

A small test track with craters and rock debris had been built at MSFC; this was used in comparing the BECO and MOLAB MTAs. In 1967, tests of the BECO MTA under one-sixth gravity were obtained through four flights on a KC-135A aircraft operating in a parabolic maneuver. The MTA was also extensively tested at the U.S. Army's Yuma Proving Ground in Arizona and the Aberdeen Proving Ground in Maryland. MSFC continued with improving the MTA during 1967 and 1968, eventually concluding that a small rover based on this design would be fully suitable for the proposed lunar missions.

In May 1969, NASA headquarters selected this type of vehicle for use in manned lunar missions, and the name Lunar Roving Vehicle (LRV) came into use. The LRV Program was approved as a MSFC hardware development, with Saverio F. Morea assigned as Program Manager. In July 1969, just before the successful Moon landing of Apollo 11, a request for proposal for the final development and building the Apollo LRV was released by MSFC, and the contract was awarded to Boeing in October.

Henry Kudish managed the Boeing effort from their Huntsville offices. As a major subcontractor, General Motors Defense Research Laboratories in Santa Barbara, California, furnished the mobility system (wheels, motors, and suspension); this effort was led by Ferenc Pavlics. The final LRV assembly and testing was at the Boeing facility in Kent, Washington. An LRV was carried on the last three Apollo missions, allowing a much larger area to be explored. Outbound, each carried an ALSEP experiment package to be set up in a remote location; on the return trip, they carried lunar rock and soil samples.

Lunar Soil Penetrometer – MSFC's Space Sciences Laboratory (SSL) was responsible for gathering and analyzing information concerning the mechanical properties of lunar soil. As the Lunar Roving Vehicle (LRV) was coming into being, it was recognized that using this vehicle would allow soil property measurements at a variety of locations for each mission. Nicholas C. Costes of the SSL believed that a soil penetrometer – a device for measuring the force required to penetrate the soil to various depths – would be beneficial, and obtained a task for Teledyne Brown Engineering (TBE, the successor to BECO as the SSL prime support contractor) to design and build such an instrument for the Apollo 15 mission.

Rolland G. Sturm led the effort at TBE, with a team devising what was called a self-recording, portable, soil penetrometer (SRP). The SRP, weighing (2.2 kg (5.0 lb), had a central shaft 109-cm (41-in.) long and 2.5-cm (1.0-in.) diameter. The shaft

Testing the TBE Soil Penetrometer

had a push-handle at the top and passed through a surface-plate at the bottom. The recorder was a copper tube around the shaft and below the handle. As the penetration was made, the depth and force were scribed along the recorder tube. Each tube could accept 12 recordings, but then could be easily changed out.

The task to TBE was assigned only a few months before the Apollo 15 scheduled launch date (26 July 1971). Thus, the SRP had to be developed quickly, and there was not time for having it fully flight qualified; it was, however, tested under one-sixth gravity through parabolic flights on a KC-135A aircraft. The SRP was likely the only item on any of the Apollo flights that flew with this just-in-time status. A second SRP was built for a 'hangar queen,' and later carried on the Apollo 16 mission launched 16 April 1972.

APOLLO MISSIONS

In the Apollo Program, each flight of a vehicle carrying an Apollo spacecraft, or a partial spacecraft, was called an Apollo Mission, and designated with an Apollo number. A few missions were unmanned, but most involved three astronauts. Two missions were launched using SATURN IB vehicles, and twelve used SATURN Vs. Most of the missions took place in the late 1960s, but the last five were in the early1970s. Time-wise, these last five missions should be in the following chapter, but, for continuity, they are included herein.

The missions were of two general types: Earth Orbit Testing and Lunar; brief descriptions follow on both types.

Earth Orbit Test Missions

The first manned test flight of the Command/Service Module (CSM) using a SATURN IB launch vehicle was scheduled for 21 February 1967. During a pre-launch test at Cape Canaveral at 6:31 p.m. EST on 17 January, a fire swept through the CM, resulting in the death of the three crew members: astronauts Virgil I. (Gus) Grissom, Edward H. White, and Roger B. Chaffee.

The exhaustive investigation of the fire and extensive reworking of the Apollo command modules by the prime contractor, North American Aviation, postponed crewed launches until NASA officials cleared them for flight, and all testing and operating procedures were reworked.

Apollo 1 – Although not a mission *per se*, this unaccomplished test flight was later designated Apollo 1. Apollo 2 and 3 remained undesignated.

It is noted that MSFC had no direct involvement, much less responsibility, in the design of the CM; therefore, it can be stated that there were no critical failures with any spaceflight hardware developed by the team led by Wernher von Braun.

Apollo 4 – While the fire investigations and changes to the CM were underway, the first test flight of a full, but unmanned, SATURN V vehicle (SA-501) was made on 9 November 1967; it was designated Apollo 4. This test flight also included the first in-flight restart of the J-2 engine on the third stage.

Apollo 5 – Saturn IB schedules had been suspended for nearly a year, and the launch vehicle that finally picked up the schedule with Apollo 5 (actually flying before Apollo 4) was designated SA-204. An investigation of testing requirements had led to a change in the payload from a CM to the first unmanned flight of a lunar module (LM), thus eliminating a full test activity and accelerating the schedule.

This mission lifted off on 22 January 1968, on a four-orbit Earth flight. This, together with other changes, placed the overall Apollo Program back on track, allowing the fulfillment of President Kennedy's national goal of a manned mission to the Moon before the end of the 1960s.

Apollo 6 – Launched on April 4, 1968, Apollo 6 was the second unmanned, Earth-orbiting test-flight of a SATURN V vehicle (SA-502). It was also the final unmanned Apollo mission. The launch phase was somewhat marred by a five-hertz longitudinal vibration (a phenomenon known as pogo oscillation) that ruptured igniter fuel lines in J-2 rocket engines on the second and third stages. Nevertheless, it made three orbits in a total flight time of 10 hours.

Apollo 7 – The first manned SATURN IB vehicle (SA-205), flew on 11 October 1968; it placed the Apollo 7 spacecraft's CSM with three astronauts into Earth orbit, where it made 163 orbits during 8 days. Apollo 7 carried out the mission that Apollo 1 had been scheduled to do. The crew was Mission Commander Walter M. Schirra, Command Module (CM) Pilot Donn F. Eisele, and Lunar Module(LM) Pilot R. Walter Cunningham.

(Apollo 8 was a lunar orbit flight and is described in the next section.)

Apollo 9 – Using a SATURN V vehicle (SA-504), Apollo 9 launched a manned CSM on 3 March 1969. (This was the first in a sometimes confusing numbering sequence, with the Apollo number five digits more than the vehicle number.) During 10 days in 152 low-Earth orbits, the three-man crew performed the first manned flight of a LM, the first docking and extraction of a LM, two spacewalks (EVAs), and the docking of two manned spacecraft, all necessary to demonstrate lunar-orbit rendezvous. Its crew was Mission Commander James McDivitt, CM Pilot David R. Scott, and LM Pilot Russell L. Schweickart. The mission was the second manned launch of a Saturn V rocket.

Lunar Orbit and Landing Missions

A flight to the Moon using a SATURN V consisted of the following: At liftoff, the five F-1 engines on the S-IC stage producing the energy equivalent of 85 Hoover Dams, burned for about 2.5 minutes, boosting the entire vehicle and payload spacecraft to near 38 miles (61 km). The first stage then separated, and the second stage (S-II) took over; the five J-2 engines burned for about 6 minutes, lifting the vehicle to near 115 miles (185 km) where the second stage was also discarded. The third stage (S-IVB), with one J-2 engine, burned for 2.75 minutes, giving the system orbital velocity of about 17,500 mph (28,200 km/hr). The J-2 was later reignited to propel the spacecraft to 24,500 mph (39,400 km/hr) – the translunar trajectory velocity. The third stage was then discarded, and the spacecraft was on its way to the Moon.

Apollo 8 – SATURN V vehicle SA-503, launched 21 December 1968, propelled Apollo 8 spacecraft with Mission Commander Frank F. Borman, CM Pilot James A. Lovell, Jr., and LM Pilot William A. Anders into 10 orbits around the Moon – the first persons to travel beyond the low-Earth orbit and the first to see the far side (commonly called the dark side) of the Moon. This was also the first manned launch from KSC.

The Apollo 8 crew made a Christmas Eve television broadcast in which they read aloud the first 10 verses from Genesis in the *Holy Bible*, starting as follows: "In the beginning, God created the heaven and the earth. . . ." At the time, this was the most watched TV program ever.

(Apollo 9 was an Earth Orbit mission and was described in the preceding section.)

Apollo 10 – Launched by SA-505 on 18 April 1969, activities of Apollo 10 continued with the efforts of Apollo 9, but in lunar orbit. The astronauts were Mission Commander Thomas P. Stafford, CM Pilot John W. Young, and LM Pilot Eugene A. Cernan. It was the first manned CSM/LM operation in lunar environments and included a simulated lunar landing. The mission lasted 8 days and included 31 lunar orbits.

Apollo 11 – On 16 July 1969, SATURN V vehicle SA-506 launched the Apollo 11 spacecraft to a landing on the Moon – the crowning achievement in the Apollo space program. Von Braun later told what he did during the final few seconds of the launch countdown – he silently recited the Lord's prayer in German: ". . . *Dein Wille geschehe, wie im Himmel, so auf Erden. . . . Denn*

dein ist das Reich und die Kraft und die Herrlichkeit in Ewigkeit. Amen."
(". . . Thy will be done on Earth as it is in heaven. . . . for Thine is the kingdom, and the power, and the glory, forever. Amen.")

The crew was Mission Commander Neil A. Armstrong, LM Pilot Edwin E. (Buzz) Aldrin, Jr., and CM Pilot Michael Collins. Armstrong and Aldrin landed on the surface and stayed over 21 hours, including about 2.5 hours extra-vehicular, while Collins circled in lunar orbit.

A while after landing, Aldrin – a devout Christian – radioed to Earth:

> This is the LM pilot. I'd like to take this opportunity to ask every person listening in, whoever and wherever they may be, to pause for a moment and contemplate the events of the past few hours and to give thanks in his or her own way.

He then took communion privately, using a wafer and small chalice of wine that he had carried.

As a world-wide television audience estimated at 600 million people watched, Armstrong stepped onto the lunar surface and said, "One small step for [a] man, one giant leap for mankind." The astronauts returned to Earth and landed safely in the Pacific Ocean on 24 July. Apollo 11 effectively ended the Space Race with the USSR and fulfilled a national goal proposed by President Kennedy in 1961.

First Step on the Moon

Apollo 12 – The second to land on the Moon, Apollo 12 was launched by vehicle SA-507 on 14 November 1969. Mission Commander Charles "Pete" Conrad, Jr. and LM Pilot Alan L. Bean spent 31 hours in lunar surface activities while CM Pilot Richard F. Gordon, Jr. remained in lunar orbit. The landing site was close to that of Surveyor 3 unmanned probe, which had landed 20 April 1967. On one of two moonwalks, the astronauts visited the Surveyor site and brought back some of the parts. A successful return to Earth was made on 24 November.

Apollo 13 – Launched by SA-508 on 11 April 1970, Apollo 13 was intended to be the third manned landing on the Moon. The crew was Mission Commander James A. Lovell with John L. "Jack" Swigert as CM Pilot and Fred W. Haise as LM Pilot. Two days after being launched and about 200,000 miles (320,000 km) along the path to the Moon, an oxygen tank exploded, crippling the SM and leaving the CM with limited-duration battery power. Mission Control guided the spacecraft into a loop around the Moon, and then placed it directly into a return-to-Earth trajectory. The crew was forced to shut down the CM completely and use the LM as a "lifeboat." Despite highly limited electrical power and suffering from a loss of cabin heat, shortage of potable water, and the need to jury-rig the carbon dioxide removal system, the crew made a safe return to Earth on 17 April.

Apollo 14 – Boosted by SA-509 on 31 January 1971, Apollo 14 was the third landing on the Moon. Its mission was much the same as originally proposed for Apollo 13. Mission Commander Alan B. Shepard, Jr. and LM Pilot Edgar D. Mitchell spent about 33 hours on the surface, while CM Pilot Stuart A. Roosa was in lunar orbit. On the surface, Shepard and Mitchell had about 9½ hours on two EVAs; 42 kilograms (93 lb) of lunar samples were collected. They returned to Earth on 9 February.

Apollo 15 – With astronauts Mission Commander David R. Scott, LM Pilot James B. Irwin, and CM Pilot Alfred M. Worden, Apollo 15 was launched by

First Lunar Excursion

SA-510 on 26 July 1971, and ended 7 August. There was much more science on this mission than on previous ones. All three astronauts had received special training in geology – that of Worden in observations from lunar orbit. The Lunar Roving Vehicle (LRV, described previously) and the Lunar Soil Penetrometer (described in the chapter that followed) were first carried on Apollo 15.

Three days were spent on the Moon, including 19 hours outside the spacecraft on lunar EVA. By using the LRV, a distance of 27.9 km (17.3 mi) was covered. Worden made 74 orbits and made many scientific measurements. While still in orbit just before the returning flight, they released a small satellite to study the plasma, particle, and magnetic field environment of the Moon and map the lunar gravity field. Apollo 15 returned to Earth 370 individual rock and soil samples, including a deep-drill core with material from 2.4 meters below the lunar surface, with a total mass of 77 kg (170 lb).

Apollo 16 – Like Apollo 15, Apollo 16 spent more time on the lunar surface, carried a Lunar Roving Vehicle and a Lunar Soil Penetrometer, and had considerable science content. It was launched by SA-511 with the crew composed of Mission Commander John W. Young, LM Pilot Charles M. Duke, Jr., and CM Pilot Thomas K. (Kem) Mattingly II. Launched on 16 April 1972, the mission lasted slightly over 11 days and concluded on 27 April. Young and Duke spent 71 hours on the lunar surface, during which they conducted three extra-vehicular activities totaling 20 hours and 14 minutes, and drove the LRV 26.7 km (16.6 mi). Mattingly spent 126 hours and 64 revolutions in lunar orbit performing observations. When they returned to Earth, they brought 95.8 kg (211 lb) of lunar samples.

Apollo 17 – This was the final mission of the Apollo lunar landing program, the sixth and last landing of humans on the Moon, and the final crewed launch of a Saturn V rocket. It was launched by SA-512 on 7 December 1972, with the crew of Mission Commander Eugene A. Cernan, CM Pilot Ronald E. Evans, and LM Pilot Harrison H. Schmitt. The mission included a three-day lunar surface stay, extended scientific capability, and the third Lunar Roving Vehicle – only two Soil Penetrometers had been built.

Apollo 17 broke all records set by previous flights, including the longest total lunar surface EVA (22 hours outside the LM); the greatest distance on the LRV (37.5 km (22.2 mi)), the largest lunar sample return (110.5 kg (243.7 lb) of lunar samples), and the longest time in lunar orbit (75 orbits). After a mission of 12 days and almost 14 hours, it returned to earth on 19 December 1972. Over four decades later, Apollo 17 remains the most recent manned Moon landing and also the last time humans have travelled beyond low Earth orbit.

Benefits of Apollo

More than $25 billion (over $100 billion in today's dollars) was spent on the Apollo Program. Most of this was used on designing and building the spacecraft, booster rockets, and other space components directly for the missions; the rest went for ground control systems, astronaut training, and other indirect expenses. However, it has been shown that if the advancements in science and technology, medicine, training techniques, and many other areas gained in the Apollo Program were independently priced, the sum would be much more – the direct economic gain from the Program was positive.

SCIENTIFIC AND ENGINEERING RESEARCH

From its beginning, MSFC had strong research projects in science and engineering. In the years of the Saturn and Apollo programs, this was important for answering questions during early developments through operations, as well as looking forward to questions in the future. Each of the missions on the lunar surface involved scientific research. Overall, there were far too many projects to include many of them in this section, but several activities are described.

Project Highwater

There were questions concerning the possible effect on radio communications and environmental condition caused by explosions of large rockets at high altitude. Project Highwater was an experiment developed by the MSFC Aeroballistics and Research Projects Divisions to try and answer these questions. In initial flights, the second stage (S-IV) of the SATURN I carried a ballast of 30,000 gallons (109,000 liters) of water. For Highwater experiments, explosive devices ruptured this stage, and the diffusion and effects of the water were monitored.

In the first test using vehicle SA-2 on 25 April 1962, the water was released at an altitude of 65 miles (104 km), eventually reaching an apex of 90 miles (144 km). On 16 November, a second test with vehicle SA-3 released the water at the flight's peak altitude of 104 miles (166 km).

In both of these experiments, within a few seconds observers saw an ice cloud form and quickly expand to an estimated several miles in diameter; lightning-like radio disturbances were recorded at ground stations; and radar on a tracking plane showed a similar disturbance. In less than a minute, all indications had disappeared. While these experiments showed the diffusion and effects of explosively released water, there were still questions concerning the propellants and oxidizers.

Project Pegasus

High-altitude experimental rockets had indicated the presence in space of micrometeoroids (small particles of rock or metal, usually weighing less than a gram). The primary purpose of Project Pegasus was to assess the potential hazard to astronauts if micrometeoroids could puncture the spacecraft skin. Three SATURN I vehicles were instrumented to study the frequency and penetration depth of impinging micrometeoroids.

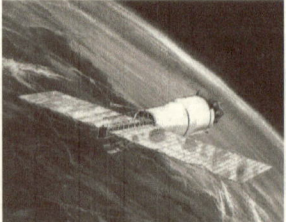

Pegasus Experiment

In a design suggested by Charles A. Lundquist of MSFC's Space Sciences Laboratory, the Pegasus detection apparatus was two "wings" with more than 1,300 ft^2 (120 m^2) of instrumented surface – 208 panels of capacitor detectors varying in thickness up to 0.016-inch (0.41-mm), some with sample protective shields to assess their effectiveness. The SATURN vehicle carried non-functional Command / Service Modules (CSM); for launching, the Pegasus wings were folded and carried within the Service Module.

Other items, including a solar power supply, signal processors, data recorders, and a transmitter, were in the CSM. After the powered second stage reached orbit, the CSM was ejected into a separate orbit. The Command Module was then opened and the Pegasus wings expanded to a width of 96 ft (29 m) – ready for micrometeoroid detection. The Space Division of Fairchild Hiller built the Pegasus equipment.

Pegasus 1, launched by SA-9, was in orbit from 16 February 1965, to 17 September 1978. Launched by SA-8, Pegasus 2 reached orbit on 25 May 1965, and decayed 3 November 1979. Pegasus 3, launched by SA-10, operated from 30 July 1965, until 4 August 1979.

In addition to micrometeoroid data, the Pegasus experiments gathered information on gyroscopic motion and orbital characteristics of rigid bodies in space, lifetimes of electronic components in the space environment, and the degrading effects of space on thermal control coatings.

Lunar Surface Experiments

Each of the six Apollo missions that landed on the Moon carried an Apollo

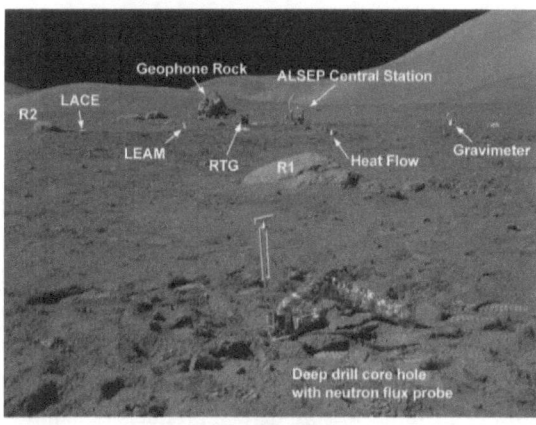

Apollo 17 ALSEP

Lunar Surface Experiment Package (ALSEP). This was composed of equipment for seven scientific experiments plus a central control station (they were controlled from the Earth), all powered with a radioisotope thermoelectric generator (RTG). The science investigators were mainly from universities, but F. Curtis Michel and Jerry L. Modisette of MSFC were co-investigators. Many other MSFC personnel and contractors also participated, particularly BECO / TBE as the prime research support contractor. The instruments were designed to run autonomously after the astronauts left and to make long-term studies of the lunar environment. The ALSEP stations were built by Bendix Aerospace.

Research Projects Laboratory

From the time of their arrival at Redstone Arsenal, Wernher von Braun and his team had an interest in starting scientific research. Ernst Stuhlinger was a particular advocate of this activity and as MSFC was formed, he was the Director of MSFC's Research Projects Laboratory (RPL, later Space Sciences Laboratory). During the 1960s, an excellent staff of scientists and engineers was assembled and engaged in a wide variety of analytical and experimental research.

Ernst Stuhlinger

Topics ranged from developing an interactive computer-programming language to high-energy astrophysics. One project involved a series of balloon-borne astronomical experiments; operating at altitudes above 80,000 feet, these payloads drifted across the United States making observations at visible, ultraviolet, and x-ray wavelengths. For many of the RPL projects, the Brown Engineering Research Laboratories was the support contractor. Other scientific projects are included in the following chapter.

INDUSTRIES AND OTHER ACTIVITIES

Prior to World War II, there were a number of industries in Huntsville dedicated to cotton-fabric manufacturing, and a few firms had come and gone that built mechanical hardware such as buggies and elementary farming equipment.   As earlier described, the Army had two major munitions manufacturing arsenals adjacent to Huntsville during WW II, but there was no local industrial infrastructure providing support.

At the close of the war, several attempts – mainly unsuccessful – were made to convert some of these munitions facilities to commercial products (chemicals and automobiles). Also, in December 1945, Nashville-based General Shoe Company (later GENESCO) opened a shoe factory in Lowe Mill – a facility dating from the fabric manufacturing days. Employing up to 800 people and operating until 1978, many U.S. soldiers in Vietnam wore boots made at GENESCO's Huntsville factory. Through the years, however, no local private industries were involved in higher levels of technology products or services.

This situation started to change in 1949 when Redstone Arsenal was designated the Ordnance Rocket Center, and two propulsion chemistry firms – Rohm and Hass and Thiokol – opened in Government-owned contractor-operated facilities. Somewhat later, two major firms – Chrysler and Boeing –

opened Huntsville operations, but these were fully dedicated to their prime contracts for missile development with most of their work performed in Government facilities on Redstone Arsenal.

In the City's fiscal year ending 30 September 1962, the Probate Office granted 2,997 business licenses; according to a brochure prepared by *The Huntsville Times*, these included the following firms and their approximate number of employees:

Boeing Company	750	Brown Engineering Company	1,900
Chrysler Company	1,900	General Analine	40
GENESCO	750	Hayes International	700
John Blue Company	350	Norton Company	50
Martin Stove Company	300	M. Lowenstein Company	1,300
P. R. Mallory	500	Olin Mathieson	150
Rohm and Hass	200	Space Craft, Inc.	110
SPACO, Inc.	350	Stauffer Chemical	120
Thiokol	2,000	Union Carbide	50
Wah Chang Corporation	75	Wyle Laboratories	75

Propulsion developers Thiokol and Rohm and Hass were in government owned / contractor operated facilities on Redstone Arsenal. Also on Redstone Arsenal were chemical manufacturers General Analine, Stauffer, and Olin Mathieson, leasing facilities held over from munitions manufacturing. Lowenstein was the only remaining fabric manufacturer; it was previously the Merrimack Mill. Shoe manufacturer, GENESCO, was in the former Lowe Mill. Union Carbide, doing business as Linde Air Products, produced liquid oxygen and hydrogen in a plant on Hobbs Island Road. Details on the other firms are contained herein.

As of the early 1960s, Redstone Arsenal (RSA) encompassed 38,300 acres, 36,459 of which the Department of the Army controlled, and the George C. Marshall Space Flight Center had under lease the remaining 1,841 acres. Approximately 15,500 acres were classified as woodlands, 9,200 were leased for agricultural use (primarily raising cattle), and 4,100 were designated as part of the Wheeler National Wildlife Refuge.

TECHNICAL INDUSTRIES

During the three decades of the Rocket City Era, many space and defense companies began operations in Greater Huntsville. In addition, a number of firms in other fields opened in the city. Huntsville's first independent, high-technology engineering and manufacturing firm opened in July 1953; this shortly became Brown Engineering Company (later Teledyne Brown Engineering). Others, both national and locally formed, followed in the later 1950s.

As the years progressed, there was an avalanche of high-technology firms starting in, or moving to, Greater Huntsville – far too many for all to be recognized in this book. In addition, many firms – including a number of large, well-known companies – opened local offices for technical representatives. A general apology is given to all who have been left out. For a good listing, but still not totally inclusive, reference is made to the periodic editions of the *Industrial Directory* of the Huntsville-Madison County Chamber of Commerce.

Company Descriptions

Brief descriptions of some of the high-technology and supporting industries in the 1949-1970 period are given in this section. Together with the name, their opening and closing (if applicable) years are designated; national firms are indicated with Huntsville or Redstone Arsenal on their name. If no closing date is given, the firm is believed to still be in operation as of 2013. For some firms, a reference to its later operation (called Updated) is noted; this is mainly where there were significant description changes during in the future eras.

Although the personnel are the most important part of any organization, and the industries in Greater Huntsville had thousands of outstanding employees, only a very small number could be cited.

AC Incorporated (1948) – Founded as a machine shop in 1948, AC Incorporated grew and added capabilities through the years to become a privately owned, ISO 9001:2008-compliant corporation serving commercial, defense, government, automobile, and aerospace markets. It manufactures missile and engine containers, dummy missiles, tent systems and other fabric and metal products; it also provides program management, inspection, and information technology services. Its headquarters and 100,000-ft^2 primary manufacturing and services facility are at 1085 Jordan Road in Chase Industrial Park. The firm also has manufacturing operations in Grove Hill, Alabama, and quality-inspection services at Lincoln, Alabama.

As of 2013, George A Smith was the President, and Daniel J. Bobrowski the VP of Operations. About 60 persons were employed at the Huntsville facilities.

Rohm and Haas - Redstone Arsenal (1949-1970) – Formed in Germany in 1907, and with a branch in Philadelphia, Rohm and Haas (RAH) gained fortune with the invention of Plexiglas in 1933, and was soon widely recognized for this and other developments in industrial and agricultural chemistry. Following WWII, RAH became an American corporation, with stock trading on the New York Stock Exchange.

From mid-1949 through 1970, Rohm and Haas conducted basic and applied research in rocket propellants, rocket systems, and related technologies at Redstone Arsenal under a Government-owned, contractor-operator, non-for-profit agreement. This began with 9 people working in former munitions-production buildings, and by 1953, this had grown to 250 scientists, engineers, technicians, and a few administrative personnel forming the RAH Redstone Research Division.

Still mainly improvising in several former munitions buildings, the headquarters and central operations were in the Josiah Gorgas Research

*Josiah Gorgas Research
Laboratory*

Laboratory. This laboratory, one of the most modern in the South, was constructed by the Government for use by Rohm and Haas in 1952; it was named after the Confederate Brigadier General who was Chief of Ordnance in the Civil War.

Initial work centered on improving small solid-rocket motors for shoulder-fired weapons. From this, the propulsion unit for the Light Assault Weapon (LAW) evolved; millions of these motors were later made by a production contractor for the Viet Nam War. Rohm and Haas was soon recognized as a major rocket-research source; they significantly contributed to the development of propulsion systems for many rockets and missiles. In many projects, there was close cooperation with personnel at Thiokol, located adjacent to Rohm and Haas.

Henry M. Shuey served as the technical lead at Rohm and Hass. The Commanding General of AMC once commented, "If Rohm and Haas had sent only Dr. Henry Shuey to Redstone, his work alone would justify every penny the Army ever paid the company." William C. Stone was another well-respected researcher. There were also many other highly qualified researchers, including specialists in mathematical modeling and instrumentation.

Rohm and Haas discontinued its operation on Redstone Arsenal at the end of 1970. Both Shuey and Stone opened propulsion consulting firms in Huntsville after Rohm and Haas closed.

Thiokol Chemical - Redstone Arsenal (1949-1996) – During WWII, Thiokol Chemical Corporation of Trenton, New Jersey, was a supplier of synthetic rubber in liquid polymer form. The Jet Propulsion Laboratory discovered that mixing an oxidizer with this polymer produced a material that burned with great intensity and produced large amounts of gas – essential properties of rocket propellants. To exploit this finding, Thiokol opened a small operation in Elkton. Maryland.

In early 1949, as the Redstone Arsenal was being reactivated as the Army's research and development center for rocketry, Thiokol agreed to move the propellant research from Maryland to facilities originally used by the Army's Redstone Ordnance Plant. William M. Mebane was the initial General Manager, and Harold W. Ritchey was hired as the first Technical Director. (Often called the Father of Solid Rocketry, Ritchey later became the President, then Chairman of the Thiokol Board.)

The first development at Thiokol's Huntsville Plant was a five-inch, end-burner solid-rocket motor that was tested in June 1949. Work was also underway on perfecting the case-bonding technique; in this, the propellant was cast vertically with a hollow core. Burning could then be simultaneous along the entire length, giving a larger thrust; the propellant also served as its own thermal insulation, allowing a much lighter outer case.

A 34-inch solid-rocket motor, nicknamed "The Thing," was built under a sub-contract from General Electric for the experimental RVA-10 HERMES (a U.S. version of the German WASSERFALL missile). In early 1951, this was successfully test fired for 41.2 seconds and a thrust of over 17 klbf (76 kN), the first

First 34-Inch Solid Rocket Motor

demonstration of a solid-propellant system for a large missile. After several more test firings, it was turned over to Jet Propulsion Laboratory and ultimately used on the MGM-29 SERGEANT missile. This led to solid-rockets ultimately becoming the standard for tactical missiles systems.

From a start of only 33 chemists, engineers, and technicians working in three buildings, employment at the Thiokol Huntsville Division peaked at 2,100 in May 1959, working in 45 main buildings spread over 900 acres. Propulsion systems were developed for the Army, Navy, and Air Force. Over the first years, missiles with Thiokol involvement included FALCON (Thiokol's first

Thiokol Area 1959

motor production), LACROSSE, BOMARC B. NIKE HERCULES, NIKE

ZEUS, PERSHING, SERGEANT, MATADOR, X-17 TEST VEHICLE, POLARIS TEST VEHICLE, LITTLE JOE, SCOUT, and BIG B (forerunner of the MINUTEMAN).

Calvin G. Wiggins was the Huntsville Division VP and General Manager during the 1990s. Thiokol abruptly closed its operations on Redstone Arsenal in 1996; there were about 400 employees at that time. Alliant Techsystems (ATK) Inc. was formed in 2001, and bought Thiokol, Inc. ATK Thiokol Propulsion maintained an engineering activity in Huntsville; this later became ATK Aerospace.

Chrysler Corporation - Huntsville (1952-2003) – Although primarily an automobile manufacturer dating from 1925, during WWII Chrysler produced tanks and mobile equipment for the military; this included the vehicles and large parabolic antennas used by the Army's SCR-584 radar, the most produced unit of its type during that period. As the REDSTONE missile was being designed at Redstone Arsenal, Chrysler was asked to provide expertise in production engineering. For this, Chrysler opened an office on Redstone Arsenal in 1952, their first activity in Huntsville.

On 11 January 1953, a contract was awarded Chrysler for the assembly of REDSTONE missiles. This was initially to work in cooperation with the Redstone Arsenal in building 12 prototypes. After successful testing, a contract for full-scale production was awarded to Chrysler in June 1955. The production was at the Navy-owned Michigan Ordnance Missile Plant in Warren, Michigan; over the next several years, a total of 101 REDSTONE missiles were built there.

The Chrysler teams at Huntsville were initially located at Service Centers on Redstone Arsenal and were later consolidated downtown at Oakwood and Meridian in the Huntsville Industrial Complex. Arleigh Trahern was the initial Engineering Manager of this activity.

In 1963, Chrysler built a 625,000-ft^2 facility at 102 Winn Drive in the Huntsville Research Park. This was primarily for working with Redstone Arsenal on the REDSTONE missile and in other Defense activities; it was managed by Arthur E. Douyard. During the 1960s, the Huntsville operations became strongly involved in electronic hardware development and manufacturing, especially for the Saturn-Apollo Program. A 176,000-ft^2 expansion to the 102 Wynn Drive facility was made in 1965. James T. Clifford was the Engineering Manager in this period. (See Chrysler Pentastar and Acuistar in next chapter.)

**Brown Engineering Company / Teledyne Brown Engineering -
Huntsville (1953)** – Since this was the first locally owned high-technology firm
in Huntsville, this description is more extensive. Its initial organization gives a
good indication of the foundation for future industry growth.

When the Ordnance Guided Missile Center (OGMC) was being formed in the
early 1950s, there was essentially no supporting infrastructure in Huntsville.
Wernher von Braun asked the Chamber of Commerce to assist in recruiting to
the city an engineering and precision manufacturing firm that could contract for
such work with the OGMC. Their search found John B. Bolton, owner of
Marietta Tool and Engineering Company in Marietta, Georgia, and provider of
support to Lockheed Aircraft. Bolton agreed to form a new firm in Huntsville,
moving his precision machine shop from Marietta and recruiting nationally for
engineers with existing security clearances for local work.

Huntsville's first high-technology firm opened 1 July 1953, with temporary
offices in the Times Building, downtown Huntsville. Aeronautical engineer

First Building

William A. Giardini was employed on the opening day
– he had previously been engaged in classified work at
the Air Force test center at Tullahoma, Tennessee.
Financed by local investors, the firm was incorporated
on 29 January 1954, under the name Alabama
Engineering and Tool Company (AETC). It soon
occupied the 13,000-ft^2 Smith Building, built for AETC
and located on the southwest corner of what is now
Memorial Parkway and Governors Drive. John E. Hatch, Jr., was the General
Manager and George F. Epps the Chief Engineer.

Rufus P. (R P.) Brown, owner of Brown's Engineering Company of
Indianapolis, Indiana, offered to merge with AETC and operate in Huntsville.
The offer was accepted, and on 29 June 1956, the corporate name was changed
to Brown Engineering Company, Inc. (BECO). In December 1957, the Board
made a decision to return the Company to local ownership and bought out R. P.
Brown's interests, but retained the corporate name.

The Army Ballistic Missile Agency (ABMA) was the main R&D operation
on Redstone Arsenal. BECO was the primary locally owned support contractor,
providing component design and manufacturing for the JUPITER missile and
analytical studies in activities such as Project Orbiter. After America's first
satellite, Explorer I, was placed into orbit by ABMA, William Giardini and four
other BECO employees were given national recognition by the American
Institute of Aeronautics and Astronautics (AIAA) for their contributions to the
program.

In February 1958, BECO moved its engineering and manufacturing operations into leased space in the Huntsville Industrial Complex (HIC), sharing the huge former Lincoln Mill textile plant with Chrysler and several other companies. On 6 May 1958, the Board appointed Milton K. Cummings (1911-1973), a local cotton broker

HIC Buildings

Milton Cummings and Wernher von Braun at BECO

and the largest investor in BECO, to serve as President. Cummings personally underwrote a new stock issue and quickly placed the Company on a sound business footing. In June 1959, Joseph C. Moquin became BECO's Executive Vice President, and the next year Raymond C. Watson, Jr., (this book's author) was employed by Cummings to establish research activities.

As the programs of NASA and U.S. Army agencies evolved in the early 1960s, so did BECO. In 1961, the Company initiated the formation of Huntsville Research Park (later named Cummings Research park and described elsewhere herein), and built a large engineering, research, and manufacturing facility on an 80-acre campus. The BECO Research Laboratories were formed in 1962,

1963 Complex in Research Park

and recruited a staff of highly qualified engineers, scientists, and technologists.

The BECO Research Laboratories were very successful from the start, setting the stage for the Company's future contracts in space and defense, as well as the formation and leaders of many future high-technology activities in Huntsville. In the mid-1960s, the following laboratories and their directors included:

• Applied Optics – Neil E. Chatterton   • Astronomy & Astrophysics – N. Frank Six   • Chemistry – Hans and Benno Walter   • Communications & Radar – Robert R. Parker   • Computer Sciences – George C. Monnig   • Control Sciences – C. S. Chen   • Geological Sciences – Ernst A. Kaarsberg   • Electromagnetics – Walter V. Bouldin   • Electro-Optics – J. William Foreman
• Flight Dynamics – Edward J. Rogers   • Fluid Mechanics – Gerald R. Guinn
• Heat Transfer – J. Wayne Littles   • Human Factors – Shiro Furukawa

• Instrumentation – Lawrence Conway • Nuclear Sciences – N. Ricky Byrn
• Plasma & Ion Physics – A. Lucile Cox • Propulsion Sciences – Hans H.
Seidel • Solid-State Physics – Harry J. Watson • Space Sciences – Ben A.
Sodek • Stress & Vibrations – James A. Baird • System Sciences – J. Richard
Fisher

All of these leaders held, or were completing, doctorates in their field.
Raymond C. Watson, Jr., was Vice President for Research, and Harry C. Crews,
Jr., served as his deputy. Roland G. Sturm, former Director of the Auburn
University Research Foundation, was an in-house consultant. Leon Deer led
new-product development, assisted by John McLean.

Program management / customer relationships were handled by Herschel A.
Matheny (Defense) and John E. White (Space). As Director of Research
Contracts, Eugene R. Andrzejewski was vital to the operations, as was the
Director of Technical Communications, Janet N. Bentley. Two persons who
provided major support to experimental research should be cited: Eugene R.
George was indispensable in apparatus development and Manfred W. Segewitz
was the mainstay of BECO activities at MSFC's Research Projects Laboratory.

Overall employment at BECO grew to about 3,500 persons. Most of the
technical personnel were assigned to the Research Laboratories or functional
departments. These departments and their managers during the mid-1960s
included the following (not all existing simultaneously):

• Aeromechanics – Warren J. Tomme • Aerospace Design – H. Dean Yantzi
• Advanced Transportation – R. Wood Calvert • Launch Operations – George
F. Norris • Commercial Systems – George H. Coyne • Electronic Support –
Robert M. Stewart • Electronic Engineering – G. Wayne Cantrell • Electronic
Systems – R. Clyde Ivy • Failure Analysis – J. G. (Pete) Belcher • Hardware
Engineering. – V. Eugene Petrie • Information Applications – Albert F. Bond
• Information Systems – Richard H. Shuford • Management Systems – Gordon
E. Darnell • Materials and Testing – I. Newton Vaughn • Materials
Laboratory – Charles E. Kaylor • Military Systems – Charles F. Ostner
• Process Engineering – John H. McLean • Propulsion – Bruce H. Neuffer

• Quality Operations – Jack Dashner • Space Systems –
Samuel D. Vollo • Space Vehicles – James L. Sanders •
Structures – James W. Fogleman • Systems Effectiveness
– Ray W. Burkett • Test Data Analysis – Anita Q. Dietzen
• Test Operations – Donald K. Martin • Vehicle Systems
– Max H. Coggins

William A. Giardini was Engineering Vice President,

William Giardini   and Jack W. Hendrix was Chief Engineer. Engineering managers were Paul H. Brown and Bruce H. Neuffer, Space Vehicles; James B. Martinelli, Equipment Systems; Charles B. Thompson and Buford Smith, Jr., Electronic Systems; and Clair E. Nicholson, Information Systems.

From its start, the Company had one of the largest and most comprehensive manufacturing operations in the region. In the mid-1960s, the manufacturing managers were Jay J. Harris, Mechanical Hardware, and M. Tyrone Samples, Electronics Hardware. David H. James managed Product Assurance; R. M. "Bucky" Walters, Quality Control; and William A. Meyer, Quality Engineering.

A major part of the engineering contracts were in support of NASA programs. William Giardini personally served as BECO's program manager for supporting the huge Saturn Vehicle activity of MSFC. Other key program managers included Billy G. Barnes, MSFC Quality Support; William J. Connell, MSFC Test Support; and Perry N. Fisher, KSC Support.

A state-of-the art Computer Center was included at BECO when the Research Park facilities were opened. Managed by Lee H. Cushman, the center supported engineering and research computing. In early 1963, BECO installed one of the first revolutionary IBM 7040 systems; this was later upgraded to the IBM 7094, and followed by an IBM 360 in 1965. As remote

IBM 7040

operations came into being, the Center supported these using a BECO-designed remote-access network, one of the first such networks in the nation.

Another major addition in the Research Park facilities was the Environmental Test Laboratory. Managed by James D. Hammons, the facility included a vacuum chamber simulating altitudes to 400,000 feet; a 10,000 pound-force, complex-motion vibrator; high-temperature thermal and humidity chambers; and many other test capabilities. This was the first full-service test laboratory of this type in the region and served the Company and outside firms for the decade.

In 1965, MSFC initiated a procurement competition for a single prime support contractor for each of the eight laboratories. Prime support contracts for both the Propulsion &Vehicle Engineering and Research Projects Laboratories were awarded to BECO, the only firm receiving two such awards. The Company also received subcontracts for support to the Test and the Quality and Reliability Assurance Laboratories.

Milton Cummings stepped down from President in April 1966, but continued as the Board Chairman; Joseph C. Moquin was named President. In April 1967, BECO was acquired by Teledyne Corporation becoming Teledyne Brown Engineering (TBE).

Joseph Moquin

Raymond Watson

In 1968, the Advanced Systems and Technologies (AS&T) Group was formed with about 1,000 personnel; Raymond Watson was the Group VP. Some of the AS&T projects were the start of TBE's long-running activities in payload integration for NASA and in ballistic missile defense and intelligence studies for the Army.

AS&T also had considerable R&D on commercial products; one project was on electronically controlled anti-skid brakes for heavy trucks and busses, and another led to the invention of the laser Doppler velocimeter – precisely non-contact measuring the speed of objects, fluids, and gasses – that ultimately led to the laser radar (lidar). Both projects had high potential for major commercial products, but, after development, were eventually terminated by corporate Teledyne because they were viewed as having high liability risks.

Throughout the Saturn-Apollo Program, BECO / TBE provided over 20 million man-hours to the lunar landing effort, called the greatest engineering achievement in history. A small but significant accomplishment was the design and development of the experimental model of what later became the Lunar Roving Vehicle (LRV). At a facility in Lewisburg, Tennessee, the Company built about 80 percent of the telemetry used with the SATURN.

In 1969, NASA presented Outstanding Public Service awards to 98 leaders from throughout industry for their contributions to the Saturn-Apollo effort; included were Milton Cummings, William Giardini, and Raymond Watson. (See Teledyne Brown Engineering, Updated, in future chapters.)

Norton / Saint Gobain - Huntsville (1955) – Norton Abrasives of Worcester, Massachusetts, was founded 1885, and is the world's largest manufacturer and supplier of abrasives for commercial applications, household, and automotive refinishing usage. In 1955, Norton opened a plant in Huntsville; located at 219 Cap Adams Road in the southeast corner of the city, it is adjacent to a railroad spur and a short distance from the Tennessee River. The Huntsville plant produces a variety of high-quality abrasive and refractory materials for industrial manufacturers worldwide. In 1990, Norton was purchased by Saint-Gobain of France, a firm dating from 1664, and now specializing in construction and high-performance materials. As of 2013, Robert S. McFadden is the Plant Manager, and there are about 140 employees.

P.R. Mallory / Capacitor Foil Plant - Huntsville (1956) – During WWII, P.R. Mallory and Company of Burlington, Massachusetts, a major producer of dry-cell batteries, used their expertise in electro-chemistry to manufacture electrolytic capacitors (then called condensers) for military products. After the war, capacitors of all types became major products of Mallory. Their electrolytic capacitors required rolls of very thin aluminum; in 1956, Mallory opened a plant in Huntsville for producing aluminum foil. Considerable electrical power was required in the manufacturing process (mainly for etching the foil surface), and the availability of low-cost power from TVA was a primary attraction of Huntsville. By the early 1960s, the plant had about 500 employees.

The foil production involves high-precision machinery, environmentally clean processes, and high levels of quality control. The 85,000-ft^2 plant at 2615 Memorial Parkway in Huntsville was, and still is, devoted to producing foil for capacitor manufacturers worldwide. The finished product is mainly rolls of 99.99-parts pure aluminum, 0.0003-inch and under in thickness. Over the years, much of the process has been automated.

Mallory sold the Huntsville operations in 1978; since then, ownership has changed a number of times. In 2002, it became Arora Technologies, Inc. (ATI), owned and operated by Mulk R. Arora, and now by his son, Rahul Arora. Although always relatively small in employment size, the firm is strongly technology based and adds another dimension to the capabilities of Greater Huntsville. ATI is the largest producer in the world of aluminum foil for electrolytic capacitors.

Ridge Manufacturing / Instruments (1959) – In 1959, D. J. Yeager founded Ridge Manufacturing, Inc. in Huntsville at 299 Cochran Road, SW. Initially, it specialized in large welding services, particularly in building non-combat military truck beds. The contract work extended into wiring harnesses and cables, and then into electro-mechanical assemblies. In 1969, the firm was reformed into Ridge Instruments, Inc., and moved to a 75,000-ft^2 facility at 7220 Governors Drive NW; an extensive environmental testing laboratory was included. In 1984, Timothy L. Yeager, son of the founder, became the Owner and President. Ridge Instruments moved to 4603 U.S. Highway 31 South in Decatur in 2004, and became a major manufacturer of TOW 2 Missile Launchers and upgrade kits.

Wah Chang - Huntsville (1960) – In 1916, mining engineer K.C. Li founded a trading company for the refinement and export of tungsten and other ores from his native China. The company was called Wah Chang, meaning "great development"; the company earned the name, pioneering the creation of new specialty metals and alloys. In 1960, Wah Chang, headquartered in Albany, Oregon, opened a plant in Huntsville on Highway 20 West. The plant soon became an industry leader in producing tungsten, tungsten carbide, and molybdenum products using powder alloy techniques. In 1962, there were about 75 employees.

In early 1967, Teledyne, Inc., of Los Angeles, acquired Wah Chang. The local operation became Teledyne Wah Chang Huntsville and was managed

separately from the Albany operation. By 1982, the Huntsville plant had about 175 employees and 90,000 ft^2 in five buildings on 21 acres at 7300/7400 Highway 20 West. In the early 1990s, Teledyne divided the operation, forming an additional firm named Teledyne

Teledyne Wah Chang
Huntsville

Advanced Materials. Together they could process tungsten and other materials from ore to powders and finished products. Bernhard Fritz Kieffer, a highly qualified metallurgist, was the overall plant manager and lead researcher.

Teledyne and Allegheny Ludlum merged in August 1996, forming Allegheny Teledyne, Inc. This merger, however, lasted only until November 1999; Allegheny Teledyne then split into three separate corporations. One of these, Allegheny Technologies Incorporated (ATI), included Wah Chang and Advanced Materals in Huntsville. Several of the other ATI companies were consolidated into what was called Alldyne Powder Technologies; later the Huntsville portion became ATI Firth Sterling. (Firth Sterling was a steel producer dating from the early 1900s and acquired by Teledyne in 1967.) (See last chapter for ATI Wah Chang Updated and ATI Firth Sterling.)

Technical Micronics Control (1960) – In 1960, Temperature Measurement Control, Inc. (TMC) was started by a local group to develop an accurate temperature monitor for supporting precision cleaning processes. Three years later, the name was changed to Technical Micronics Control, Inc., reflecting a broadening of products in the chemical cleaning field. George F. Epps was Chairman, and TMC facilities were on Highway 20 near Madison. By 1975, TMC had 45 employees and built a facility on six acres in the CRP at 210 Wynn Drive. They won a long-term waste management contract from the Army Missile Command and, over the next decades, made environmental services the firm's primary business. Doris R. Humphrey, who joined TMC as the bookkeeper, eventually became President and CEO. As of 2013, TMC continues with industrial and government contracts for waste management, environmental consulting, and chemical laboratory services; Michael W. Davidson is the President.

Hayes International - Huntsville (1960-1976) – Headquartered in Birmingham, Alabama, Hayes International Corporation was formed in 1951. Based at the Birmingham Airport, it was originated primarily for maintaining and upgrading commercial and military cargo aircraft. In the early 1960s, Hayes began offering the services of its engineering department to private and governmental clients, and opened an office in Huntsville pursuing on-site engineering support to the newly formed Marshall Space Flight Center. Over the following years, Hayes supplied engineering services in vehicle development and system testing; at times, several hundred employees were involved. Essentially all of the work was performed at MSFC on Redstone Arsenal, but Hayes had some activities and offices off-site in the Huntsville Industrial Complex at 204 Oakwood Avenue. In 1988, Hayes changed its name to Pemco Aeroplex, and later to Alabama Aircraft Industries; however, there was no work in Huntsville under these names.

Atlantic Aerial Surveys / Atlantic Technologies (1961-2011) – In the early 1960s, the Architect and Civil Engineering Department of Brown Engineering Company (BECO) recognized the value of aerial photography in developing accurate city and county maps. Raymond J. Meehan was hired to provide a capability in aerial photogrammetry, and Atlantic Aerial Surveys, Inc. was formed as a subsidiary in 1961; facilities were at 803 Franklin Street. BECO was acquired by Teledyne in 1967, and the new company, Teledyne Brown Engineering (TBE), was required to divest all of its subsidiaries. Atlantic Aerial Surveys became an independently owned firm, with Ray Meehan as President.

Through the 1970s and 1980s, Atlantic Aerial Surveys provided services in remote sensing, surveying, and consulting. In 1993, the firm broadened its activities to include computer-based geographic information systems (GIS) and was renamed Atlantic Technologies. R. Steven Denny was President, their facilities were at 208 Exchange Place, they owned a small fleet of aircraft, and there were about 60 technical specialists; a capability in airborne laser survey (LiDAR) information was added.

In 2001, Atlantic Technologies was acquired by Canadian firm MacDonald Dettwiler, and then by Aeroquest International in 2010. In 2011, the Huntsville operation became a part of Magnolia River Services, headquartered in Hartselle, Alabama. Magnolia River Geospatial, Inc. under Mark W. Brooks, Vice President, Technology, is located at 4975 Bradford Drive.

Space Craft, Inc. / SCI Systems (1961-2001) – A physicist by education, Olin B. King (1934-2012) came to Huntsville in 1957. Initially employed by RCA, he soon joined the Army Ballistic Missile Agency (ABMA) where he worked on the Nation's first satellites. Following ABMA into the formation of NASA's MSFC, there King saw the need for a local firm devoted to aerospace electronics and formed Space Craft, Inc. (SCI) in 1961.

SCI was started in the garage of King's home; the original objective was to design and build spacecraft for private buyers. An initial success was a contract from Johns Hopkins University; for this, a facility was rented and a staff employed. Electrical engineer A. Eugene Sapp, Jr. joined SCI in 1962, and was thereafter the closest associate of King. It was soon recognized that SCI could not compete in selling spacecraft *per se*, and attention was given to subcontract work on government programs.

SCI 8600 S Mem. Parkway

This redirection led to SCI opening a manufacturing plant at 8600 South Memorial Parkway as the Saturn-Apollo Program was getting underway at MSFC. Before the end of 1963, there were about 250 employees in 40,000 ft^2 of facilities. For over a decade, SCI primarily served as a subcontractor to Boeing and other prime contractors to NASA, the U.S. Air Force, and the U.S. Navy, building subsystems and electronic assemblies. Each SATURN V vehicle carried over 400 SCI-built assemblies. The company's longest-running contract was for hardware on the TITAN III launch vehicle; starting in 1966, it extended into the 21st century. (See SCI Continued in next chapter.)

Parker-Hannifin Corporation - Huntsville (1961) – In 1918, engineer Arthur L. Parker established the Parker Appliance Company in Cleveland, Ohio, to develop pneumatic braking systems for trucks and buses. After major growth during WWII, in the post-war period the company had operations across the nation, and then worldwide. The Hannifin Corporation was acquired in 1957, and the firm's name was changed to Parker Hannifin Corporation

Parker Hannifin opened a branch operation in Huntsville in 1961; three years later, a 130,000-ft^2 plant was occupied at 9400 South Memorial Parkway. This eventually became the company's Instrument Products Division (IPD), a global leader in the design, manufacture, and distribution of critical-flow and ultra-purity components. At its peak in the 1990s, Parker Hannifin had over 400 employees in the Huntsville operations. In this period, Charles L. Neidhardt was the General Manager, and Joseph D. Servick the Manufacturing Manager.

In 2003, the IPD relocated to a new plant at 1005 Cleaner Way, just off Triana Boulevard in the Chelsea Industrial Park. As a manufacturer of precision hardware, the IPD was a victim of foreign competition; local employment decreased to about 100. H. Craig Beckwith was General Manager at this time. A new Instrumentation Engineering Center was added in 2011. One of the first new products was the Parker THM Analyzer, an analytical system that quickly measures trihalomethane concentrations in water at ppb levels; this was developed by Kazi Z. Hassan. As of 2013, the Division General Manager was Tony Wyszkowski.

SPACO, Inc. (1961-1974) – With the formation of NASA's Marshall Space Flight Center (MSFC) in 1960, there was a need for support contractors; in 1961, John E. Hatch, Jr., led in the formation of SPACO, Inc., to provide such services. With Hatch as the initial President, George F. Epps was hired to develop and run the Engineering Department; both Hatch and Epps had previously been officers with Brown Engineering Company.

For the next several years, SPACO provided engineering, research, manufacturing, and test support to the various laboratories at MSFC. Some of the work was done at the NASA facilities, but there were also full capabilities within SPACO's facilities at 3016 University Drive. William Neely was the Director of Manufacturing. Considerable work by SPACO was in Human Factors; as a part of this, a "light switch" was developed, whereby the motion of the eye could be used to close or open a switch. SPACO used this light switch in an in-house project in designing a control system for wheel chairs.

In 1965, NASA directed MSFC to obtain a prime support contractor for each of its laboratories; major competitions were held. Through an extensive proposal effort led by George Epps, SPACO bid and won prime for the Quality & Reliability Assurance Laboratory. With Epps serving as the Program Manager, this contract continued with about 300 personnel until 1971. During this time, SPACO was bought by John B. Lampe, and eventually closed in about 1974.

IBM Federal Systems - Huntsville (1961-1975) – Computer systems began their emergence in the 1950s; before the end of that decade, IBM had become the world's leading technology firm, dominating the design and building of computers – often called artificial brains. In heavy-lift boosters, the rocket engines provided the power, but complementary brains were also necessary. When ARPA was reviewing the proposed design for the booster that could deliver payloads to Earth orbit or launches into space, IBM was turned to for providing the brains – the Instrumentation Unit (IU).

In 1957, IBM formed the Federal Systems, an organization devoted to applying the company's technical expertise and manufacturing to programs of the Federal Government. For the SATURN, IBM offered to not only participate with MSFC in the design of the IU, but the Federal Systems would open a plant in Huntsville for building this highly complex electronic system. IBM was awarded the IU prime contract, and the Federal Systems Huntsville Facility at 100 Sparkman Drive in the Huntsville Research Park was opened in 1964.

Unlike other firms described in this section, the IBM Huntsville Facility was a single-product plant – they built nothing but IUs for SATURN IB and V vehicles. These are described in the relevant sections. Clinton H. Grace was the IBM Huntsville Facility manager. When at peak, this had about 2,000 employees.

At the end of the Apollo program, the facility remained in operation and supported Skylab, the Nation's first Space Station , built from and launched by the last SATURN V. Also, both onboard and ground computers and software from IBM's facility in Owego, New York, helped in the scheduling and execution of Skylab activities. The last Saturn IB flight was launched on 15 July 1975, carrying a three-man crew on a six-day mission to dock with a Soviet Soyuz spacecraft. The IBM Huntsville Facility was closed after this; ultimately, the buildings were purchased by Olin B. King and given to the University of Alabama in Huntsville.

Boeing - Huntsville (1962) – William E. Boeing took his first plane ride in July 1915; this was with barnstormer Terah Maroney, a son-in-law of Madison County's William L. Quick, builder of the first airplane in Alabama. Boeing, then head of a timber industry and owner of a shipyard in Seattle, Washington, immediately became fascinated with aircraft. He incorporated Pacific Aero Products Company on 15 July 1916, and with Conrad Westervelt, a marine engineer who had some knowledge of aeronautics, built their first floatplane. In 1917, the firm's name was changed to the Boeing Airplane Company. Wong Tsoo, one of the few aeronautical engineers in the nation, was retained to design new planes. As WWI got underway, the company received from the U.S. Navy its first production order: 50 Model C seaplanes that had been designed by Tsoo. By mid-1918, Boeing had 337 employees.

In the 1920s Boeing became the nation's leading producer of military aircraft. After Charles A. Lindbergh made the first solo, nonstop, trans-Atlantic flight in 1927, there was a tremendous surge of interest in commercial aviation. In 1929, Boeing built the 12-passenger, three-engine Model 80 biplane, the first plane built by the Company specifically as a passenger transport. By 1930, with 800 employees, Boeing was one of the largest aircraft manufacturers in the country.

During the Great Depression, Boeing Aircraft survived by building mail-carrying planes and operating mail and cargo services. In 1934, William Boeing left the Company and Claire Egtvedt became Chairman and President. Egtvedt believed that the Company's future lay in large bombers developed in tandem with equally large passenger airplanes. In WWII, Boeing produced America's most important heavy bombers, the B-17 and the B-29. The new Boeing B-29 Superfortress entered combat in 1944; it was the largest bomber to go into production during the war and had the longest range.

With the end of WWII, Boeing factories shut down and 70,000 people lost their jobs. Over the next decade, Boeing had little success in new aircraft for either military or commercial markets, and the Company augmented aircraft building with systems management – providing technical direction for advanced weapons systems. In late 1958, Boeing was awarded a contract for the assembly and test of MINUTEMAN missiles (LMG-30). This was then the Company's largest, longest, and most complex military contract, and established Boeing as a manager of large, technology-based systems.

As MSFC began production of the Saturn V booster for the Apollo Program, Boeing was the successful bidder for the S-IC stage, doing the engineering in

Engineering Design Before Computers

Huntsville, manufacturing at Michoud near New Orleans, and testing at the Mississippi Test Facility. After signing the contract in December 1961, Boeing quickly organized a large staff of engineers and support personnel in Huntsville. By the summer of 1962, Boeing had some 1,000 employees on site at MSFC and at the HIC

Building. In 1964, Boeing built a complex at 220 Wynn Drive in the Cummings Research Park (CRP). At a maximum, there were about 4,500 Boeing employees working on the Saturn Program, mainly in MSFC facilities.

Boeing - Wynn Drive

General Electric - Huntsville (1962-1993) – When Redstone Arsenal was made the U.S. Army's Ordnance Rocket Center and Wernher von Braun's team at Fort Bliss, Texas, transferred to Huntsville in 1950, there were 102 employees of General Electric (GE) at White Sands who were also transferred to Redstone Arsenal. The GE group was working on the HERMES missile, but this project was terminated in a few months and the GE personnel mainly went to Valley Forge. Although this was, in fact, the first, GE activity in Huntsville, it did not stay and take roots.

In 1962, while the Apollo program was getting underway, General Electric returned to Huntsville as GE Aerospace under a contract from NASA headquarters to the corporate Space Division. As the efforts increased the personnel were consolidated into a new facility at 4000 South Memorial Parkway. The facility was built with support of the City of Huntsville and encompassed 124,000 ft^2 in two main buildings plus warehouses, and had up to 1,600 employees. GE Aerospace designed and fabricated equipment for the pre-launch test and launch control of the SATURN IB and V systems. James E. Keister served as the General Manager, and Leroy Softer the Engineering Manager. Frederick Schultz led in developing an outstanding capability in general-purpose test equipment.

Going into the 1970s, there was a significant decrease in business and GE moved to smaller facilities, eventually occupying a building on Leeman Ferry Road. At the lowest point, the only contract work was data systems analysis for MSFC. There was, however, a good capability in automated test equipment, and business in this area was pursued. The turning point was the awards by U.S. Army Missile Command for new automated test equipment for the Land Combat Support System and the Improved Hawk Program.

By the early 1980s, there was considerable recovery in contracts. With Lowell A. Kelly as Manager of Huntsville Programs and Dennis Singletary as Engineering Manager, GE moved to office and manufacturing facilities at 110 Wynn Drive in Cummings Research Park. In 1984, a major contract from the U.S. Navy for a computer-aided support system led to regaining almost 300 employees.

In 1993, GE Corporate made a decision to close existing Huntsville Operations and moved all contract work to Daytona Beach, Florida.

Wyle Laboratories - Huntsville (1962) – Wyle Laboratories was established in El Segundo, California, by Frank S. Wyle in 1949. Within a short time, the company expanded to mechanical, electrical, and pneumatics testing for prime aircraft contractors. In 1956, Wyle designed and offered the services of a low-frequency, high-force electrohydraulic actuator called Hydrashaker for large-mass test specimens.

To support the emerging Saturn Booster Program at the Marshall Space Flight Center, Wyle established an operation at Huntsville in 1962; a new facility was built at 7800 Highway 72 West. Under contract to IBM, Wyle performed flight qualifications testing on the Saturn 1B instrument unit, including low frequency, 200,000-force-pounds sine-wave and random vibration tests.

Wyle Laboratories 1962

In 1968, Wyle Huntsville began a major combined environmental test program for Rockwell/NASA. This involved static structural load-testing for the SIVB forward skirt, Instrument Unit, Space Lunar Adaptor (SLA), Lunar Excursion Module (Descent and Ascent stages), Service Module, and Dummy Control Module.

Wyle Huntsville diversified in 1970, with programs for DoD, DoT, and DoE using transfer of aerospace, engineering and test technology. One major contract was for the engineering and development of a Rail Dynamics Laboratory to support the DoT Ground Transportation Test Center in Colorado. In 1972, General Electric awarded Wyle Huntsville a contract to develop a system to simulate nuclear blast shock effects in testing antiballistic missile support equipment.

In 1977, Wyle Huntsville entered Space Shuttle development by designing and developing a special-purpose test facility for simultaneous combined environments (acoustic, mechanical loads, high temperature, and cryogenic cooling) to qualify ablative materials for the external tank.

In its two decades of operation, Wyle Huntsville became one of the most complete independent testing laboratories in the nation. Covering 93 acres with 163,000 sq. ft. in buildings, it had facilities and equipment large enough to accommodate systems and subsystems that otherwise could only be tested at specialized government or OEM's facilities. Their environmental and electromagnetic interference (EMI) and electromagnetic compatibility (EMC) resources include more than 50 environmental and simulation test chambers. Keith B. Wilson was the General Manager.

Northrop Corporation - Huntsville (1963) – John K. (Jack) Northrop formed Northrop Aircraft, Inc., at Hawthorne, California, in 1939. Through WWII, the still relatively small company built military aircraft, the P-61 Black Widow night fighter being the best known. Upon the successful flight of their XB-35 Flying Wing in 1947, public attention was drawn to Northrop. With expanding objectives, the firm was renamed Northrop Corporation in 1959. Shortly thereafter, Northrop Services, Inc. (NSI) was formed as a subsidiary for pursuing non-hardware business; branches were planned for a number of locations across the U.S., including Huntsville, Alabama.

Northrop opened a corporate office in Huntsville at 600 Boulevard South in 1962. Also in that year, Northrop purchased land on what was at that time a field road in Huntsville Research Park (forerunner of Cummings Research Park). A 15,000-ft^2, 2-story building at what became 6000 (later 6025) Technology Drive was opened for NSI in 1963. MG (Ret) John A. Barclay was named the General Manager of NSI-Huntsville, and Patrick R. Odom was hired to lead the technical activities. Essentially all initial work at NSI-Huntsville was performed for the parent Northrop; gradually, however, contract studies were obtained from NASA at MSFC and the Army at MICOM.

Lockheed Corporation - Huntsville (1963) – Allan and Malcolm Loughead (Lockheed) formed the Alco Hydro-Aeroplane Company at San Francisco, California, in 1912. This converted to the Lockheed Aircraft Company of Burbank, California, in 1926, and then became Lockheed Corporation in 1934. At the beginning of World War II, Lockheed – under the guidance of Clarence (Kelly) Johnson, one of America's best known aircraft designers – built a highly unusual twin-engine, twin-boom interceptor aircraft, the P-38 Lightning. Later, Johnson began a secret facility that became known as the Skunk Works; his well-known products included the P-80 Shooting Star, the U-2, and the SR-71 Blackbird.

In the early 1960s, a subsidiary, Lockheed Missile and Space, Inc. made the decision to open a facility in Huntsville. A 57-acre plot was purchased in the

Lockheed 1963

northwest corner of the newly opened Huntsville Research Park. A building was started at 4800 Bradford Drive in May 1963, and in November, Lockheed's Research Park facility was opened. James S. Ferrior, III, was assigned as the Manager of Lockheed M&S Huntsville, remaining there until 1980. The company began work in Huntsville as a NASA contractor, mainly in MSFC on-site support of SATURN development, but in the later 1960s the work became primarily in missile defense.

AVCO / AVEX - Huntsville (1963-1999) – Aviation Corporation Inc. was formed in Wilmington, Massachusetts, in 1928; it later shortened its name to AVCO. In 1963, as NASA-MSFC was getting underway, AVCO set up a small manufacturing operation in Huntsville. By the mid-1970s, this had grown to where a large facility was opened at 4807 Bradford Drive, CRP. AVCO was purchased by Textron in 1984; then the Huntsville Electronics Division was purchased by a private investor, J. M. Huber Corporation of New Jersey, in 1985. The firm's name became AVEX – keeping it close to AVCO. With operating headquarters in Huntsville, AVEX expanded both nationally and internationally; near the end of the 1990s, it had about 800 Huntsville employees. In 1999, AVEX was acquired by Benchmark Electronics.

Computer Science Corporation - Huntsville (1966) – Computer Science Corporation (CSC) was founded in California by mathematician Roy Nutt and businessman Fletcher Jones in 1959. In 1961, it obtained its first government contract – with NASA's Jet Propulsion Laboratory, and in 1962, opened its first service bureau – offering "computing time" over telephone lines. In 1965, two large divisions of ITT were acquired, making CSC the largest software services company in the United States. The corporate headquarters of CSC were in El Segundo, California,

In June 1966, CSC won a five-year contract for supporting computer services at NASA's Marshall Space Flight Center. Their first office was opened in Huntsville at 4835 University Square; Ralph Rodney, Jr. was the Director of Program Development. Since then, CSC has had a continuous presence in Huntsville, providing services to federal, health care, and commercial activities. By the early 1990s, CSC had about 100 personnel in Huntsville; Glennis Williams was the resident VP.

Toroid Corporation (1966) – In 1966, Paul and Anne Paelian started a business in Huntsville winding toroid coils for load cells – force-monitoring devices widely used in rocket development and other mechanical and structural applications. The Toroid Corporation was formed and a facility built at 225 Winn Drive in what is now Cummings Research Park. Toroid specialize in highly accurate electronic weighing systems, and their catalog of load cells and strain-gauge type load pins shows hundreds of standard items. As of 2013, Anne Paelian continues as the firm's Owner and President, with Paul Paelian, Jr., as Vice President.

Battelle - Huntsville (1967) – Battelle Memorial Institute is the largest nonprofit research and development organization in the world. Headquartered in Columbus, Ohio, since its founding in 1929, Battelle serves the national security, health and life sciences, and energy and environmental organizations, both governmental and commercial. The Institute was founded on industrialist Gordon Battelle's vision that business and scientific interests can go hand-in-hand as forces for positive change.

Recognizing that this city was destined to be a leading science and engineering center, Battelle opened a Huntsville office in 1967. From the start, Battelle Huntsville had a small technical staff and facilities to provide direct-access services to local activities, but it has primarily served as a conduit to the outstanding capabilities in the Columbus and other large laboratories. As of 2013, Battelle was located at 7047 Old Madison Pike with Bobby L. Evars the Site Manager.

Automatic Electric - Huntsville (1967-1986) – Aaron Strowger patented the automatic telephone exchange in 1891; this involved rotary dialing and a stepping-relay switch to connect telephone users without the need for operator assistance. In 1901, the Automatic Electric Company was formed, buying the Strowger patents and beginning the manufacturing of automatic telephone equipment. During the first half of the 1900s, Automatic Electric was second only to the Western Electric (half owned by Bell Telephone), and the largest supplier to independent telephone companies. Automatic Electric became an independent division of General Telephone and Electronics (GTE) in 1955.

GTE opened an Automatic Electric plant at Huntsville in 1967. Located at 13000 Memorial Parkway in the Farley area on the southern end of the city, the plant covered 550,000 ft^2 of space; it was inaugurated by Robert J. Gressens, President of Automatic Electric. The plant manufactured business and consumer telephones, as well as coin-operated instruments; there was also a telephone refurbishment operation. Initially, there were about 1,000 employees. By 1974, the plant was manufacturing between 45,000 and 55,000 telephones per week; of these, about 45 percent were touch-tone sets and the others several types of rotary-dial instruments. Employment peaked at about 3,000.

The most popular set was the Styleline; this had a push-button dialer, with duel-tone multi-frequency (DTMF) operation. With DTMF, each button pressed gives two simultaneous and distinct audio frequencies; this requires eight separate tone oscillators, all mounted on a tone-dialer chip. The eight frequencies must conform to an international standard, and must be stable over wide variations in temperature. Power for the tone-dialer chip comes from the connected telephone line (around 45-V DC); thus, no battery or outside power supply was required, but a resistor divider was needed to reduce the voltage and protect the transistor. As dialing tones were generated, the chip had to switch off the set's microphone, otherwise there could be outside-sound interference.

The DTMF unit was the most expensive part of the telephone, and the plant had an electrical design group that, in addition to other activities, continuously experimented with circuit-cost reduction. Bruce H. Bodin was the managing engineer of this group, and made many improvements in the DTMF technology. Product designers included Ronald C. Ware, Ronald M. Kenny, and Richard B. Kostey.

By 1984, the Huntsville plant was the only GTE facility making telephones. The plant, however, was closed in 1986, as domestic labor and production costs rose sharply against overseas competitors; at closing time it had 800 permanent and 300 temporary employees.

PPG - Huntsville (1969) – Pittsburgh Plate Glass, Inc. was formed at Pittsburgh, Pennsylvania, in 1883. Later called PPG Industries, for many years it has been a leader in manufacturing glass and related products. It is now the world's largest supplier of sealants, coatings, aircraft maintenance chemicals, transparent armor, transparencies, and application systems, serving original equipment manufacturers and maintenance providers for the commercial, military, regional jet, and general aviation industries.

In 1969, PPG Industries opened a plant in Huntsville to develop and manufacture aircraft windshields. Located at 1719 Highway 72 East in what is now the Chase Industrial Park, it was a part of the PPG Aerospace Division.

PPG Huntsville

George A. Ouskarich was the Plant Manager, and there were about 500 employees. Within a few years, the Huntsville plant became the largest producer of aircraft windshields and windows in the world.

Typical of their defense work is a five-year contract with the Defense Logistics Agency's Defense Supply Center to supply windshields for Sikorsky UH-60 Black Hawk helicopters and variants operated by the U.S. Army, U.S. Navy, and U.S. allied forces. PPG has produced heated-glass windshields for the Black Hawk helicopter and its variants since the 1970s.

An example of commercial work is the production of windshields and side cockpit windows for the Eclipse 550 private jet from Eclipse Aerospace. These are glass-faced acrylic windshields that combine the chemical, abrasion, and craze resistance of glass with the lightweight benefits of acrylic, and resist the impact of a 4-pound (9-kg) bird at 350 knots (400 mph). Veronica Frain is the PPG program manager, aerospace transparencies.

In addition to aircraft transparencies, the Huntsville plant has cooperated with PPG's transparency manufacturing facility in Sylmar, California, in developing a variety of ballistic transparent armor – glass and polycarbonate – for military ground vehicles. The Huntsville plant is FAA-approved for overhauls and repairs of framed window assemblies.

From its beginning, the Huntsville plant had also been engaged in highly specialized research and testing concerning impact resistance; they are well known for their "bird gun," simulating the impact of a bird during flight at relative speeds of Mach 1 (740 mph). Luis E. Deganis is a lead researcher at this facility.

At the start of the 21st century, the Huntsville facility was called PPG Aerospace - Transparencies. As of 2013, Brian K. Pollock was the Plant Manager and there were about 750 employees.

Phoenix Industries (1969) – Since 1969, the Huntsville Rehabilitation Foundation – a 501(c)(3) organization – has done business as Phoenix Industries of Huntsville; this is the totally self-supporting industrial side of the Foundation's comprehensive community-based vocational rehabilitation programs. (See Non-Profit Organizations in Chapter V for details.)

M&S Computing (1969) – M&S Computing was founded by James W. Meadlock, Nancy Broadwater Meadlock, and Terry T. Schansman in 1969; they and two other initial personnel – Robert B. Thurber and Keith H. Schonrock – had previously been with IBM, developing software for the Saturn vehicles Instrument Unit. In a short time, James F. Taylor, Jr., joined the staff. All but Schansman stayed with the company for many years and held key positions. Working in a small rented facility, M&S Computing's initial contracts were with ABMA in developing hardware and software systems that applied digital computing to real-time missile guidance issues. (See M&S Computing / Intergraph – Updated in next chapter.)

Universal Data Systems (1969-1978) – As computers emerged in the early 1960s, data communication over telephone lines between computers used a modem (short for modulator and demodulator). The first modems involved use of a standard called "serial synchronous data transmission" (SSDT), which required that the sending computer be perfectly synchronized with the receiving computer. In the early 1970s, AT&T's Bell Laboratories created a new standard, "asynchronous serial data transmission" (ASDT), that enabled computer transmitters and receivers operating at different speeds to successfully communicate data to each other. Since AT&T owned the telephone lines and limited the attachment of devices, users of these standards were forced to rent modems from AT&T.

Researchers at the Stanford Research Institute developed a device that converted digital pulses into audio signals and vice versa. Placing a telephone handset onto the twin cups on this device (called an acoustic coupler) allowed data transmission over telephone lines without a direct electrical connection, thus circumventing AT&T's monopoly over the communications sets.

While working at Space Craft, Inc. (SCI) in Huntsville, Mark C. Smith realized that the merger of telephone and computer technologies would revolutionize the transmission of data. In 1969, Smith founded Universal Data Systems (UDS) in Huntsville, primarily to make moderately priced data communications equipment for the growing U.S. computer industry. In a short time, UDS built a 110,000-ft^2 engineering and manufacturing facility at 5000 Bradford Drive in the Huntsville Research Park. When the acoustic coupler came into being in the early 1970s, UDS – as well as several other firms – took on this product. These early acoustic couplers might allow transmission rates of 300 baud (~bits/second), but 150 baud was more typical.

During the first part of the 1970s, the ARPANET – forerunner of the Internet – was slowly being developed. This was a system of small computers called Interface Message Processors (IMPs - later called routers), modems, and telephone lines interconnecting mainframe computer centers at government and university research centers. Later in the 1970s, this technology expanded and UDS obtained a considerable share of the market.

To gain an entry to this market, electronics giant Motorola acquired UDS in 1978. Smith became a Motorola vice president, and the Huntsville operation continued as Motorola UDS. Meanwhile, AT&T was pursuing a major law suit against Carterphone, one of the leading suppliers of acoustic couplers.

HUNTSVILLE / CUMMINGS RESEARCH PARK

With increasing contract work from MSFC as well as the Army, Brown Engineering Company (BECO) soon outgrew its facility in the Huntsville Industrial Complex. In early 1961, a search turned up a large tract of undeveloped land on the western edge of the city. Located adjacent to land that had recently been acquired by the University of Alabama for developing a Huntsville Center and within a few miles of MSFC and Army agencies on Redstone Arsenal, this area was ideal for establishing a high-technology research park. BECO purchased 360 acres of this land in mid-1961, and soon started on the first building. At that time, the only access was via a dirt farm road; this later became Sparkman Drive.

BECO executives Milton K. Cummings and Joseph C. Moquin proposed that the City of Huntsville zone the area as a research park district. Dean Y. Matthews, head of City Planning, visited Research Triangle in North Carolina; impressed with what he found and visualizing a similar activity in Huntsville, he recommended that the City Council replicate this with a park district of 3,000 acres. Charles H. Younger in the City Attorney Office is credited with developing the extensive zoning that controlled the future use of this land.

In 1962, the land was officially designated Huntsville Research Park. Holding 80 acres for its own central campus, BECO sold other land at cost to firms such as IBM and Northrop, and donated a right-of-way to the city for thoroughfares and utility stations.

The 3,000 acres was in two distinct parts, divided by the Research Drive (now Research Park Boulevard), a main thoroughfare into Redstone Arsenal. A 1,000-acre eastern part contained the original acreage purchased by BECO, and, to forestall land speculation, the remainder was bought by Research Park

Foundation (RPF), a new 501(c)(6) not-for-profit land-holding organization. (The RPF later became the University of Alabama Huntsville Foundation.) Lockheed was one of the first land-buyers from the RPF, and built their facility in 1963 – second only to BECO.

The 2,000 acres in the western part of the Research Park was held by the RPF under rights-of-first refusal, and remained undeveloped for a decade. It was first opened as what was called Research Park West in 1982; Dynetics, Inc. was the first occupant.

Dedication of CRP, Senator John Sparkman and Mrs. Vastus Ivy Cummings on right

Renamed Cummings Research Park (CRP) in 1973, honoring Milton Cummings who had died early that year, it is now commonly called the second largest park of its type in the Nation – only Research Triangle Park (RTP) in North Carolina is larger than CRP. However, since RTP is spread over areas in two counties and intermingled with many non-technical firms, and CRP is fully concentrated with strong restrictions as to the type of occupants, there is justification to credit CRP as being the largest contiguously formed research park of its type in America, possibly in the World. In 1997, CRP was voted as the Most Outstanding Science Park in the World by the Association of University Research Parks.

U.S. SPACE & ROCKET CENTER

The U.S. Space & Rocket Center in Huntsville is one of the nation's best technology museums. Operated by the government of Alabama, it is located just south of the Cummings Research Park at Exit 15 on Interstate 565. It contains over 1,500 permanent rocketry and space-related artifacts, physically showing a large portion of Greater Huntsville's technological evolution. Commonly called the Space Center, it opened in 1970, just after the second manned mission to the lunar surface. Originally named the Alabama Space & Rocket Center, it was renamed when it became listed on the National Register of Historic Places.

Earlier, the facilities of the U.S. Army and Marshall Space Flight Center on Redstone Arsenal had included exhibits centering on the rockets developed by the Army. Wernher von Braun was a major leader in promoting a statewide bond referendum to finance construction of a new museum; the referendum passed in November 1965. Redstone Arsenal donated 35 acres of land, and Huntsville architect David Crowe designed the initial building with 22,000 ft^2

(2,000 m^2) of exhibit space. An Omnimax (now called IMAX Dome) theater was installed in early 1984.

Edward O. Buckbee was selected by von Braun as the founding Director; remaining until 1994, he deserves full credit for leading the first quarter century of growth.

Full SATURN IB and V systems are the most noted items on display; Buckbee worked with von Braun in obtaining these vehicles from NASA in 1969. The SATURN I was placed erect in the outside display; the SATURN V was initially laid out horizontally with the stages separated. For 10 years, this was the only SATURN

Edward Buckbee

V on national public display. Other outdoor displays included a German V-1 flying bomb and a V-2 missile, a lunar landscape with lunar lander mockup, an F-1 rocket engine standing 18.5-feet (5.6-m) high, and a Pershing missile system poised as if ready for launch.

In the hundreds of inside exhibits are the actual Apollo 16 spacecraft that went to the Moon (this and a Moon rock are the highest

U.S. Space & Rocket Center

regarded artifacts); an engineering prototype of Skylab – America's first space station; a test model of the Lunar Roving Vehicle; and Project Mercury and Gemini simulators showing the cramped conditions endured by the first

Americans in space. The Center is the resting place of Miss Baker, a squirrel monkey who flew on a suborbital test flight of the JUPITER rocket in May 1959; Miss Baker lived in a specially designed habitat at the Center from 1971 until she died in 1984.

Redstone Arsenal eventually deeded a total of 450 acres to the Center, a large

Space & Rocket Center Interior

part of which was in turn leased to the City of Huntsville for the Botanical Gardens and recreational areas. Some of the new area was used for two camp programs: the U.S. Space Camp and the Aviation Challenge; these provide residential and day educational programs for children and adults – particularly teachers – using simulators, lectures, and training exercises.

In 1999, a full-scale model of the SATURN V rocket was erected in front of the main building, joining an actual Lockheed SR-71 Blackbird – the fastest air-breathing manned aircraft in the world – on loan from the U.S. Air Force. (See Chapter V for U.S. Space & Rocket Center Update.)

HIGHER EDUCATION

During the first decades of the 1940s, The University of Alabama (UA, in Tuscaloosa), as well as Alabama Polytechnic Institute (API, later becoming Auburn University) gave courses in Huntsville under the Engineering, Science, and Management War Training (ESMWT) program. Both schools ended Huntsville courses at the close of the war in 1945.

At this time, there were three other higher-education institutions in the Greater Huntsville area. Alabama A&M College, a historically Black school located just northeast of the city limits, had undergraduate offerings in certain areas of science, but the programs were primarily for teacher training and agriculture. Founded in 1896, Oakwood College was owned and operated by the Seventh-day Adventist Church and located in the northwest part of the City; Oakwood began offering bachelor's degrees in 1945, but its initial courses gave little support to the local technology needs. Dating from 1822, and the oldest college in Alabama, Athens College was located in Athens, about 20 miles northwest of Huntsville; owned and operated by the Methodist Church, its bachelor's programs at that time were primarily in education and liberal arts.

Brief information on Alabama A&M College is contained in this section. More information on all three of these institutions is given in the next chapter.

University of Alabama – Huntsville Center

As previously noted, the University of Alabama (UA) gave ESMWT courses in Huntsville in the early 1940s. On 15 January 1950, the UA returned to Huntsville, opening the Huntsville Extension Center (commonly called the Huntsville Center). Huntsville attorney Patrick W. Richardson had lobbied the UA president for this, and is generally given primary credit for obtaining the Center.

This extension activity began before Redstone Arsenal was designated the Ordnance Guided Missile Center and before the arrival in Huntsville of the rocket team from Fort Bliss; thus, the first offerings did not have a technical emphasis. John R. Morton, then Director of Adult Education at the UA in Tuscaloosa, served as the initial Director. The UA Center classes and offices used space in West Huntsville High School (predecessor of Butler High School) on the corner of Clinton and Fifth Avenue (later part of Governors Drive).

By mid-1951, over 5,000 persons were working at Redstone Arsenal, and there was a demand for broader types of courses. In September, UA's Graduate School received a Government contract to give engineering, physics, and mathematics courses for military and civil-service employees. Called Redstone Arsenal Graduate School, this used classrooms of the UA Center, but otherwise the instructional operations were independent. The contract continued until 1954, and at that time the graduate courses were opened to the public and given by the UA Center.

Foundation – In 1956, the Huntsville Industrial Sites, Inc. (HIS), was formed by 30 local investors. They purchased land on the northeast corner of Redstone Arsenal that was made surplus by the construction of a new Highway 20 to Decatur, and quickly sold most of it for considerable profit. This began transactions on industrial land that continued for several years. Although organized as a for-profit corporation, the purpose of HIS was to build funds that could ultimately be used for developing the UA Center into a university. Later, HIS reconstituted into a 501(c)(6) not-for-profit, land-holding entity, and still later it converted to the University of Alabama Huntsville Foundation.

Academic Programs – In early 1960, MG John Medaris, commander of the recently formed Army Ordnance Missile Command, addressed the Alabama Legislature, saying that if the State wanted technology-oriented industrial growth, the UA Center in Huntsville should be expanded to a full-scale university. By then, the University Center, directed by Phillip M. Mason, had grown to 1,100 students (mainly part-time). The Legislature allocated funds to UA for the Center's campus.

Land initially purchased was located between U.S. Highway 72 (a portion of which was renamed University Drive) and Athens Pike (an extension of Clinton Avenue), and bordered on the west by an unpaved farm road (soon named Sparkman Road and later Sparkman Drive). Bonds were issued for an initial

building; this opened on the northwest corner of the campus in January 1961, and was named Morton Hall (honoring the Center's first Director).

There were a few full-time faculty members, but most of the instruction was by an adjunct

Morton Hall

faculty drawn from local government agencies and industrial firms. (As an adjunct faculty member, this book's author taught one of the first courses given on this new campus.) A library for the Center was highly important, and this was started by Christel Ludewig McCanless; the

Redstone Scientific Information Center (RSIC) donated about 1,000 books. Most students could also have access to the huge RSIC collections of books and journals on Redstone Arsenal. In addition to Phil Mason, the administration included Nan G. Hall, the first Registrar.

It must be noted that racial segregation still existed at this time in Alabama; Blacks, even if employed as professionals at Redstone Arsenal, could not attend the UA Center. The campus of Alabama A&M College was only a few miles from the new UA Center campus, and it would have been to the long-term benefit of both schools if integrated programs could have been offered from the beginning. Segregation ended with Governor George C. Wallace's "stand in the schoolhouse door" at the UA in Tuscaloosa in 1963. It is noted, however, that even earlier Blacks had quietly been admitted to Huntsville UA Center classes.

During the 1960s, there were significant changes in Huntsville's higher education. As the UA Graduate Center increased in enrollment, the offerings were greatly expanded – particularly in engineering, physics, mathematics, and chemistry. Although the degrees were awarded through the Tuscaloosa Graduate School, master's degrees in these fields could be completed in Huntsville as of 1964, and undergraduate degrees were added in 1965. Cooperative doctoral programs were initiated between Huntsville and Tuscaloosa; thesis and dissertation research was often conducted at the student's place of employment. The first master's degrees were awarded through Huntsville in 1964, followed by undergraduate degrees in 1968

In 1969, the Board of Trustees of the University of Alabama established the University of Alabama System, with three independent, autonomous operations at Tuscaloosa, Birmingham, and Huntsville. The University of Alabama in Huntsville (UAH) became fully operational the following year. The evolution of UAH starting in 1970 is given in the next chapter.

Research Institute – In June 1960, as a follow-up to General Medaris's plea for educational expansion, a group of officials from Redstone Arsenal asked the UA President to add a research activity in Huntsville. In October, an interim Research Institute was started, using loaned personnel from the Tuscaloosa campus. On 24 January 1961, MSFC funded the first work of this Institute, and a beginning operation started in the existing teaching facilities. On 21 June 1961, Wernher von Braun addressed a joint session of the Alabama Legislature. He cited the University Center, the Research Institute, and the potential Research Park as vital elements in the future economic development of Alabama, and suggested that special funding be allocated to fully develop the Research Institute.

The Legislature responded by unanimously voting $3 million (about $14 million present dollars) for the UA Research Institute. A bond issue was approved 3-to-1 by Alabama voters on 5 December 1961. Huntsville and Madison County provided funds to buy 200 acres of additional land on the south end of the existing University Center campus. A building was started, providing space for both the Research Institute and additional engineering offices and classrooms. Initially, John F. Porter, Jr. led the central research activities, then Rudolf Hermann was recruited as the Director. Hermann had built hypersonic wind tunnels for von Braun at Peenemünde, and recently had been a professor of Aeronautical Engineering at the University of Minnesota. The Research Institute officially opened in 1963, temporarily operating in space at the Research Laboratories of Brown Engineering Company.

Alabama A&M College

Through the efforts of former slave William Hooper Councill, the Huntsville Normal School was organized in 1875, providing the first opportunity in higher education for Madison County's Black population. In that era, a "normal school" was an institution for training elementary and high school teachers, instructing the students in teaching standards or norms – thus, the name. Huntsville Normal School's first formal program, started in 1878, was in Industrial Education. Under the second Morrill Act of 1890, the School became a Land Grant Institution and moved to its present location at Normal, Alabama (adjacent to the northeast city limits of Huntsville) in 1891.

Courses in agriculture were added, and, in cooperation with Alabama Polytechnic Institute (now Auburn University), a research farm was operated. Upon earning the status as a junior college in 1919, the name was changed to the State Agricultural and Mechanical Institution for Negroes. Courses at the senior college level were added in 1939, and the first graduating class received bachelor's degrees in 1941. The name was changed to Alabama A&M College in 1949, and the College became a fully accredited member of the Southern Association of Colleges and Secondary Schools in 1963.

As a part of its agriculture and teacher education programs, Alabama A&M had longstanding offerings in mathematics, chemistry, physics, and other disciplines that benefitted the areas technology evolution. Until the early 1960s, however, there was segregation

Alabama A&M University

of the races and the College had very few advanced courses that attracted the few African-American engineers, scientists, and technologists in the local professional community. The College was elevated to University status in 1969. (The evolution of AAMU starting in 1970 is given in the next chapter.)

SPACE CITY USA

Space City USA was announced to the public in January 1964 – Huntsville was going to have an amusement park that would rival Disneyland. A group of local investors had raised capital to start the project; it was estimated to cost $5 million (about $35 million in today's dollars) and bring an estimated 1.2 million visitors annually to Huntsville. In the planning stage for three years, construction had already started and the park would open in 1965.

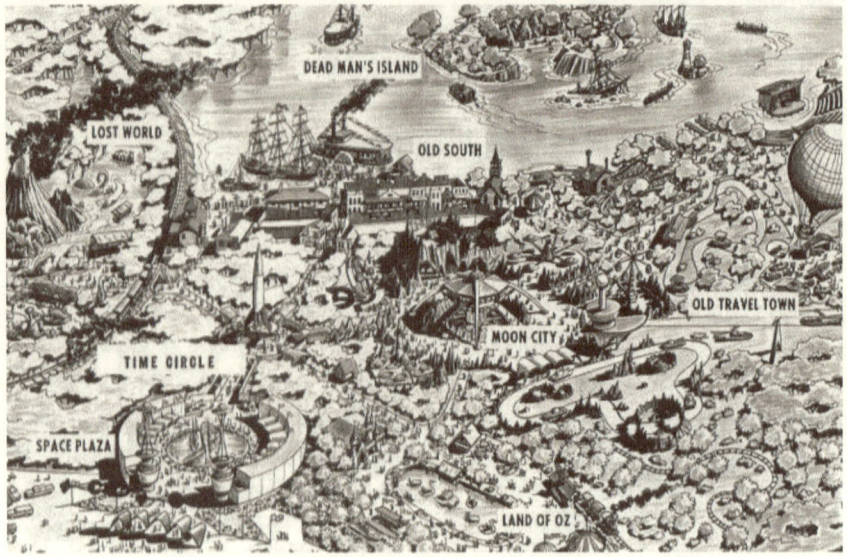

Planned Layout for Space City USA

Space City would be located on an 850-acre site surrounding Lady Ann Lake on Highway 20 between Huntsville and Decatur. Entry to the park would be through a "time machine" into themed area including Lost World, Moon City. Land of Oz, and Old South. There would be a skyway ride, a glass-bottom boat, a jet-car ride, and a flying saucer; 24 park rides in total would offer something for everyone. Running around the area would be a totally rebuilt 1897 steam locomotive with five passenger cars. A 150-acre commercial area with shops, a 10-story hotel, and restaurants was to be on Highway 20.

Space City USA initially incorporated with 24 stockholders. Engineering and land grading began in early 1964. A few of the buildings were constructed, an iron structure for the volcano was erected, and a section of 36-guage track was layed. The 1897 H.K. Porter Locomotive, named the *Melodia B*, arrived in June. It seemed that the Space City park development was fully underway.

Challenges developed, however, and progress on the theme park began to stall. By early 1967, the park opening date had been delayed several times and it appeared that financing had run out. A lawsuit by investors followed, and the rapidly decaying park was put on the auction block in October.

By 1970, essentially no visible signs of the grand theme park remained. Space City USA was one of the biggest commercial things that never happened in Greater Huntsville.

AGRICULTURE IN ROCKET COUNTRY

Differing from the other portions of this chapter, this section covers the period extending somewhat beyond 1970; this is due to the incremental periods of availability of data.

While Huntsville was becoming known as the Rocket City, the surrounding Madison County was still highly involved in agriculture. A very telling comparison of these two activities is found in the equipment involved; while the rocket-development facilities at Redstone Arsenal had the latest tools and machines, much of the farming in the county remained largely by hand or using simple horse/mule-drawn equipment. This was, however, changing. As of the 1945 census, equipment on the average farm in Alabama was valued at only $320; less than one percent of the farms had a tractor, but 10 years later the number of tractors on farms exceeded the number of horses and mules.

In the State of Alabama as a whole, farm numbers in Alabama decreased from 211,500 to 60,000 during the period 1950-1974; at the same time, the average size of a farm increased from 99 acres to 198 acres. In 1950, 31.4 percent of Alabama's population lived on farms compared with only 4.6 percent in 1970. The single proprietorship continued to be the dominant form of legal organization for Alabama farms; in 1974, 92.4 percent of all farms with sales of $2,500 and over were owned by individuals or families. Tenant farming declined from 41.5 percent in 1950 to 6.8 percent in 1974. Between 1950 and 1976, the average value of land and buildings per Alabama farm increased from $4,800 to $81,200.

Nationally, the farm output per man-hour rose at a faster rate than the industrial output per man-hour. In 1950, the U.S. farm output per man-hour was 35 percent of the index for 1967, and the industrial output per man-hour was 67 percent. By 1973, the farm labor index had increased to 133, while industrial labor index had increased to 114. The increased productivity of farm labor was, to a major extent, due to farming mechanization. From 1960 to 1970, Alabama cotton producers switched from 98 percent hand-picked cotton to 98 percent machine-picked cotton

Cotton, the historic "King" of Alabama agriculture, has greatly declined in importance since 1950. During this period, harvested cotton dropped from a high of 1.6-million acres in 1953, to a low of 400-thousand acres in 1975 – about a 300 percent decrease. Alabama's percentage of national cotton production decreased from 5.7 percent in 1950 to 3.8 percent in 1975. Similarly, corn production dropped from 2.0 percent to 0.6 percent in this period. Soybean production, however, went from 0.6 percent to 2.1 percent.

Crop yield – pounds or bushels of a crop per acre harvested – is an important measure in agriculture. Modern advances in machinery, fertilizers, seed varieties, and pesticides have enabled farmers to produce greater levels of agricultural output. Fertilizer application on Alabama farmland increased on the average from 261 pounds per acre in 1950 to 409 pounds per acre in 1975; anhydrous ammonia was increasingly used as a cheap source of nitrogen.

In insect control, DDT was initially extremely effective, but boll weevils in the U.S. developed resistance by the mid-1950s. Methyl parathion, malathion, and pyrethroids were subsequently used, but environmental and resistance concerns arose as they had with DDT and control strategies changed. In 1971, DDT production was prohibited by the new Environmental Protection Agency.

The control of weeds has been inherent to cotton farming since its beginnings. Through the years, this control was through the labor-intensive process of hoeing. The chemical glyphosate was discovered to be a highly effective herbicide by Monsanto in 1970; brought to market under the trade name Roundup in 1973, this essentially ended hand weed control.

The following are included in the official agricultural statistics for Madison County in the years shown:

	1949	1979
Number of Farms	5,004	1,213
Total Farm Area, Acres	415,332	209,745
Percent of County	80	40
Average Farm Size, Acres	83	173
Cotton, Acres	109,400	45,200
Cotton, Bales	50,800	41,400
Cotton, Yield (lb/acre)	223	440
Corn, Acres	56,400	11,800
Soy Beans, Acres	5,400	88,100

Although the total area in the County devoted to farming about halved in this period, and there were 76 percent fewer farms, these farms were more than double in size; this shows the increased importance in mechanized farming. While the acreage in cotton was reduced by more than 60 percent, the corresponding drop in cotton lint (bales) was less than 20 percent – the yield was about doubled. In this period, Madison County's acreage in soy beans increased by a factor of over 16, while corn acreage decreased by a factor of almost 5. Richard O. Magnusson was the County Agent for many years starting in 1955.

The Tennessee Valley Research and Extension Center (formerly known as the Tennessee Valley Substation) at Belle Mina in adjacent Limestone County has been the area's center for cotton research since 1929. During the 1950s and 1960s, the Center gave attention to cotton mechanization work under the leadership of Tom E. Corley. (The Agriculture Engineering Building on the Auburn University campus was named to honor Corley in 1998.)

KOREAN WAR

After the surrender of Japan at the close of WWII, the country of Korea was divided at the 38th parallel, with the north called the Korean People's Republic (KPR) supported by the Soviet Union and the south the Republic of Korea (ROK) supported by the United States. On 25 June 1950, the KPR, using Russian-made tanks and other weapons, attacked the ROK. On 27 June President Harry S. Truman declared that the U.S. would assist the South Korean régime, and the United Nations immediately did likewise.

American troops arrived in Korea on 30 June. What is generally called the Korean War followed; although 17 nations were in the UN effort, most of the troops and equipment were from America. An armistice was eventually signed 27 July 1953, leaving Korea still divided. At maximum, the U.S. had 327,000 persons in combat; the Department of Defense acknowledges that almost 40,000 of its servicemen died, either in battle or of other causes. There were 21 persons from Madison County among the casualties.

At Redstone Arsenal, production of munitions had ended in 1946. When the Korean War began, ammunition production was resumed by the Munitions Production Division; about 39 million rounds of artillery munitions were produced from July 1951 through July 1955. Products included mortar shells, 3.5-inch rockets, 105- and 155-mm shells, and 81-mm illuminating shells. Except for the munitions production, there was little local involvement by Army organizations at Redstone Arsenal during the Korean War.

VIETNAM WAR

To assist in expelling invading North Vietnamese troops, President Dwight D. Eisenhower sent 685 military advisors to South Vietnam in May 1960. The conflict accelerated later in the year, and by 1962, President John F. Kennedy had 10,000 "advisors" there. This eventually reached a maximum of 536,000 in 1968. Although the U.S. had never declared war, President Richard M. Nixon's representative, Henry Kissinger, was a party to a peace agreement signed 27 January 1973. In March, the last U.S. troops left South Vietnam. The capture of Saigon by the North Vietnamese Army marked the end of the war in April 1975. During this war, 58,159 Americans died in combat or from other war-related causes, 1,719 remain missing, and some 303,635 suffered wounds; 1,207 of the casualties were from Alabama.

In support of the war, stocks of virtually every fielded missile system managed and supported at Redstone Arsenal were sent to Southwest Asia. Very few of these systems, however, were actually used in the conflict – they were primarily designed for a different type of warfare and appropriate targets were seldom presented. Civilian personnel from MICOM made trips to Vietnam to see first-hand how the equipment might be improved to fit the environment and operational conditions. Huntsville industries were awarded contracts to build special-purpose equipment and facilities.

Selected Bibliography / Additional Information for Chapter III

"25 Years Since," Special Supplement showing the impact on Huntsville by rocket development; *Huntsville Times*, 3 November 1974

Akens, David S., "Historical Origins of George C. Marshall Space Flight Center"; NASA Marshall Space Flight Center, 1960

Akens, David S., "Saturn Illustrated Chronology: Saturn's First Eleven Years, April 1957 through April 1968"; NASA Marshall Space Flight Center, 1971; http://history.nasa.gov/MHR-5/contents.htm

"Army Ballistic Missile Agency," *Wikipedia*; http://en.wikipedia.org/wiki/Army_Ballistic_Missile_Agency

Bergaust, Eric, *Rocket City U.S.A.*; Macmillan, 1963

Bilstein. Roger E., *Stages to Saturn: A Technological History of the Apollo/Saturn Launch Vehicles*; NASA History Series, SP 4206, Government Printing Office, 1980

Cagle, Mary T., "History of Redstone Arsenal," Historian's Office, Redstone Arsenal (undated)

"Chronology of the Development of Ballistic Missile Defense, 1955 to 1969," Ballistic Missile Defense Command, July 1970

Dunar, Andrew J. and Stephen P. Waring, *Power to Explore: A History of Marshall Space Flight Center*, 1960-1990, National Aeronautics and Space Administration, 1999

Dornberger, Walter, *V-2 – Der Schuss ins Weltell [The Shot into Space]*, Viking Press, 1952; English Version 1954

"Early Rocketry," The Military Standard; http://themilitarystandard.com/missile/earlyrocketry.php

"Explorer I and Jupiter C," Department of Astronautics, National Air and Space Museum, Smithsonian Institution; http://history.nasa.gov/sputnik/expinfo.html

"Ernst Stuhlinger," Wikipedia; http://en.wikipedia.org/wiki/Ernst_Stuhlinger

Garber, Steve, "Sputnik and The Dawn of the Space Age," NASA History Web; http://history.nasa.gov/sputnik

George, Lance, "Space City U.S.A.- A Never-Completed Space Theme Park," *Huntsville Rewound*, http://huntsvillerewound.com/HSVspacecityUSA.htm

"History of Redstone Arsenal," The Military Standard; http://themilitarystandard.com/missile/redstone.php

Huntsville / Madison County Chamber of Commerce, "Industrial Directory," 1970

Klee, Ernst, and Otto Merk, *The Birth of the Missile: The Secrets of Peenemüde*; E. P. Dutton, 1965

"Lunar Roving Vehicle," *Wikipedia*; http://en.wikipedia.org/wiki/Lunar_Roving_Vehicle

Lundquist, Charles A., *Transplanted Rocket Pioneers*, M. Louis Salmon Library, University of Alabama In Huntsville, 2014

"Marshall Space Flight Center," *Wikipedia*; http://en.wikipedia.org/wiki/Marshall_Space_Flight_Center

"Milton K. Cummings," *Wikipedia*, https://en.wikipedia.org/wiki/Milton_K._Cummings

Mitchell, Eddie, "Apogee, Perigee, and Recovery: Chronology of Army Exploitation of Space"; Rand Corporation, Report ADA254602, 1989

"National Aeronautics and Space Administration," *Wikipedia*; http://en.wikipedia.org/wiki/NASA

"Operation [Project] Paperclip," *Wikipedia*, http://en.wikipedia.org/wiki/Operation_Paperclip

Ordway, Frederick I, III, and Mitchell R. Sharpe, *The Rocket Team: From the V-2 to the Saturn Moon Rocket—The Inside Story of How a Small Group of Engineers Changed World History*, Crowell, 1979

"Project Horizon; Vol. I: Summary and Supporting Considerations," United States Army; http://www.history.army.mil/faq/horizon/Horizon_V1.pdf

"Redstone Arsenal," *Wikipedia*, http://en.wikipedia.org/wiki/Redstone_Arsenal

"Redstone Arsenal Historical Information," AMCOM (Comprehensive Information); http://history.redstone.army.mil/index.html

Stuhlingler, Ernst, Frederick I. Ordway, III, Jerry C. McCall, and George C. Brown, eds, *Astronautical Engineering and Space: From Peenemüde to Planetary Space*; McGraw-Hill, 1963

U.S. Census of Agriculture, published every five years by U.S. Department of Agriculture's National Agricultural Statistics Service; http://agcensus.mannlib.cornell.edu/AgCensus/homepage.do;jsessionid=E763E7 5D11EAB5599E4159EF707035AC

Vaughan, Otha ("Skeet"), "Lunar Driver Simulation History," http://www.knology.net/~skeetv/SimHist3.html

von Braun, Wernher, and Frederick I. Ordway, III, *History of Rocketry and Space Travel*, Crowell, 1966

von Saurma, Ruth, and Walt Wiesman, "Dr. Wernher von Braun and the German Rocket Team," Alabama Space Science Commission, 1990

Ward, Bob, *Dr. Space – The Life of Wernher von Braun*, Naval Institute Press, 2005. ISBN 1-59114-926-6

Watson, Raymond C., Jr., "Fifty Years of Technical Leadership – A Brief History of Teledyne Brown Engineering," Special 50th Anniversary Document, 2003

Wicks, T. Gary, *Huntsville Air and Space*; Arcadia Publishing, 2010. ISBN 978-0-7385-6607-8

Chapter IV

DIVERSIFICATION ERA

Having developed the powerful SATURN V rocket system and used it to safely initiate the exploration of the Moon, and, at the same time in the Rocket City Era, developed the rocketry-based technologies that might defend the Nation from intercontinental ballistic missiles, capabilities in Greater Huntsville began to be diversified.

The Diversification Era – in this book considered the three decades from 1970 through 1999 – started with the continuation of the Apollo Program and the opening of America's first ballistic missile defense system, but the momentum of both space and defense was soon lost because of declines in the U.S. economy and decrease in public interest.

Nevertheless, much of importance in Army and NASA organizations, as well as industries, occurred during this period. In the previous era, a huge talent pool of engineers, scientists, and technologists was developed in Greater Huntsville; using this capability and the outstanding facilities of government agencies and industries, significant changes occurred in technology during the Diversification Era.

By the early 1970s, MSFC leaders realized that propulsion work alone was unlikely to sustain the Center. During the decades of the Diversification Era, MSFC developed specializations in many other areas that made it the most diversified of NASA's field centers. A similar situation existed with the local Army activities, particularly in expanded missile defense research and development, and the overall mission was broadened to include aviation. In the commercial sector, significant diversification, especially in electronics and vehicle manufacturing, also occurred.

Although Huntsville increased in population by only 15 percent in this era, Madison County grew 48 percent – a significant shift took place from growth inside the Huntsville city limits to the surrounding areas; the nearby town of Madison became the fastest growing city in the State.

This chapter is divided into three sections: Space Activities, Defense Activities, and Industries and Other Activities. Each of these is time-sequenced, overlapping throughout the era.

SPACE ACTIVITIES

With the successful mission of Apollo 11 to the Moon, there had been near perfect flights of fifteen SATURN IBs and five of the giant SATURN Vs. To many Washington politicians and much of the American public, it appeared that the Marshall Space Flight Center and its director, Wernher von Braun, were no longer needed. One of the first official acts of President Richard Nixon was forming a Space Task Force to establish goals and recommend a long-term space program for America. In September 1969, they submitted a plan with three options: (1) a $10 billion per year program involving a manned Mars exposition and including space stations in Earth and lunar orbits; (2) a $8 billion per year version of Plan 1; and (3) a $4 to $6 billion per year program involving an Earth space station and a linking space shuttle.

NASA Administrator Thomas O. Payne, an appointee of President Nixon, invited Wernher von Braun to leave MSFC and join him in charting a future course for the space program. Realizing that there was little left in a role at Huntsville, von Braun transferred to Washington as NASA Associate Administrator for Future Programs in February 1970. In March, President Nixon announced that he would follow a limited version of the Task Force Option 3, starting the space shuttle but delaying a decision on the space station. Disappointed by this, Payne resigned in July, leaving von Braun to fend for himself. After two frustrating years, von Braun resigned in June 1972, taking the position of EVP for Engineering and Development with Fairchild Industries in Germantown, Maryland.

Five years later at age 65, Wernher von Braun died on 16 June 1977. His simple gravestone reads: WERNHER VON BRAUN 1912-1977 Psalms 19:1. That scripture is: "The heavens declare the glory of God; and the firmament showeth his handiwork." At his memorial service, former NASA Administrator James C. Fletcher, quoting the Hebrew prophet Joel, called von Braun one of the "few men [who] arise in each century who 'see visions' and 'dream dreams' that give hope and spiritual nourishment to us all. . . . Such men cling to this vision despite all efforts to destroy it."

When von Braun left MSFC, his long-time Deputy Eberhard Rees was appointed as Director. The following were MSFC Directors and their dates through the end of this century:

Wernher von Braun	1 July 1960 - 27 January1970
Eberhard F. M. Rees	1 March 1970 - 19 January 1973
Rocco A. Petrone	26 January 1973 - 15 March 1974
William R. Lucas	15 June 1974 - 3July 1986

Thomas Jack Lee (acting)	3 July 1986 - 29 September 1986
James R. Thompson, Jr.	29 September 1986 - 6 July 1989
Thomas Jack Lee	6 July 1989 - 6 January 1994
Gene Porter Bridwell	6 January 1994 - 3 February 1996
Jerroll Wayne Littles	3 February 1996 - 3 January 1998
Carolyn S. Griner (acting)	3 January 1998 - 11 September 1998
Arthur G. Stephenson	11 September 1998 – 20 May 2003

APOLLO APPLICATIONS PROGRAM

As the Apollo landing missions were being planned in the early 1960s, NASA headquarters began a parallel plan for use of the Apollo hardware in other applications; the name soon became the Apollo Applications Program (AAP). Included in the plan was an Apollo Lunar Base with a two-man excursion team staying on the surface for up to 200 days, a Manned Venus Flyby, an earth-orbiting Space Station, a Solar Observatory, a Grand Tour of the Outer Solar System, and a series of Mars Lander probes. Congress, however, having fully funded a lunar exploration saw this as sufficient space activities and was not interested in the AAP. As planning for the AAP progressed, it turned to much less costly applications of the surplus SATURN vehicles. Wernher von Braun assigned Arthur L.H. Rudolph to represent MSFC in AAP planning.

Skylab and ATM

In 1959, the Army's plan for Project Horizon included the use of spent SATURN vehicles linked together to form an Earth-orbiting workshop. The AAP planners picked up on this and in December 1965, MSFC was authorized to have the Orbital Workshop as a formal project. In 1966, George E. Mueller, NASA Associate Administrator for Manned Space Flight, assigned MSFC the responsibility for the development of the orbiting space station hardware, as well as overall systems engineering and integration. In receiving this assignment, MSFC was diversifying – getting away from just rocket propulsion.

The Orbital Workshop was designed using the hull of a SATURN V third stage; this would be fully refitted on the ground and boosted into orbit by a full SATURN V. The Orbital Workshop was renamed Skylab in February 1970; two were built – one for orbital flight and the other for testing and mission simulation at MSFC. SATURN IB vehicles with their CSMs would be used to ferry three-man crews to dock with the Skylab. Leland F. Belew served for eight years as the overall Skylab Program Director.

The Solar Observatory, originally planned as an AAP mission using an Apollo spacecraft, was allowed to continue as an adjunct to the Orbital Workshop. Renamed the Apollo Telescope Mount (ATM), the project was assigned to MSFC in 1966, but the two developments were kept as separate projects; Rein Ise was the ATM Project Manager. The ATM included eight major instruments for observations of the Sun at wavelengths from extreme ultraviolet to infrared.

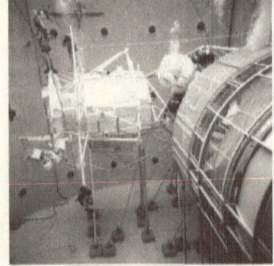

At MSFC, a 75-foot (23-m) diameter 40-feet (12-m) deep tank containing 1.3 million gallons (4.9 million liters) of water was opened for testing and mission simulation in March 1968. Called the Neutral Buoyancy Simulator (NBS), this was used by engineers and astronauts to simulate the weightlessness (or near zero-g) environment of

Working on Skylab in NBS

space. The NBS was large enough to accommodate full-size mockups of space hardware and structures submerged within it. There were 71 observation portholes, 36 containing high-intensity lamps for illuminating the underwater area. Initially conceived by Charles R. Cooper of MSFC's Test Laboratory, Robert J. Schwinghamer, Manager of the Materials and Processes Laboratory, led in developing the NBS. The facility was declared a National Historic Landmark in 1985.

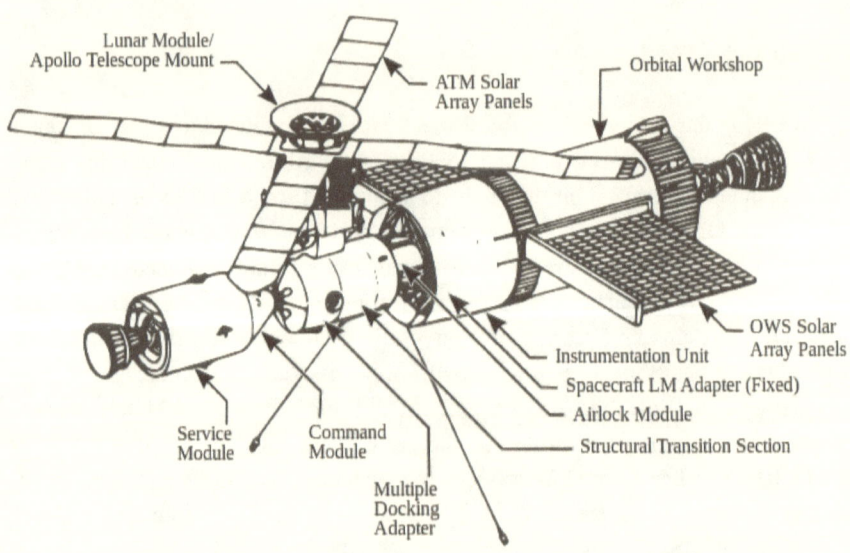

Components of Skylab and ATM

On 14 May 1973, the 77-ton (70,000-kg) Skylab, America's first Space Station, was launched into a 235-nautical-mile (435-km) orbit by a SATURN V vehicle (the last to be flown), with Skylab replacing the upper stage. Severe damage occurred during Skylab launch and deployment, with the loss of its micrometeoroid shield/sun shade and one of its main solar panels. Repair plans and tools were quickly made, and on 25 May the first crew was launched and, in the first-ever American spacewalk, partially corrected the Skylab damage.

The first crew remained on Skylab for 28 days, returning on their CSM. Two more crews were launched on 28 July and 16 November, remaining in orbit 59 and 84 days, respectively. The last Skylab crew returned to the Earth on 8 February 1974. *Skylab with Missing Solar Panel*

In three missions, Skylab, with the ATM, logged about 2,000 hours on some 300 scientific and medical experiments.

Concept Verification Test

Experiments conducted on Skylab led to an important ground-based activity at MSFC called Concept Verification Test. In one activity during 1974, an all-female crew performed a five-day exercise to test the feasibility of experiments proposed for future missions. Ann F. Whitaker, Carolyn S. Griner, and Mary H. Johnson worked eight-hour days in materials science experiments under conditions – except zero gravity – simulating those of a future space station. The objective was to determine practical applications of the experiments and to identify integration and operational problems that might later occur.

Apollo–Soyuz

Although not a mission of MSFC, the Apollo-Soyuz Test Project should be noted. Conducted in July 1975, this was the first joint U.S.- Soviet space flight, and the last flight of a SATURN IB vehicle as well as an Apollo spacecraft. The primary purpose was as a symbol of the policy of détente that the two superpowers were pursuing at the time.

The Soyuz spacecraft was launched by a SOYUZ rocket. The docking adapter – allowing the Apollo and Soyuz spacecraft to be coupled – was jointly designed by NASA's JSC and the Soviet Academy of Science, and then built in the U.S. During the flight, the Apollo carried three astronauts and the Soyuz had two cosmonauts. Launched into 123.9-nautical mile (229.4-km) circular orbits on 15 July 1975, the spacecraft were maneuvered into a successful docking. During several days of joint activities, the five men practiced docking and undocking procedures and carried out five joint experiments.

This was the last manned U.S. space mission until the first Shuttle orbital flight in April 1981.

SPACE SHUTTLE

At the time, the Space Shuttle was likely the most complex spacecraft ever built. It was developed as a reusable space-launch vehicle, serving NASA, the DoD, the telecommunication industry, and even foreign users. Although MSFC was not responsible for developing the centerpiece – the Orbiter Vehicle (OV) – it was responsible for all of the rocket propulsion elements: the OV's three main engines, the External Tank (ET), and the two Solid-Rocket Boosters (SRBs). MSFC was also responsible for Spacelab, the research facility carried in the Shuttle's cargo bay on certain flights. Alex A. McCool, Jr. was Manager of MSFC's Space Shuttle Projects Office. The propulsion project managers were James R. Thompson for the engine, George B. Hardy for the solids, and James B. Odom for the external tank.

The Space Shuttle used a computerized fly-by-wire digital flight-control system. Developed by Honeywell, Inc. in Clearwater, Florida, it was composed of four interrelated subsystems, each addressing a different mission phase: ascent, descent, on-orbit, and aborts.

A total of five Space Shuttles were built for orbital flight: *Columbia, Challenger, Discovery, Atlantis,* and *Endeavor.* In addition, *Enterprise,* a full OV without engines, was built for testing in the atmosphere. These tests were conducted at Edwards Air Force Base in California during 1977, being taken aloft by a Boeing 747 Shuttle Carrier Aircraft (SCA). In 1978, *Enterprise* was brought by the SCA to MSFC where it underwent thorough vibration testing.

At MSFC for Ground Testing

Shuttle Propulsion

From the start of the program in 1972, the management and development of Space Shuttle propulsion was a major activity at MSFC. The SRB motors were built by the Wasatch Division of Morton Thiokol Corp., at Brigham City, Utah. The SRBs were parachuted into the Atlantic after each flight, then recovered and refurbished by United Space Boosters Inc. at KSC. The external tank was built by Martin Marietta at its Michoud facility in New Orleans, Louisiana; it was not recovered and reused. The Space Shuttle main engines were built by Rockwell's Rocketdyne Division at Canoga Park, California.

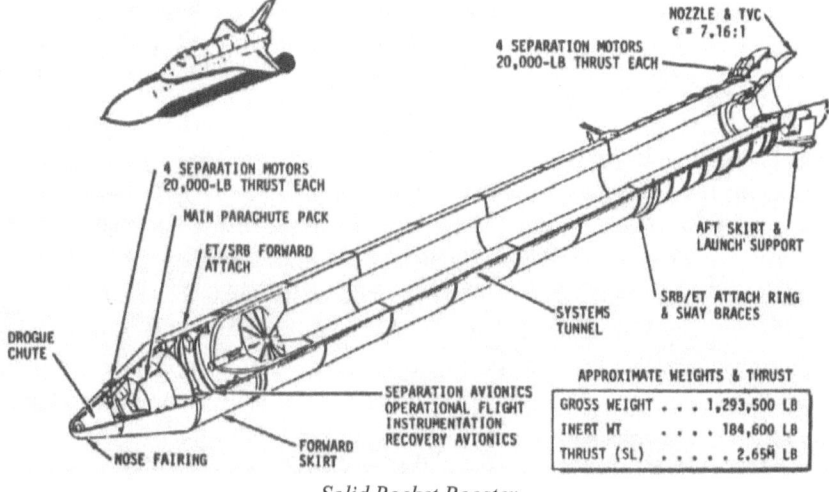

Solid Rocket Booster

The boosters with the three main engines generated a combined thrust of 30.16 MN (6.78 Mlbf), sufficient to insert a Space Shuttle cargo of 24,300 kg (53,600 lb) into a low-Earth orbit (LEO), or 3,800 kg (8,400 lb) to a geostationary transfer orbit (GTO). For readers interested in aeronautical performance numbers, the mass ratio was 15.4, and mass fraction was 0.935.

MSFC engineers participated in tests related to launch plans for the first Space Shuttle. Prior to each later Shuttle launch, personnel in the Huntsville Operations Support Center monitored consoles to evaluate and help solve

Shuttle with External Fuel Tank and Solid Rocket Boosters

any problems at the Florida launch that might involve Shuttle propulsion.

Space Shuttle Flights

All orbital flights of the space shuttles originated at KSC in Florida. Initially, the landing site was at Edwards Air Force Base in California (one landing was at White Sands, New Mexico). Starting in 1984, the prime landing site was the Shuttle Landing Facility at the KSC, a landing strip constructed for this purpose. Thereafter, landings were at Edwards only when KSC was closed by the weather. The Boeing 747 SCA was used to ferry the Shuttle between locations. A detailed listing of Space Shuttle Flights is given in *Wikipedia*.

On 12 April 1981, the *Columbia* made the first orbital test flight of a full Space Shuttle with two astronauts aboard. Designated STS-1 (Space Transportation System-1), this verified the performance of the entire system including the astronauts. (It is noted that the STS numbers were sequential based on the planned, not actual, launch dates.) STS-2 followed on 12 November; this also used *Columbia* and was primarily to demonstrate safe re-launch of a Shuttle System. Two more test flights (STS-3 & 4, both with *Columbia*) were made during 1982; STS-4 carried a small DoD payload. STS-5, still with *Columbia*, launched 11 November was the first operational mission; it carried four astronauts and two commercial satellites were deployed.

STS-6, launched 4 April 1983, was the first flight for Space Shuttle *Challenger*. From then through April 1984, there were six more flights using *Challenger* or *Columbia*, but the flight numbers were changed to a different series starting in February 1984. On STS-51J in October 1985, Space Shuttle *Atlantis* was first launched.

STS-51L – which would have been STS-25 in the original numbering series – with Space Shuttle *Challenger* was launched 28 January 1986. One-minute, 13-seconds into flight, the entire *Challenger* was enveloped in a fireball and broke into several large segments, killing the seven astronauts. Subsequent analysis of the high-speed tracking films and telemetry signals indicated that a leak occurred in a joint on one of the SRBs, the escaping flame impinged on the surface of the ET; there followed a complex series of very rapid structural failures, and in milliseconds the hydrogen and oxygen streaming from the ruptured tank exploded.

The basic cause of the disaster was determined to be an O-ring failure in the right SRB; cold weather was a contributing factor. The redesign effort, directed by MSFC, involved an extensive test program to verify that the SRBs were safe. There were no Space Shuttle missions in the remainder of 1986 or in 1987. Flights resumed on 19 September 1988, with sequential numbering starting with STS-26.

After September 1988 and through December 1999, there were 70 additional Space Shuttle flights – a total of 96 flights since the start in April 1981. Authorized to replace *Challenger*, *Endeavor* was the fifth and last Space Shuttle to be built; it first flew in May 1992. After the turn of the century, there were 39 Space Shuttle flights (STS-97 through STS-135); these are briefly described in the next chapter.

Missions and Payloads

The Space Shuttles had a wide variety of missions and payloads – from scientific research equipment to classified military satellites. For all missions, a crew of between three and seven astronauts was fully involved; these included a mission commander, a pilot, and one or more mission specialists and/or payload specialists.

On some missions, the Space Shuttle served as the launch platform for planetary spacecraft. An example of this was the Magellan that was launched by STS-30 on 4 May 1989, and sent on a 15-month loop around the Sun and then on to orbit Venus for four years in radar mapping; for launching this spacecraft, MSFC was responsible for a two-stage, solid rocket – called an Inertial Upper Stage – for propelling the spacecraft from LEO to planetary velocity.

In addition to their primary payload, most Shuttle flights also carried equipment for the crew to perform on-board research. This equipment was accommodated in two forms: on pallets or other arrangements in the Shuttle's cargo bay, or within a reusable laboratory called Skylab. All such experimental payloads were under the responsibility of MSFC. James A. Downey, III was Manager of the Payloads Project Office.

Pallet Experiments – Space Shuttle pallet experiments involved a very wide spread of types and complexity; many of them were in fluid physics, materials science, biotechnology, combustion science, and commercial space processing. Under a long-running contract, Teledyne Brown Engineering (TBE) integrated

TBE's Experiment Bridge

many of these pallet experiments; for holding them in the Shuttle, TBE developed an aluminum bridge that fit across the cargo bay. This could carry 12 standard canisters holding isolated experiments, particularly those under the Getaway Special (GAS) program. (The GAS flights were made available at low cost to colleges and universities, American industries, individuals, and friendly foreign governments.)

Some pallet experiments constituted the full payload; an example of this is the Astronomy Laboratory-1 (ASTRO-1), developed by Goddard Space Flight Center and flown on STS-35 in December 1990. The payload included an X-ray telescope and three ultraviolet (UV) telescopes. The displays for the on-board scientist-astronaut to aim the UV telescopes failed; an astronomer in the MSFC Payload Control Center then did the telescope pointing.

Another full-payload was the Atmospheric Laboratory for Applications and Science (ATLAS 1), carried on STS-45 and launched in March 1992. This had 12 instruments from the U.S., France, Germany, Belgium, Switzerland, The Netherlands, and Japan. Experiments were conducted in atmospheric chemistry, solar radiation, space plasma physics, and ultraviolet astronomy.

Spacelab – In addition to the pallet experiments, many other experiments were flown and performed using Spacelab. This was a reusable laboratory with multiple components, including a pressurized module, an unpressurized carrier, and other related hardware. Under a program managed by MSFC, ten European nations jointly designed, built, and financed the first Spacelab through the European Space Research Organisation (ESRO). James A. Downey, III, was the MSFC Project Manager. In addition, Japan funded a Spacelab for STS-47, a dedicated mission. Over a 15-year period, Spacelab flew on 22 shuttle missions, the last in April 1998.

Spacelab 1 was flown on STS-9, launched November 28, 1983. A Shuttle *Columbia* flight, this was the first with six astronauts, including two Payload Specialists from the ESRO. There were 73 experiments carried out in astronomy and physics, atmospheric physics, Earth observations, life sciences, materials sciences, and space plasma physics

Spacelab in Shuttle Cargo Bay

Another Spacelab, U.S. Microgravity Laboratory 1 (USML-1), was launched in June 1992 on STS-50, the first Extended Duration Orbiter. During 14 days, 31 microgravity experiments were completed in round-the-clock operations. Following this, USML-2 was launched in October 1995 on STS-73 with an MSFC scientist, Frederick W. Leslie, as an on-board Payload Specialist.

In early 1990, MSFC's new Spacelab Mission Operations Control Center took over the responsibility for controlling all Spacelab missions. This replaced the Payload Operations Control Center formerly located at the JSC from which previous Spacelab missions were operated.

OTHER MAJOR SPACE PROGRMS

The advent of the Space Shuttle made possible several major space programs in which MSFC had significant responsibilities. These were the International Space Station, the Hubble Space Telescope, the Chandra X-Ray Observatory, and the Compton Gamma-Ray Observatory; the latter three were part of NASA's series of Great Observatories.

Space Station Freedom / International Space Station

A manned space station had long been an objective of visionaries. In his widely read *Collier's Magazine* 1953 article, Wernher von Braun described this to be a huge wheel, rotating to produce gravity-like forces on the occupants. In Project Horizon, prepared by the U.S. Army in 1959, a space station was proposed to be built by assembling spent boosters. In 1973, MSFC used a modified stage of SATURN V to build Skylab; with Skylab's success, President Ronald Reagan announced plans to build Space Station Freedom in 1984. Luther B. Powell was MSFC's initial Program Manager for Freedom; James M. McMillion was the Chief Engineer and Robert J. Naumann the Project Scientist.

The USSR had been first with their *Salyut* series started in 1971, and continued with Mir in 1986. By the start of 1990, planning was underway for America's *Freedom*, the Soviet/Russian *Mir-2*, the European *Columbus*, and the Japanese *Kibō*. In 1992, with the Cold War over, American President George H. W. Bush and Russian President Boris Yeltsin agreed to cooperate on space exploration. In November 1993, plans were announced for *Freedom, Mir-2*, and the European and Japanese modules to be incorporated into a single *International Space Station* (ISS). Boeing began as NASA's prime contractor for U.S. hardware in January 1995.

The ISS would be composed of a number of modules, sharing electrical

power from large arrays of solar power cells. The first module, *Zarya* from Russia, was delivered to orbit by a PROTON rocket on 20 November 1998. On 4 December, the first American component, *Unity*, a connecting module, was carried up by Space Shuttle *Endeavour* on flight STS-88; it was then joined with *Zarya* to form an embryonic ISS.

Unity Module Under Constructed

Unity (also called *Node 1*) was built by Boeing in MSFC facilities. Additional building supplies were carried aboard STS-96 to the ISS in May 1999. (See next chapter for ISS updated.)

Hubble Space Telescope

In the early years of NASA, the Orbiting Solar Observatory was launched and was followed by the Orbiting Astronomical Observatory. Between 1968 and 1972, these carried out ultraviolet observations of stars and showed the value of space-based astronomy (ultraviolet radiation cannot pierce the Earth's atmosphere). This led to the planning of the Large Space Telescope (LST) that would be launched and maintained from the forthcoming space shuttle.

This occurred in the era of budget cuts and as MSFC was phasing out of the Saturn project – MSFC needed new, diversified work. Wernher von Braun had personally stated, "That's the type project that I would like to see Marshall do."
The astronomy community – especially Lyman Spitzer, Jr. – and the National Science Foundation pressed for a major program in this area; Congress finally funded LST in 1978, with an intended launch date of 1983. In the LST studies, Charles L. Wyman led a group in the MSFC Astrionics Laboratory that laid the foundation of the optical, structural, and control technologies that ultimately were used for the LST.

MSFC was given responsibility for the design, development, and construction of the telescope, while Goddard Space Flight Center (GFC) was to control the scientific instrument and the ground-control center. As the Project Scientist, MSFC brought on board C. Robert O'Dell, then chairman of the Astronomy Department at the University of Chicago. Jerry W. Richardson was the initial Project Manager.

The telescope assembly was designed as a Cassegrain reflector with hyperbolic mirror polished to be diffraction limited; the primary mirror had a diameter of 2.4 m (95 in). Mirrors had been made by Eastman Kodak and Perkin-Elmer, but it was the latter that did the final polishing and assembly. Although it had been partially tested, adequate end-to-end testing had not been available. (It later turned out that Perkin-Elmer in their Danbury, Connecticut facility had this capability, but it was tied up on the highly classified KH-9 Hexagon "Big Bird" satellite development.) Neither did MSFC have a facility for complete testing of the mirror assembly, so the telescope could not be totally checked until launched and placed in service.

The LST was named the Hubble Space Telescope in 1983. In the development at MSFC, Fred S. Wojtalik was the Project Manager, Jean R. Olivier the Chief Engineer, and Martin C. Weisskopf the Project Scientist . In the final development, James B. Odom was the Project Manager. After much delay in the development and the loss of Challenger, Shuttle Discovery launched the Hubble telescope on Mission STS-31 on 24 April 1990.

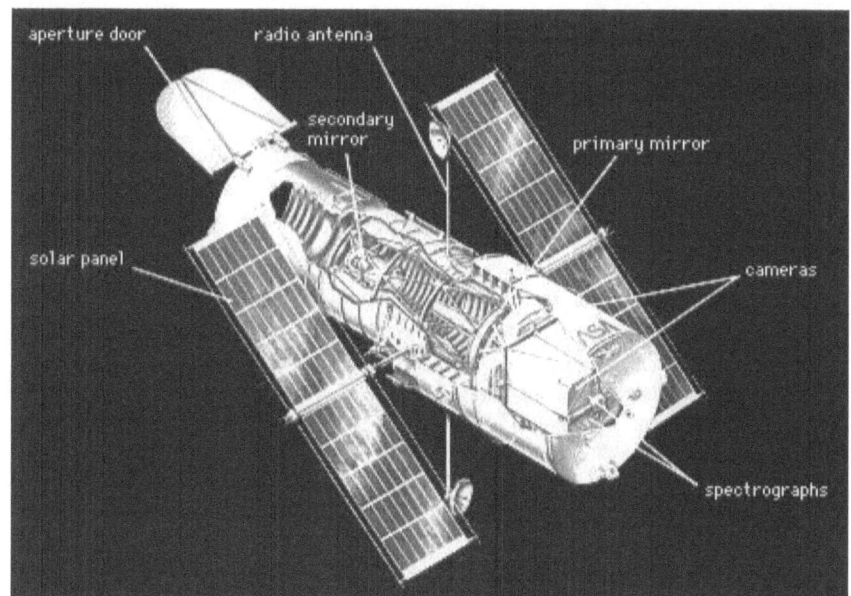

Components of Hubble Space Telescope

It was soon realized that the optical performance was not as expected; analysis of the images showed that the primary mirror had been ground to the wrong shape, resulting in spherical aberration. Specifically, it was eventually determined that the conic constant of the main mirror as built was -1.01390 ± 0.0002, instead of the intended -1.00230.

Fortunately, the *Hubble* telescope had been designed to allow in-space maintenance; thus, in December 1993, mission STS-61 carried astronauts to the Hubble to make corrections. A second repair mission, STS-82, was made in February 1997, and a third, STS-103, in December 1999. For these repair missions, the astronauts practiced the work in MSFC's Neutral Buoyancy Facility, simulating the weightless environment of space.

Through the 1990s, the *Hubble* provided images that had never before been seen. During the next decade, two additional repair missions were made, eventually bringing the telescope to even better that its initially intended performance.

Chandra X-Ray Observatory

Activities in high-energy astronomy began at MSFC in the 1960s. Scientists at Ernst Stuhlinger's Research Projects Laboratory, supported by the TBE Research Laboratories, made measurements aboard balloon-borne instruments.

They then turned to developing MSFC's first scientific satellites, the High-Energy Astronomy Observatory (HEAO, described later). Even before HEAO-2 (the Einstein Observatory) was launched, MSFC began preliminary studies for a larger X-ray telescope.

In 1976, to support this effort, an X-Ray Test Facility (described later) was constructed at MSFC for verification testing and calibration of X-ray mirrors, telescope systems, and instruments. With the success of HEAO-2, MSFC was given responsibility for the design, development, and construction of what was then known as the Advanced X-ray Astrophysics Facility (AXAF). The Smithsonian Astrophysical Observatory partnered with MSFC on this effort. Charles L. Wyman was the AXAF Chief Engineer and Martin C. Weisskopf was the Chief Scientist.

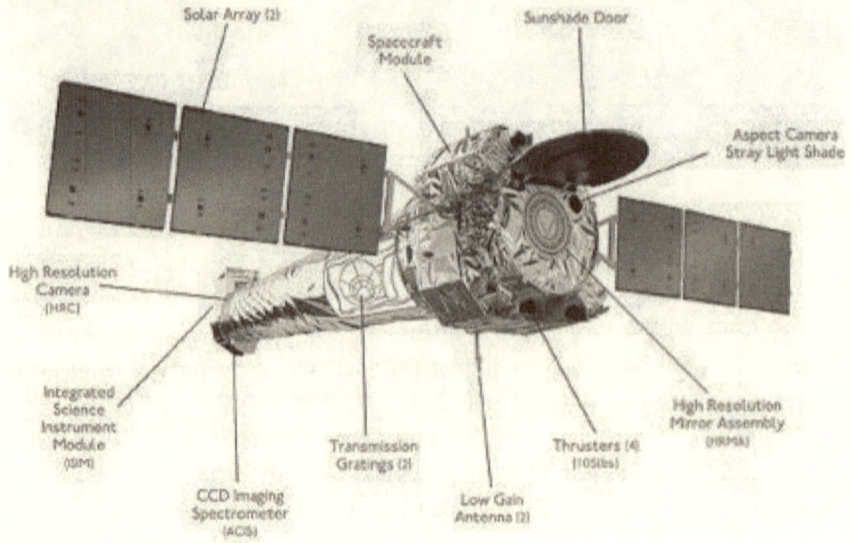

Components of XAXF - Chandra X-ray Observatory

Work on the AXAF continued through the 1980s. Following a review in 1992, four of the twelve planned mirrors and two of the six scientific instruments were eliminated. The planned circular orbit was changed to an elliptical one, reaching one-third of the way to the Moon at its farthest point. This placed the spacecraft above the Earth's radiation belts for most of its orbit, but eliminated the possibility of improvement or repair using the Space Shuttle.

In honor of Nobel-prize winning Indian-American astrophysicist Subrahmanyan Chandrasekhar, AXAF was renamed *Chandra X-ray Observatory* in 1998. It was launched on 23 July 1999, by the Shuttle *Columbia* (STS-93). An Inertial Upper Stage booster, adapted by MSFC, was used to transport *Chandra* to its high orbit. With a mass of about 22,700 kg (50,000 lb), this was the heaviest payload ever launched by a Shuttle. It initially had an expected life of five years, but was extended to 15 years or longer.

Compton Gamma Ray Observatory

The *Compton Gamma Ray Observatory* (CGRO) was launched 5 April 1991, on Shuttle flight STS-37. The CGRO was 14 years in development; TRW was the prime contractor. At 17,000 kg (37,000 lb), it was the heaviest astrophysical payload ever flown at that time.

Gamma radiation (rays) is the highest energy-level of electromagnetic radiation, having energies above 100 keV and thus frequencies above 10 exahertz (10^{19} Hz). This radiation is produced by sub-atomic particle interactions, including those in certain astrophysical processes. The continuous flow of cosmic rays bombarding space objects, such as the Moon, generate this radiation; Gamma rays also result in bursts from nuclear reactions. The CGRO was designed to image continuous radiation and to detect bursts.

MSFC was responsible for the Burst and Transient Source Experiment (BATSE) using the CGRO. This detected sudden changes in gamma count-rates lasting 0.1 to 100 s; it was also capable of detecting less impulsive sources by measuring their modulation using the Earth occultation technique. In nine years of operation, BATSE triggered about 8000 events, of which some 2700 were strong bursts that were analyzed to have come from distant galaxies.

The CGRO was not designed for on-orbit repair and refurbishment. Therefore, after one of its gyroscopes failed, NASA decided that a controlled crash was preferable to letting the craft come down on its own at random. On 4 June 2000, CGRO was intentionally de-orbited, with final debris falling into the Pacific Ocean.

OTHER RESEARCH ACTIVITIES

Activities in high-energy astronomy began at MSFC in the 1960s. Scientists at Ernst Stuhlinger's Research Projects Laboratory, supported by the TBE Research Laboratories, made measurements aboard balloon-borne instruments, then turned to developing MSFC's first scientific satellites.

High-Energy Astronomy Observatory

Successful experience with balloon-borne high-energy experience led MSFC to consider a satellite devoted to high-energy astronomy. A project was approved by NASA Headquarters, and, in early 1969, TBE's Research Laboratories was tasked with the conceptual analysis and preliminary design (Phase A) for a High-Energy Astronomy Observatory (HEAO). Coordinated for TBE by Perry F. McDonald, the study included spacecraft design, analysis of high-energy instrumentation, and computer simulations.

With highly satisfactory results from the study, MSFC issued an RFP for the Phase B design of two 9,700-kg (21,400-lb) HEAO satellites. Contracts were awarded to Grumman and to TRW in May 1970, and their designs completed in April 1971. Fred A. Speer was the HEAO Project Manager for MSFC; technical leadership was by Thomas A. Parnell and Martin C. Weisskopf in the Space Science Laboratory.

In July 1971, MSFC issued an RFP for building and testing the two HEAO satellites; the contract was awarded to TRW in June 1972, with the first launch to be on a TITAN III in 1975. The two satellites would carry 13 experiments. In January 1973, NASA budget cuts forced a delay in HEAO. In the ensuing

HEAO Configuration

delay, the project was redefined, changing to three smaller satellites to be launched by ATLAS-CENTAUR rockets starting in 1977. Scientists from across the Nation were principal investigators, and the instruments were from a variety of sources. During 1974, MSFC used balloon flights to test developmental models of the HEAO instruments.

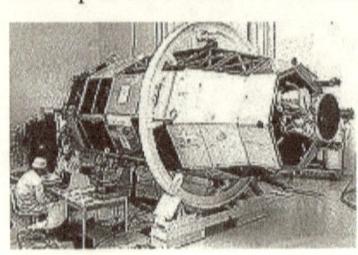
HEAO A in Construction

Designated HEAO-A, B, and C, (later 1, 2, and 3) the spacecraft were about 18 feet (5.5 m) in length, weighed between 6000 and 7000 lb (2700 and 3200 kg), and carried some 3000 lb (1400 kg) of experiments for X-ray and gamma-ray astronomy and cosmic-ray investigations. HEAO-1 was launched 12 August 1977, and operated until September 1979. On 13 November 1978, HEAO-2, also called the Einstein Observatory, was launched and operated for 2.5 years. The last, HEAO-3, was launched 20 September 1978, and returned data for 20 months. The project provided revolutionary insights into celestial objects by studying their high-energy radiation from space.

Other Space Science Research

Three other space science research projects of MSFC in this period should be noted: the Laser Geodynamics Satellite (LAGEOS), Gravity Probes, and Lightning Sensors. Although relatively wise (as compared with those previously discussed), they will be described in some detail to show the complex nature of advanced scientific research.

Laser Geodynamics Satellite – In the original Apollo Applications Program, MSFC and the Smithsonian Astrophysical Observatory did studies of a satellite that could provide an orbiting benchmark of maximum accuracy. Like most AAP proposals, this was not taken up. In 1974, NASA Headquarters asked MSFC to reexamine such a satellite with mirrors to reflect laser light for benchmarking; the Laser Geodynamics Satellite (LAGEOS) was then developed. Donald R. Bowden was the Task Leader, Lewis L. McNair the Chief Engineer, and Joseph L. Randall and James L. Zurasky developed the optics.

LAGEOS-1

LAGEOS-1 is a 2.0-ft (60-cm) diameter, 900-lb (408-kg) sphere covered with 422 prismatic mirrors. Its circular orbit is at 5,900 km (3,700 mi) with an inclination of 109.8 deg. Measurements are made by transmitting pulsed laser beams from 34 ground stations to the satellites; the beams are reflected by the satellite mirrors and the travel times are precisely measured, permitting ground stations in different parts of the Earth to measure their separations to an accuracy better than 10^{-8} (one inch in thousands of miles). Built by Bendix Aerospace Systems, LAGEOS-1 was launched by a DELTA-1 rocket on 4 May 1976. A second sphere, LAGEOS-2, was built by the Italian Space Agency and deployed from shuttle STS-52 on 23 October 1992; it has an orbital inclination of 52.6 deg. The satellites have extremely long life expectancies.

Gravity Probes – General relativity is the geometric theory of gravitation conceived by Albert Einstein in 1916. Through the years, many tests were proposed to provide experimental proof of this theory, but not until the space age were such tests feasible. At the Smithsonian Astrophysical Observatory (SAO), R. F. C. Vessot designed the Gravitational Redshift Space-Probe experiment to examine a predicted relationship between gravity and electromagnetic radiation. Leonard I. Schiff of Stanford University outlined an experiment called Motion of a Gyroscope According to Einstein's Theory of Gravitation. Specifically, this was to verify frame-dragging and geodetic effects, phenomena predicted by general relativity. It was later realized that George E. Pugh of MIT had earlier proposed a similar experiment.

Proposals from SAO and Stanford were submitted to NASA Headquarters, where they were accepted and assigned the names Gravity Probe A (GP-A) and Gravity Probe B (GP-B), respectively. These efforts, however, had little in common except in their names.

In GP-A, an atomic hydrogen, cryogenically cooled, maser oscillator was carried in a 220-lb (100-kg) spacecraft. The object was to measure the effect of gravitational potential on the frequency of the maser. General relativity theory predicts that decreases in gravity at increasing altitudes would cause the maser frequency to increase, and vice versa on the return flight. At an altitude of 10,000 km (6,200 mi), the theoretical prediction was that the frequency should be 4.5 parts in 10^{-10} higher than the frequency from an identical maser on the Earth. Measurement of this very small change involved the microwave transmission of the maser signal to a ground station, where it was compared with the signal from an identical maser. All effects – such as the Doppler shift – other than the gravitational inducement had to be discounted. The maser frequency was stable to one part in 10-15.

The GP-A spacecraft was launched by a SCOUT D rocket at Wallops Island, Virginia, on 18 June 1976. The launch was nearly vertical to give maximum change in local gravity; the total flight duration was 115 minutes. The observed gravitational effect on frequency matched the prediction to an accuracy of about 70 parts per million. Available MSFC documents do not name the project manager or other participants.

For GP-B, NASA grant funded Stanford for the project between 1964 and 1977 – the longest running NASA grant ever awarded; in 1965, MSFC was

brought in for collaboration, then was given full responsibility in 1971. Planning for a flight program began under contract in 1977, MSFC performed a Phase A study in 1980, and Stanford followed with a Phase B study in 1982. The resulting spacecraft was too large and the program too expensive; thus, Stanford and MSFC

Gravity Probe B

collaborated in a new study leading to the start of spacecraft design in 1984.

Lockheed-Martin was selected to build the spacecraft and integrate the experimental hardware. The GP-B spacecraft centered on a 650-gallon (2,500-liter) dewar of superfluid helium, maintaining a temperature of under 2 K (−271 °C, −456 °F); this housed a reference telescope and four gyroscopes. The gyros were a nearly perfect spherical rotating mass made of fused quartz covered with an extremely thin layer of niobium superconducting material. The spinning superconductor layer generated a magnetic field precisely aligned with the rotation axis; their spin axes could then be sensed by monitoring this magnetic field. To eliminate friction, the rotor assembly was centered by the electric field from six electrodes. After the initial spin-up to 4,000 RPM by a jet of helium, the polished gyroscope housing was evacuated to an ultra-high vacuum to further reduce drag on the rotor.

GP-B Dewar

In operation, the reference telescope was sighted on HR8703 (also known as IM Pegasi) – a bright star close to the ideal positions near the celestial equator of the sky coordinates; this star also emits relatively strong radio signals. In polar orbit, with the gyro spin directions also pointing toward HR8703, the frame-dragging and geodetic effects came out at right angles, each gyroscope measuring both.

There were a number of nominal project managers at MSFC and several different Principal Investigators at Stanford, the last being C. W. Francis Everitt. In 1995, MSFC named Rex D. Geveden the official Program Manager and Anthony T. Lyons the Chief Engineer, positions held through the GP-B launch in 2004. Starting in 1998, a GP-B Science Advisory Committee composed of seven eminent scientists was established; the Committee met 18 times, the final meeting to examine the end report.

After 40 years of development (almost all under NASA funding), the spacecraft was successfully launched from Vandenberg AFB by a DELTA-2 rocket on 20 April 2004, at 9:57:23 AM PDT – it had only a one-second launch window because of the precise orbit required by the experiment. In a few hours it was placed into a 642-km (400-mi) polar orbit with a very small eccentricity (0.0014). The mission lasted 16 months.

Preliminary results of the data analysis were released in 2007; the data appeared to confirm Einstein's predicted geodetic effect to a precision of better than 1 percent. NASA funding and sponsorship of the program ended in September 2008, but alternative funding was obtained from King Abdulaziz City for Science and Technology in Saudi Arabia to allow continuation of data analysis. On 4 May 2011, it was succinctly stated in *Physical Review Letters* 106 (22) that the data from GP-B indeed confirms the two predictions of Einstein's general theory of relativity.

Lightning Sensors – Lightning has long been the subject of scientific investigations. In America, Benjamin Franklin wrote about lightning in 1849, then conducted a classical kite experiment that showed that lightning was, indeed, an electrical discharge. Two of the originators of radar – Robert Watson Watt in Great Britain and Basil F. J. Schonland in South Africa – earlier developed instruments for detecting and locating distant lightning strikes.

Two lightning sensors, the Optical Transient Detector (OTD) and the Lightning Imaging Sensor (LIS), were both developed as in-house projects at MSFC. Rex D. Geveden was the Project Manager, and the hardware was built by Lockheed Martin and Kaiser Electro Optics.

Optical Transient Detector

Optical Transient Detector (OTD) was a compact unit containing optical and electronics units that could detect momentary changes in a scene, indicating a lightning occurrence. The OTD was designed for satellite containment; the total weight was 18 kg (40 lb). The camera – a 120x120 charge-coupled device array sampled at about 500 frames/sec – had maximum sensitivity around 770-nanometers (in the near infrared), allowing operation in both day and night conditions. In the electronics, a real-time event processor was used to determine when a lightning flash occurs, even in the presence of bright sunlit clouds. Cloud-to-ground, intracloud, and cloud-to-cloud discharges are all detected. The lens system had a 100-degree field-of-view, sweeping a wide viewing path along the Earth. A point on the Earth or a cloud was within view for 80-90 seconds, sufficient to estimate the flashing rate.

The OTD was carried as a secondary payload on a PEGASUS, an air-launched rocket of Orbital Sciences. Launched on 3 April 1995, the OTD was placed into an Earth orbit of approximately 710-km (446-mi) altitude, with an inclination of 70 degrees. With that orbit, and OTD's wide 100-degree field of view, it was able to survey virtually all areas of the globe where lightning normally occurs. OTD was expected to be in operation for only two years, but continued to function until beyond year 2000.

Lightning Imaging Sensor (LIS) is a follow-on to the OTD and uses essentially the same general elements, but the array (camera) is about three times more sensitive. It is one instrument on the Tropical Rainfall Measuring Mission (TRMM) satellite; this was launched into a 350-km (200-mi) orbit with an inclination of 35 degrees on 28 November 1997. This is optimum for monitoring thunderstorm activity in the Earth's tropics region. The data are transmitted on a daily basis to a ground station in Fairmont, West Virginia, and then sent on to the Global Hydrology and Climate Center in Huntsville for processing, analysis and distribution to the scientific community.

Flight Robotics Laboratory

In the 1980s, considerable attention was given at MSFC to techniques for automated rendezvous and docking of spacecraft. A special laboratory for testing was developed in Building 4619. Officially called Flight Robotics Laboratory (FRL), it was best known as the Flat-Floor Facility and classified the best of its type in the World.

A floor area of 44 by 80 feet was precision-poured with an epoxy material, giving a surface that was flat to within seven thousands of an inch over large areas. Test vehicles – called test sleds -- weighing up to 400 pounds would by suspended by air bearings fed from high-pressure sources, typically on the vehicle itself. The test vehicle could then be moved essentially without friction, simulating zero gravity in two dimensions.

Flight Robotics Laboratory

The laboratory had an overhead gantry capable of carrying up to 1000 pounds. With a robotic arm on the target vehicle, the combination of the target and gantry provided eight degree-of-freedom motion. In addition a dynamic lighting system could simulate the motion and brightness of the Sun The FRL provided sophisticated real-time simulation capabilities for research in the human/system interactions of remote systems.

DEFENSE ACTIVITIES

From 1949, when Redstone Arsenal was assigned the responsibility for all missile development in the U.S. Army, there had been a single organization for the overall coordination of activities. Since 1962, this had been the Army Missile Command (MICOM). As the Diversification Era started, the activities in ballistic missile defense (BMD) had increased to where a second command was formed. Also in this Era, certain other local defense activities evolved that were under other commands. By the start of the next century, the number of DoD personnel had more than doubled.

BALLISTIC MISSILE DEFENSE

As described in the previous chapter, America's ballistic missile defense (BMD) was originated by Army organizations at Redstone Arsenal. As the BMD activities evolved, so did the responsible organizations. In March 1969, these had become the Safeguard System under the Safeguard System Command (SAFSCOM).

BG Ivey Drewry, then CG of SAFSCOM and the 'father' of Huntsville's BMD activities, retired in July 1969; BG Robert C. Marshall was then named CG and served until April 1973. Oswald H. Lange, an original member of the Wernher von Braun rocket team, was the SAFSCOM Chief Scientist.

America's First BMD System

The primary mission of SAFSCOM was to implement the Safeguard System; working closely with the Corps of Engineers (CoE) -Huntsville, sites were started near Malmstrom AFB, Montana, and Grand Forks AFB, North Dakota. COL Hartsell H. Northington, P.E., of the CoE-Huntsville was in charge of the actual installation activities.

The Safeguard System was composed of SPRINT (low-altitude missiles), SPARTAN (high-altitude missiles), Missile Site Radar (MSR), long-range Perimeter Acquisition Radar (PAR), and a Missile Defense Center that included a centralized Data Processing Center (DPC). On 6 August 1969, the U.S. Senate approved (by only two votes) the Phase I deployment of the system, authorizing the commencement of construction work on two sites: Malmstrom AFB, Montana, and Grand Forks AFB, North Dakota.

From the start of BMD developments, Bell Telephone Laboratory and its sister firm, Western Electric Company (BTL/WECo), had served as the system prime contractor. BTL/WECo informed the government that they would withdraw from defense work in early 1971. This not only affected the Safeguard hardware development but also the engineering analysis and system integration activities. SAFSCOM initiated a large procurement for a System Engineering and Technical Assistance Contractor (SETAC), attracting many major defense firms.

Led by R. Stephen McCarter, previously head of BTL's Radar Research Department, many of the BTL engineers and scientists joined Teledyne Brown Engineering (TBE) in Huntsville. The SETAC award was made to TBE in May 1971, to a large extent based on the former BTL personnel.

The United States and the USSR had begun Strategic Arms Limitation Talks (SALT) in November 1969. By 1972 an agreement had been reached to limit strategic defensive systems. Each country was allowed two sites at which it could base a defensive system, one for the capital and one for ICBM silos. The ABM Treaty was signed on 26 May by the President of the United States, Richard Nixon, and the General Secretary of the Communist Party of the Soviet Union, Leonid Brezhnev; the U.S. Senate ratified the Treaty on 3 August 1972.

The Kwajalein Test Site (later renamed Kwajalein Missile Range) in the South Pacific had been transferred from the U.S. Navy to the U.S. Army in 1964. A facility for testing subsystems for Safeguard was set up on Meck Island, a part of the Kwajalein Test Site. The facility consisted of a prototype

MSR, SPRINT and SPARTAN missile sub-systems (described earlier), and a DPC.

During the R&D phase, a series of test processes were implemented on Meck, each with increasing complexity and more stressing system objectives. The Safeguard System test program, which began at Kwajalein in 1970, was completed in August 1974. Of the 54 tests conducted, 47 were successful, 2 partially successful, and only 5 were classified as failures. In a majority of the

Dual Sprint Fired From Meck

tests, only software changes were involved.

The ABM Treaty was amended in July 1974, permitting only a single defensive site per party with a total of 100 interceptor missiles. The sites selected were Moscow for the USSR and the North Dakota Safeguard Complex for the U.S. Construction continued at what was called the Stanley R. Mickelsen Safeguard Complex near Grand Forks AFB; this finally reached initial operational condition in April 1975, and was declared fully operational on 1

Mickelsen Complex

October with 70 SPRINT and 30 SPARTAN missiles. However, the U.S. had a fully authorized and operating BMD system for only one day!

On 2 October 1975, the U.S. House of Representatives voted to shut down the Safeguard BMD system. In February 1976, the Grand Forks site was placed into caretaker status, and then closed two years later. In general, the U.S. public was 'war-weary' from Viet Nam and was not interested in missile defense. The PAR was transferred to the U.S. Air Force, where it was operated as part of its space-track and early-warning system.

Consolidated BMD / BMDO

In May 1974, as the sole Safeguard System site neared completion, the Secretary of the Army realigned all BMD efforts under one organization: the Ballistic Missile Defense Organization (BMDO). A field-operating agency of the Chief of Staff U.S. Army (CSA), BMDO was located in the Huntsville Research Park, with a BMD Program Manager and Program Office in the Washington, D.C., area. William A. Davis, Jr., was the civilian Director of BMDO-Huntsville. LTG Walter P. Leber briefly served as the first BMD PM; then MG Robert C. Marshall served as BMD PM between 1974 and 1976.

BMDO was responsible for implementing the Safeguard System, and also had the more general role of conducting research and development in advanced ballistic missile defense including Theater Missile Defense (TMD); it also managed the Kwajalein Missile Range (KMR). In 1994, BMDO gained public attention when, in cooperation with NASA, it launched a space probe, *Clementine*, to the Moon.

New BMD Organizations

Also in the reorganization of May 1974, the SAFSCOM in Huntsville was renamed the Ballistic Missile Defense Systems Command (BMDSCOM), and the Ballistic Missile Defense Advanced Technology Center (BMDATC) was formed, replacing parts of the original ABMDA. BG John G. Jones commanded BMDSCOM between 1975 and 1977; for the last year he was also the BMD PM and was named the U.S. Army Project Manager of the year. From 1977 to 1979, MG Stewart C. Meyer served simultaneously as BMD PM and commanded both BMDSCOM, and BMDATC. William A. Davis, Jr., served as the civilian Director of BMDATC from 1971 to 1975; Davis then became Deputy BMD PM. James D. Carlson succeeded Davis as Director of BMDATC.

Filling out the 1970s, BMDSCOM also continued the development in a new program called Site Defense – a SPRINT-based system for protecting single strategic sites. All activities were conducted within the bounds of a 1974 Congressional ban on prototyping that limited research and development to the subsystem and component levels. The BMDATC / BMDSCOM explored future technologies such as sensors, missiles, and software. Research in non-nuclear options – kinetic-kill technology and directed-energy weapons – was begun.

Optical techniques for discrimination were given much attention; the Optical Signatures Code (OSC) was developed by contractor TBE and distributed to military and industrial organizations. OSC is a set of programs that, among other functions, models ballistic and satellite trajectories and calculates infrared and visible optical signatures of hard-bodies. A number of years in evolution, the full OSC matured under the leadership of Jerry C. Edwards. Being continuously updated, the OCS is still an industry standard.

Homing Overlay Experiment

With concerns about BMD programs using interceptors with nuclear warheads, consideration turned to hit-to-kill methods, where an interceptor missile would destroy an incoming ballistic missile by colliding with it. In 1979, to examine the feasibility of the so-called Kinetic Kill Vehicle (KKV), the BMDATC initiated major missile research activity called the Homing Overlay Experiment (HOE) – Homing meant that the missile had its own seeker, and Overlay was the term for exo-atmospheric interceptions). Edward L. Wilkinson was the Program Manager.

The KKV, integrated by Lockheed, had an infrared seeker, guidance electronics, and a propulsion system. It extended a 13-ft (4-m) diameter structure similar to an umbrella skeleton to enhance its physical cross section; this device was intended to destroy the ICBM reentry vehicle on collision. The target was a MINUTEMAN-boosted re-entry vehicle launched from Vandenberg AFB 4,500 miles (7,200 km) away on a trajectory so that it would impact north of Kwajalein Atoll in the central Pacific. The ARPA Lincoln C-Band Observables Radar (ALCOR) on Roi-Namur acquired and tracked the target before

Kinetic Kill Vehicle used in the HOE

the HOE was launched, about 20 minutes into the target flight. The kill vehicle was launched from Meck Island by a two-stage MINUTEMAN booster.

After three test failures, the HOE was successful on 10 June 1984, intercepting and destroying the target RV with a closing speed of about 13,600 mi/hr (6.1 km/s) at an altitude of more than 100 mi (160 km). This successful experiment was highly important in BMD evolution; it showed that it would be possible to develop a BMD system against ICBMs using the hit-to-kill concept, leading to the eventual elimination of nuclear warheads on interceptors.

In addition to the BMDATC and Lockheed Missiles and Space Company, team members sharing the success included Teledyne Brown Engineering, Nichols Research Corporation, Honeywell International, Rocketdyne, Space Vector Corporation, McDonnell Douglass, Aeromet Incorporated, MIT Lincoln Laboratory, and the Sandia National Laboratory.

STRATEGIC DEFENSE INITIATIVE

For many years, the strategic offense of the United States had been the doctrine of Mutual Assured Destruction (MAD) – both America and the USSR

had more or less the same capability for retaliating should either nation launch a nuclear attack. The Strategic Defense Initiative (SDI) was proposed by President Ronald Reagan on 23 March 1983, to use ground-based and space-based systems to protect the United States from attack by strategic nuclear ballistic missiles.

President Reagan Announces the Strategic Defense Initiative

The initiative focused on strategic defense rather than the prior strategic offense doctrine of MAD. The ambitious initiative was widely criticized as being unrealistic, even unscientific; it was derided in the mainstream media as "Star Wars." There was also the question of the ABM Treaty signed with the USSR in 1972; this Treaty, however, concerned defending specific sites, and since the SDI was for widespread protection, it did not apply.

In 1984, the Strategic Defense Initiative Organization (SDIO) was set up within the Department of Defense to oversee the SDI. Space and BMD programs from throughout the DoD and other agencies were brought together in the SDIO. The master plan included space-based chemical lasers, ground-based laser weapons, space-based neutral particle beam weapons, and nuclear (X-ray) directed energy; much of the concept was originated by renowned scientist Edward Teller.

One of the first efforts of SDIO was to solicit proposals from industry to become centers of technical excellence for SDI; Teledyne Brown Engineering (TBE) was one of 10 firms selected nationwide for this 1985 award. A TBE team led by Walter M. Langley developed concepts for a long-range research and development approach to the SDI; the product was mainly technical and

cost trade-offs of conceivable approaches and laid a foundation for SDI activities throughout the life of the program.

An early system concept centered on space-based weapons. It soon became obvious to the SDIO that space-based systems would not be practical for many years, primarily because of their enormous prime-power requirements.

Early SDI Space-Based Concept

(Teller proposed thermo-nuclear detonations in space to generate this power.)

Attention then turned to ground-based directed-energy systems with their output beams reflected toward targets by space-based mirrors (large mirrors in geosynchronous orbit directing the beam to smaller battle mirrors in low-earth orbit). The directed-energy sources being examined included lasers (free-electron, chemical, eximer, and X-ray – the last type being an original concept of Edward Teller), and particle beams (subatomic protons, electrons, or heavy ions).

Preliminary studies indicated that the prime power – extremely high for very brief surges – might be supplied by either superconducting magnetic energy storage (SMES) devices or the national electrical power grid. SMES development was estimated to take 10 or more years – an unacceptable time. In 1987, an in-depth analysis led by Raymond Watson (this book's author) at the Applied Research Center of Huntsville's Southeastern Institute of Technology showed that the commercial power grid could not provide the necessary surge. These findings, plus many others, ultimately led to the dismissal of giant directed-beam weapons.

SDIO sensor research continued; this encompassed visible light, ultraviolet, near- and mid-infrared, and radar technologies. Some of the projects included Boost Surveillance and Tracking System (BSTS), Space Surveillance and Tracking System (SSTS), and Brilliant Eyes, a simpler derivative of the SSTS for detecting theater ballistic missiles.

STRATEGIC DEFENSE COMMAND

The various elements of the Ballistic Missile Defense Organization (BMDO), including BMDSCOM and BMDATC, were merged in 1985, creating the U.S. Army Strategic Defense Command (USASDC). With headquarters at Arlington, Virginia, LTG Jay M. Garner was the initial Commanding General of USASDC; he was succeeded by LTG Robert D. Hammond, serving between 1988 and 1992.

The USASDC had a major field office in Huntsville, led by BG Grayson D. Tate, Jr. Here responsibilities were expanded into new areas of research, particularly anti-satellite missile systems and innovative sensors. Of the 12 major components examined for the SDI program, the USASDC managed or contributed to nine of these.

During 1986, the concept of employing "brilliant technology" became popular in the USASDC. Ultimately called Brilliant Pebbles (another Teller concept), this involved thousands of interceptors, each with miniature sensors and computers, capable of independent operations against whatever comes within their field of view. After much study, it was realized that this was impractical.

An ongoing USASDC project was the High Endoatmospheric Defense

Interceptor (HEDI), a missile system that might hit and destroy incoming ballistic missiles in the final portion of their trajectory – a kinetic-kill technology. A test missile was put together by Lockheed Martin using off-the-shelf components and a Ka-band radar seeker. In the last few seconds of the flight, 216 small thruster/attitude-control rockets would fire as needed to direct the missile to the target. Alan D. Sherer was the Project Manager.

HEDI SDI Missile

On 21 May 1987, at White Sands Missile Range, the HEDI successfully made a direct hit on an incoming LANCE missile at 16-kft (4.9-km) altitude. Like the Homing Overlay Experiment of 1984, this was a milestone in hit-to-kill development; it also demonstrated the guidance accuracy of a small, agile, radar-homing vehicle.

In the late 1980s, USASDC began to move in new directions. Continuing with the ground-based directed energy concept, the Kinetic Energy Antisatellite Project Office was formed under BG J. Morgan Jellett in 1989. In October 1990, as part of an effort to centralize laser research, the High-Energy Laser Systems Test Facility (HELSTF) at White Sands was transferred to the USASDC from the Army Materiel Command. The USASDC mission was further enhanced in January 1991, when the Command was assigned all Theater Missile Defense functions.

HELSTF

On 13 December 2001, President George H. W. Bush announced that the U.S. was pulling out of the 1972 ABM Treaty. On 1 June 1992, President Bush established the U.S. Strategic Command (USSTRATCOM), one of the eight Unified Commands (Army, Marines, and Navy, and Air Force) of the Department of Defense (DoD). This new Unified Command, headquartered at Offutt Air Force Base south of Omaha, Nebraska, is responsible for both early warning of and defense against missile attack, as well as long-range strategic attacks. Within this, the Army's responsibilities included land-based missile defense.

In May 1993, the DoD changed the SDIO to the Ballistic Missile Defense Organization (BMDO – not to be confused with the BMDO of the 1970s). Although no full SDI system was ever built, many technologies that it evolved led to those in the MDA's activities. Also, much credit is given to the SDI for ending the Cold War, primarily through initiating a defense system that the Soviet Union could not afford to either overcome or match.

The Missile Defense and Space Technology Center (MDSTC) was formed in Huntsville to manage technology-based research and development for the BMDO. J. Richard Fisher served as Executive Director of the MDSTC and pioneered much of the work in ballistic missile target discrimination and sensor phenomenology. He initiated the Army's development of directed energy for ballistic missile defense which ultimately became the THEL system. The MDSTC programs included communications; Reconnaissance, Surveillance, and Target Acquisition (RSTA); fire support; position location, and navigation; and computational sciences for the Army in Space.

The USASDC was reorganized to focus elements upon specific needs and missions. As part of this decision, several missile and radar projects were transferred from the USASDC to a newly created Program Executive Office (PEO) for Global Protection Against Limited Strikes (GPALS, later renamed PEO for Air and Missile Defense). The PEO was assigned the mission of developing and deploying viable national missile defense against strikes of 200 or less and theater missile defense systems.

Projects under the new PEO included the Ground Based Interceptor (GBI), the High-Endoatmospheric Defense Interceptor (HEDI), the Theater High-Altitude Area Defense (THAAD), the ARROW defensive missile, the Ground Based Radar (GBR), and the Extended Range Interceptor (ERINT).

The ERINT project was a follow-on to the HEDI hit-to-kill technology and demonstration of 1987. Also built by Lockheed Martin, the ERINT had a more powerful rocket motor, was 15.7-ft (4.79-m) long, and weighed about 700 lb (320 kg). In a test at WSMR in November 1993, the ERINT successfully intercepted a STORM target missile (STORM was a low-cost target missile that directly simulated a theater ballistic missile). The ERINT was subsequently selected as the Patriot Advanced Capability-3 (PAC-3) missile.

ERINT Hit

ARMY MISSILE COMMAND

While all of the activities at Redstone Arsenal in ballistic missile defense were underway, the Army Missile Command (MICOM) continued with its own missions. MICOM had been created in August 1962, taking over programs, facilities, and personnel of ARGMA and the remnants of ABMA. As earlier described, MG Francis J. McMorrow was MICOM's first Commanding Officer; he established the Directorate of Research and Development housed in a large facility that, after McMorrow died unexpectedly, was named the Francis J. McMorrow Missile Laboratories (commonly called the McMorrow Labs). In

1972, this organization was named the Missile Research, Development, and Engineering Laboratory (MRDEL). Over the next several years, the Laboratory was considerably expanded; this included the transfer of most research and engineering development activities from throughout MICOM into the facility.

Split Commands

In 1976, the U.S. Department of the Army underwent a major reorganization. In this, the Army Materiel Command (AMC), to which MICOM reported, was reorganized into the U.S. Materiel Development and Readiness Command (DARCOM). In January 1977, MICOM was abolished and its facilities, personnel, and programs were divided between two new commands: the Missile Research and Development Command (MIRADCOM) and the Missile Materiel Readiness Command (MIRCOM), both at Redstone Arsenal.

MIRADCOM managed missile research, development, testing, and initial acquisition, while MIRCOM managed launch, fire-control, and ground support equipment, as well as readiness of deployed missile and rocket systems. MIRADCOM also handled all aspects of foreign sales of Army missiles. The MRDEL was also divided into the Engineering Laboratory and the Technology Laboratory. These split organizations functioned until October 1979, when the two commands officially combined and MICOM was reestablished.

The Commanding Generals throughout this period and their associated dates were as follows:

MG Francis J. McMorrow	June 1962 - Aug. 1963	MICOM
MG John G. Zierdt	Sept 1963 - June 1967	MICOM
MG Charles W. Eifler	July 1967 - Sept. 1969	MICOM
MG Edwin I. Donley	Nov. 1969 - Oct. 1973	MICOM
MG Vincent H. eifler	Nov. 1973 - Oct. 1977	MICOM
MG George E. Turnmeyer	Oct. 1975 - Jan. 1977	MICOM
MG George E. Turnmeyer	Jan. 1977 - Sept. 1977	MIRCOM
BG Grayson D. Tate, Jr.	Jan. 1977 - July 1977	MIRADCOM
MG Charles F. Means	July1977 - Aug. 1978	MIRADCOM
MG S. Louis Rachmeler	Sept. 1977 - June 1979	MIRCOM
BG Frank P. Ragano	Aug.1978 - June1979	MIRADCOM
MG S. Louis Rachmeler	July 1979 - July 1980	MICOM
MG Robert L. Moore	July 1980 - July 1983	MICOM
MG Jerry Max Bunyard	July 1983 - Aug. 1985	MICOM
MG Peter G. Burbules	Aug. 1985 - May 1986	MICOM
MG Thomas D. Reese	May 1986 - July 1988	MICOM

MG August M. Cianciolo	July 1988 - Oct. 1989	MICOM
MG William S. J. Chen	Oct. 1989 - July 1992	MICOM
MG Joe W. Rigby	July 1992 - June 1994	MICOM
MG James M. Link	June 1994 - July 1997	MICOM

Research, Development, & Engineering

In November 1980, the Army Missile Laboratory (AML) was formed as the

single, central research and development laboratory of MICOM. William C. McCorkle, Jr. was appointed to the dual role of Technical Director of MICOM and Director of the Army Missile Laboratory (AML). McCorkle had come to Redstone Arsenal in 1957, just after earning a doctorate in physics, and was subsequently involved with missile-related research and development on virtually every Army missile and rocket system during this period.

William McCorkle   Systems Simulation and Development, led by Maurice M. Hallum, III, was one of the most active units in the early 1980s. In 1985, the AML was renamed the Missile Research, Development, and Engineering Center (MRDEC), starting a basic designation that exists until today.

For the remainder of the 1980s and through most of the 1990s, MRDEC added facilities and functional organizations. Included were Directorates for Engineering, Test & Evaluation, Propulsion & Structures, Software Engineering, and Weapon Sciences. There were special facilities for radar and optical sensors, prototype development, and small missile testing. Richard G. Rhodes served as Associate Director during this time.

Through much of this period, a special development project was the Fiber Optic Guided Missile, or FOG-M. An internally funded activity strongly supported by McCorkle, this missile would allow an infantryman to attack a tank from the top, where the armor is barely an inch thick. Although proven in tests, support by the DoD for advanced development was not forthcoming.

While MRDEC had highly applied activities such as FOG-M, there was also special research in relatively basic sciences. Charles M. Bowden, a Research Scientist at MRDEC, received international recognition for his studies in nonlinear and quantum optics; he organized a number of scientific meetings held at Redstone Arsenal and published many papers and several books. After his death in 2004, the Charles M. Bowden Physical Sciences Complex at MRDEC was named in his honor.

Army Missiles

Three general types of fielded missiles were under the responsibility of MICOM in this period: air defense (CHAPARRAL, HAWK, STINGER, and PATRIOT); anti-armor (DRAGON, TOW, HELLFIRE, and SHILLELAGH); and artillery (HYDRA-70, Army TACMS). MICOM was also responsible for the Multiple Launch Rocket System (MLRS) and Avenger self-propelled mobile launchers. JAVELIN was added to the anti-armor list just before the century ended.

In addition, MICOM supported other systems such as the Ground/Vehicle Laser Locator Designator (G/VLLD), the Mast Mounted Sight (MMS), the M-901 Improved TOW Vehicle (ITV), and the Forward Area Alerting Radar (FAAR).

Only one rocket remained under another Command: the Picatinny Arsenal in New Jersey was responsible for the 2.75 Inch Rocket System. In September 1979, after supplying 35 million 2.75 Inch rockets to the U.S. military and about 30 allies, the Picatinny PM for this system was closed and the mission transferred to MICOM. It is noted that the HYDRA-70 system – widely used in Army Aviation – uses the 2.75 Inch Rocket as its base.

Many of these MICOM-managed missiles were fielded and have already been described; others were under development or final testing and are now described.

STINGER – Originally started by General Dynamics as REDEYE II in 1967, further development led to the name FIM-92A STINGER in 1972. This is a man-portable, infrared-homing, surface-to-air missile, with an effective range of 4.8 km (3.0 mi) and a speed of Mach 2.2 (750 m/sec). In addition to being shoulder-launched, it can be fired from Avenger, MLRS, and other mobile launchers, as well as the APACHE helicopter.

STINGER Missile

Manufactured by Raytheon, it went into service in 1981. An improved FIM-92B STINGER was available in 1983; other variants have followed. STINGER missiles have been adopted by 30 countries.

ATACMS – While the STINGER was the smallest of MICOM's missiles, the Army Tactical Missile System (MGM-140 ATACMS) was the largest in that period. Earlier designated M39, the MGM-149 is 4.0-m (13-ft) long, 0.61-m (2.0-ft) diameter, and weighs 1,670 kg (3,690 lb). Its maximum range is 165 km (103 mi), and the flight ceiling is 48 km (155 kft). It has an inertial guidance

ATACMS Missile

system aided by GPS. After launch, fins with a span of 1.4 m (4.6 ft) are deployed. The warhead carries up to 950 anti-personnel and anti-materiel submunitions. Initially designed by Ling-Temco-Vought in 1986, it was produced by Lockheed Martin and first fielded in 1991. ATACMS is primarily fired from the MLRS mobile launcher.

HELLFIRE – This anti-armor missile was initially called Helicopter Launched, Fire-and-Forget, but changed to the acronym HELLFIRE (AGM-114). Lockheed Martin designed the missile starting in 1974, and began production in 1982. It soon became the 100-pound-class precision air-to-ground weapon for the United States and many other nations. It is 1.6-m (5.3-ft) in length, 17.8-cm (7.0-in) in diameter, and weighs 47 kg (103 lb). Its maximum operational range is 8 km (5 mi). The AGM-114 has a millimeter-wave radar

HELLFIRE System

seeker and semi-active laser homing, requiring no operator guidance after launch. The warhead has 9.1 kg (20 lb) of high-explosive, anti-armor charge. Primarily for use on the APACHE helicopter (it can carry 16 of the missiles), there are a number of variants for other applications.

HYDRA-70 Rocket Pod

HYDRA-70 – Although in service with the U.S. Navy since 1948, and tested at MICOM since the 1960s, the HYDRA-70 free-flight aerial rocket only came into MICOM responsibility when the Army Aviation and Troop Command was absorbed in 1997. The 6.2-kg (13.6-lb) folded-fin rocket has a universal 2.75-inch motor with 6.29 kN (1.41 klbf) thrust that gives a speed of 739 m/sec (2.42 kft/sec) and an effective firing range of 8 km (25 kft). Warheads include impact-detonating, remote-set, and air-burst options. Several types of launchers for rotary and fixed-wing aircraft commonly hold 7 or 19 rockets. The HYDRA-70 is used by all four of the U.S. Military Services.

JAVELIN – As technologies evolved, the man-portable anti-tank weapon reached a new high with the development of the JAVELIN FGM-148. Designed in a joint effort by Texas Instruments and Martin Marietta (now Raytheon and Lockheed Martin) starting in 1989, it has a carry-weight of 22.3 kg (49.2 lb) with the reusable command launch unit (CLU) of 6.4 kg (14.1 lb). It is a lock-on-before-launch and fire-and-forget system. The gunner centers the target on the display of the CLU, switches to the missile's independent IR seeker, and pulls the trigger. The on-board imaging seeker and autopilot directs the missile path; for an armored-vehicle target, this is normally a

JAVELIN System

top-down approach. The effective firing range is 75 to 2,500 m (570 to 8,300 ft), and the peak altitude is 150 m (500 ft). The warhead is a tandem shaped charge weighing 8.4 km (18.5 lb). For the safety of the gunner, a small launch motor ejects the missile from the launcher, and the solid-propellant flight motor ignites after a delay to clear the operator. Production by Raytheon and Lockheed Martin started in 1996.

PATRIOT – Under the FABMEDS program, studies had begun on an updated surface-to-air supplement / replacements for MIM-23 HAWK in 1959. This became AADS-70 in 1962, and then SAM-D in 1964. All of these were study programs – no full hardware resulted. They all had the ability to counter aircraft threats, and also had a limited capability against short-range ballistic missiles. In May 1967, Raytheon was selected as the prime contractor for SAM-D; test missiles were available in November 1969, and engineering development came in 1973.

In early 1974, a major new requirement was imposed: the system must use Track-Via-Missile (TVM) guidance. With TVM, the missile itself generates the tracking information using an on-board radar; with its being closer to the target, this would not only enhance accuracy but also increase the discrimination between decoys and real threats. This change slowed the engineering schedule. TVM was demonstrated in tests in 1975, and full-scale development finally started in January 1976.

In May, the missile was renamed PATRIOT, derived from the name of the radar component of the system. (The AN/MPQ-53 radar system is known as the Phased Array Tracking Radar to Intercept On Target, a bacronym for PATRIOT.) Testing took place in the remainder of the 1970s, and in October 1980, Raytheon was finally awarded a production contract. The system was designated MIM-104A. The missile weighed 900 kg (2000 lb) and had a speed of Mach 5; the Thiokol TX-486 motor delivered 107-kN (24-klbf) thrust, giving it a range of 70 km (43 mi) and a maximum altitude greater than 24 km (80 kft).

During the early 1980s, an upgrade – labeled Patriot Anti-tactical ballistic missile Capability, phase 1 (PAC-1) – was made to the system software,

PATRIOT System

changing the ground radar's upward search pattern from low to the horizon (best for detecting incoming aircraft) to near vertical (best for a counter to the steep parabolic trajectory of inbound tactical missiles and artillery rockets). The missile system with software PAC-1 upgrade was tested at WSMR in 1986, knocking a LANCE missile off course and making what was classified as a "mission kill."

Later in the 1980s, the system was modified to improve its capability for intercepting inbound ballistic missile; this led to the PAC-2 missile and system upgrades. These included further optimization of the radar search algorithms, as well as a change in the missile's warhead (significantly increasing the size of blast fragments as well as the fuse timing). With these changes, the system was designated MIN-104-C.

The AN/MSQ-104 Engagement Control Station is the brain of the system, and is the only manned portion of the operation. At the center is the AN/MPQ-53 phased-array radar, emitting a random beam that is very difficult to jam. PAC-2 was first tested in 1987, and reached Army units in 1990, just in time for use in the Persian Gulf War. There, Patriot was first recognized as a successful ABM system and proof that ballistic missile defense was indeed possible.

AN/MPQ-53 Radar

REDSTONE TECHNICAL TEST CENTER

The missile flight test ranges and rocket motor test stands – both dating from the late 1950s and the ABMA – formed the core facilities of the Test and Evaluation (T&E) Directorate of MRDEC. Through the years, the T&E Directorate greatly expanded, gaining the capability to conduct tests on a wide variety of missile components and subsystems. In October 1990, the T&E Directorate of MRDEC was transferred to the jurisdiction of the U.S. Army Test and Evaluation Command (ATEC), headquartered at Aberdeen Proving Ground, Maryland.

Test Center with Tank Target

Under BRAC 1995 (described later), the merging of the Aviation and Troop Command (ATCOM) and MICOM to form the Aviation and Missile Command (AMCOM), together with the establishment of PEO

267

Aviation, brought more capabilities under RTTC. The activities were particularly enlarged by aircraft test work. The RTTC test area covered about 14,000 acres; in addition, special test capabilities were in over 100 buildings across Redstone Arsenal.

In October 1999, a new era in modern military testing began as Army developmental and operational testing consolidated under one command, the U.S. Army Test and Evaluation Command (ATEC) in Alexandria, Virginia.

MISSILE INTELLIGENCE AGENCY

On 17 June 1970, the MICOM Missile Intelligence Directorate was made a direct report of the Army Materiel Command (AMC); it was renamed the Missile Intelligence Agency (MIA), and was led by a U.S. Army Colonel. In 1975, COL Jack D. Wilson was assigned as the MIA Director and Rankin A. Clinton, Jr. was the Deputy; when Colonel Wilson retired in 1983, Clinton was named the first civilian Director. The Army Intelligence Agency (AIA) was formed in 1984, and MIA became a unit of the AIA; Clinton served concurrently as the MIA and AIA Director until he retired in 1985. MIA had about 450 civilian and military personnel at that time.

In 1985; R. Glenn Priddy was made MIA Director, and served until 1990; he was followed by Joseph W. Holmes who served for four years. MIA became a unit of the Defense Intelligence Agency in January 1992; at that time it was renamed the Missile and Space Intelligence Center (MSIC), and was no longer just an Army organization. From 1994 to 1998, AF Col. Fredrick J. Driesbach was Director; he was followed by AF Col. John T. Wigington who served to 2000.

During much of this period, the MIA/MSIC functioned with the following divisions: Tactical Air Defense Missile Systems, Land Combat Missile Systems, and Strategic Air Defense Missile Systems; there was also the Systems Simulation & Science Applications Division. The division names indicate the areas of foreign weaponry for which the organization had intelligence responsibility. Major technical competencies included missile propulsion, guidance and control, and structures, as well as electro-optics, radars, computers, and telecommunication.

In 1999, MSIC occupied a new 200,000-ft^2 facility, the Richard C. Shelby Center for Missile Intelligence (Building 4545 on Fowler Road); at that time, there were about 650 personnel.

CORPS OF ENGINEERS – HUNTSVILLE DIVISION

In 1967, the U.S. Army Corps of Engineers (USACE) activated a Huntsville Division with its sole mission being the Sentinel (later Safeguard) ballistic-missile defense program. Facilities were at 110 Wynn Drive, in a wing of what was widely called the BMD Building. The Stanley R. Mickelsen Safeguard Missile Complex in North Dakota – the only operational BMD system deployed by the United States – was designed and constructed by the Huntsville Division, During the mid-1970s, the Huntsville Division transitioned from a design and construction organization, with a few missions dominated by the missile programs, to a diversified, high-technology engineering and design and procurement organization. Its activities included installation support, energy, ordnance and explosives, chemical demilitarization, and engineering and environmental programs. In 1995, the Division was named the U.S. Army Engineering and Support Center; that same year, the Center moved into new offices in northwest Huntsville at 4820 University Square. The Center had about 800 employees.

BRAC 1995 – A HISTORICAL CHANGE

Base Realignment and Closure (BRAC) is a process by a United States federal government commission to increase United States Department of Defense efficiency by planning end of the Cold War realignment and closure of military installations. Commissions in 1988, 1991, and 1993, had essentially no effect on DoD agencies in Huntsville. However, BRAC 1995 – which in overall had 29 closures and 11 realignments – led to the most significant diversification in the history of Redstone Arsenal. Since 1949, this had been the Center for Army Missiles; it would now equally become the Center for Army Aviation.

The Aviation Troop Command (ATCOM) in St. Louis, Missouri, would be closed, with the Aviation Research, Development, and Engineering Center; Aviation Management; and Aviation Program Executive Office moved to Redstone Arsenal and integrated with MICOM. It is noted that eventually some 65 percent of the civil service personnel at St. Louis elected to transfer to Huntsville, a record in such activities. The most senior of these was Paul Bogosian, the Director of the Aviation PEO; Bogosian was nationally recognized as a leader in Army Aviation and the principal architect of the move of ATCOM to Huntsville. The Secretary of Defense was required to implement the recommendations not later than 15 September 2011, but those affecting Huntsville were made prior to the turn of the century.

AVIATION AND MISSILE COMMAND

On 17 July 1997, the changes of BRAC 1995 began to be implemented. After 35 years of being the dominant organization on Redstone Arsenal, MICOM became the Aviation and Missile Command (AMCOM). The last MICOM Commander, (then) MG James M. Link, turned the AMCOM Command over to MG Emmitt E. Gibson, the first Commander. On 1 October 1998, AMCOM assumed operational control of Corpus Christi Army Depot (CCAD) in Texas, which maintained aviation systems, and Letterkenny Army Depot (LEAD) in Pennsylvania, which maintained missile systems. Both depots – previously parts of the U.S. Army Industrial Operations Command – remained at their respective locations, but answered directly to AMCOM. The MRDEC was expanded to become, on 1 October 1999, the Aviation and Missile Research, Development, and Engineering Center (AMRDEC).

Aviation Systems Added to Responsibilities

With the merging of the Missile and Aviation Commands, the existing aviation systems came under the responsibility of AMCOM and the Aviation PM (later PEO Aviation). These are mainly helicopters with system names after Native American tribes. Initial systems included the following:

AH-64 Apache - The AH-64 Apache is a four-blade, twin-engine attack helicopter with tandem cockpit for a two-man crew. It has a nose-mounted sensor suite for target acquisition and night vision systems, and is armed with a 30-mm (1.18-in) M230 Chain Gun carried under the aircraft's forward fuselage. It has four hardpoints mounted on stub-wing pylons, typically carrying a

Apache

mixture of AGM-114 Hellfire missiles and Hydra 70 rocket pods. It was developed by Hughes Helicopters, and first flew in 1975. Starting in 1997, the Apache was being produced by Boeing; as of 2011, over 1,200 had been produced.

Black Hawk

UH-60 Black Hawk - The UH-60 Black Hawk is a four-bladed, twin-engine, medium-lift utility helicopter designed and produced by Sikorsky Aircraft. The prototype was the winner of a fly-off in 1976, and entered service with the U.S. Army in 1979. First used as the Army's tactical transport helicopter, this was followed by the fielding of electronic warfare and special operations variants. About 4,000 have been built.

CH-47 Chinook - The CH-47 Chinook is an American twin-engine, tandem rotor heavy-lift helicopter. Its primary roles are troop movement, artillery placement, and battlefield resupply. It has a wide loading ramp at the rear of the fuselage and three external-cargo hooks. It has a carrying capacity in excess of 8,000 lb (3,640 kg), and a top speed of 170 knots (196 mph, 315 km/h). They are generally armed with a 7.62-millimeter M60 machine gun on both sides for self-defense. The design began with the Vertol Corporation (earlier Piasecki Helicopter), which was acquired

Chinook

by Boeing in 1960, and then completed and first flown by Boeing Vertol in 1961. Produced by Boeing Rotorcraft Systems, over 1,200 have been built. The Chinook helicopter has been sold to 16 nations; the largest user, the U.S. Army, plans to operate the CH-47 until 2038.

OH-58 Kiowa - The OH-58 Kiowa is a family of single-engine, single-rotor, military helicopters used for observation, utility, and direct fire support.

Kiowa

Developed and manufactured by Bell Helicopter, the OH-58 has been in continuous use by the U.S. Army since 1969. The latest model, the OH-58D Kiowa Warrior, entered service in 1985; it is primarily operated in an armed reconnaissance role in support of ground troops. Each of the helicopter's two pylons can be armed with two Hellfire missiles, seven Hydra 70 rockets, two air-to-air Stinger missiles, or one .50-calibre fixed forward machine gun. The OH-58D has a Mast Mounted Sight, supplied by DRS Technologies, positioned above the rotor system with a gyro-stabilized platform containing equipment that gives the aircraft the capability of target acquisition and laser designation in both day or night, as well as in limited-visibility and adverse weather.

RAH-66 Comanche - The RAH-66 Comanche was an advanced five-blade, armed reconnaissance and attack helicopter being developed by Boeing-Sikorsky. In 1991, following a study concerning requirements for helicopters then in service, the Boeing-Sikorsky team was chosen to produce prototypes. The Comanche would incorporate stealth technologies, featuring a number of designs previously untried. It was to employ advanced sensors in its reconnaissance role, and was intended to designate targets for the AH-64 Apache. The aircraft was also armed with missiles and rockets to destroy armored vehicles. Two RAH-66 prototypes were built and were flight tested from 1996 to 2004. After nearly $7 billion had been spent, the RAH-66 program was canceled in 2004.

Comanche

RQ-7 Shadow - The RQ-7 Shadow is an unmanned aerial vehicle (UAV) used for reconnaissance, surveillance, target acquisition, and battle damage assessment. Launched from a trailer-mounted pneumatic catapult, it is recovered with the aid of arresting gear. Its gimbal-mounted, digitally stabilized, liquid-nitrogen-cooled electro-optical/infrared (EO/IR) camera relays video in real time via a C-band line-of-sight data link to the ground control station. Originally developed in Israel by the Israeli Aircraft Industries (IAI) from their RQ-2 Pioneer, it is now built by AAI Corporation in Hunt Valley, Maryland. The AAI Shadow was first flown in 1991, and adopted by the U.S. Army in 2001.

Shadow UAV

Sparkman Center – Headquarters

The Mobile District Corps of Engineers had contracted to build a complex of six buildings in September 1992. This project was the first phase of a master plan to build a complex of nine buildings. In August 1994, the first six buildings were completed and the complex was dedicated to Senator John J. Sparkman. In April 1996, the seventh building was opened. Two additional buildings were funded by BRAC 1995 to house Army personnel relocating from St. Louis; these were occupied in late 1997, and named the Heflin Center for Alabama Senator Howell T. Heflin, recognizing his life-long dedication to the defense of this country and support to Redstone Arsenal.

Located at the intersection of Martin and Patton Roads, the Sparkman Center was built for the headquarters for AMCOM and the Logistics Support Activity (LOGSA – described in next chapter). The Center, the first portion of which opened in 1996, was the first new administrative building to be constructed on Redstone Arsenal by the Army since 1960. The Center is a campus-style complex, consisting of 992,390 ft^2 of space in 9 structures: one 5-story, one 4-story, six 3-story, and one 1-story. The complex can provide facilities for over 3,000 people.

Sparkman and Heflin Centers

Defense Acquisition University

The Defense Acquisition University (DAU) is a Department of Defense training establishment formed in October 1991. The DAU was created to provide the DoD Acquisition, Technology, and Logistics workforce with a professional career path and consistency of training content and opportunity. It started as a loose consortium of existing training departments at various commands, then in the late 1990s, this was formalized with a central structure that operated more like a corporate university. The DAU focused on case-based instruction and emphasizes the use of distance learning, mission assistance, and knowledge sharing. With its headquarters at Ft. Belvoir, Virginia, several regional operations were begun, including one at Redstone Arsenal.

U.S. ARMY SPACE AND MISSILE COMMAND

The first Army organization at Colorado Springs, Colorado, was opened in 1984; this was for liaison with the U.S. Air Corps Space Command. In 1987, this liaison activity enlarged to become the Army Space Agency (ASA), the foundation of the Army's operational capability in space.

In April 1988, ASA was reorganized and renamed the U.S. Army Space Command (ARSPACE), headquartered in Colorado Springs. The Army's Defense Satellite Communication System (DSCS) platform and payload control mission came under ARSPACE. The mission of ARSPACE grew in importance up to and during the Persian Gulf War; there the Army found space increasingly needed in conducting its operations.

U.S. Army Strategic Defense Command (USASDC, described earlier) and ARSPACE were merged in 1992, creating the U.S. Army Space and Strategic Defense Command (SSDC). As the focal point for the Army's space and missile defense, SSDC continued providing support to the SDIO (BMDO in 1993), and matrix support to the PEO GPALS.

On 1 October 1997, the Department of the Army created the U.S. Army Space and Missile Defense Command (SMDC), finally rejoining the space and missile activities. This basically brought the Army to the position of the pre-1960s, before Wernher von Braun and the space capabilities were separated from ABMA to form NASA's Marshall Space Flight Center. (See next chapter for the further evolution of SMDC.)

PERSIAN GULF WAR

The Persian Gulf War (August 1990 to February 1991, code named Operation Desert Shield), was waged by Coalition U.N. forces from 34 nations against Iraq, responding to Iraq's invasion and annexation of Kuwait. On 29 November 1990, the U.N. Security Council passed a resolution that gave Iraq until 15 January 1991 to withdraw from Kuwait. The Persian Gulf War began with an extensive aerial bombing campaign on 17 January 1991. In the air war, the Coalition flew over 100,000 sorties, widely destroying military and civilian infrastructure.

Iraq had AL-ABBAS missiles (upgraded Russian SCUD B missiles). When preparation for the war started, engineers at MICOM quickly designed changes to the PAC-2 PATRIOT missile systems, specifically to destroy AL-ABBAS missiles. The new version of the PATRIOT had a more powerful warhead and larger steel fragments to strike its targets. Produced by Martin Marietta, these were deployed to Coalition locations in Saudi Arabia and to Israel.

During the war's seven weeks, 88 SCUD missiles were fired against Coalition military targets and against Israel. (Iraq had hoped to provoke a military response from Israel, expecting other Arab states to then withdraw from the Coalition rather than fight alongside Israel.) This war was the U.S. Army's first occasion to use anti-missile missiles in warfare, and MICOM personnel went to the area to gain first-hand operational data for future developments.

The war ended with the destruction of most of the Iraqi forces and equipment as they attempted to escape along the highway returning to Baghdad. Unfortunately, the U.N. did not follow into Iraq, and the cease-fire terms were such that they led to a second Persian Gulf War 12 years later.

OTHER ACTIVITIES

In addition to the previously described Space and Defense Activities, in this Era Greater Huntsville had many other diversifying activities that highly influenced its technological evolution. The industrial base greatly expanded, as did the educational institutes. Several unusual research facilities were established, and agriculture significantly changed. All of these are considered in this section.

OUTSTANDING RESEARCH FACILITIES

Over the years, many excellent research facilities had been developed by the Army and NASA on Redstone Arsenal – some of which have previously been noted – but two are world-class and will be given special attention. In addition, Greater Huntsville has two outstanding computer facilities that will also be described.

Hypervelocity Research Center

Hidden away on the west side of Redstone Arsenal is a large building that houses one of the most unusual, and least known, research facilities in Greater Huntsville. The central experimental apparatus is a hypervelocity light-gas gun that can shoot the largest projectile at the highest velocity in the world. The facility

Hypervelocity Gun Building

also contains two other light-gas guns with lesser capabilities. Originally used in the early 1990s to develop methods of discriminating between ICBM reentry warheads and decoys based on wake characteristics, and later to test the effects of kinetic-energy weapons in the Strategic Defense Initiative, it now serves as a test range for the Aerophysics Research Center of the University of Alabama in Huntsville.

Technically described, the largest of the two-stage light-gas guns uses up to 100 pounds of gunpowder to fire a piston down a 125-foot long, 10-inch diameter pump tube containing hydrogen gas. When the pressure reaches a predetermined level, it ruptures a diaphragm that allows the hydrogen to expand into a smaller, interchangeable 2- to 6-inch diameter, 75-foot long launch tube containing a projectile carried on a sabot.

Large Two-Stage Light-Gas Gun

Projectiles may be up to 10 kg (35 ounces) in mass, and the velocity is typically between 2,000 to 5,000 meters per second (4,500 to 13,000 miles per hour). With the sabot stripped off, the projectile enters a 10-foot diameter, 41-foot long impact chamber containing an atmosphere from sea level to that of 40 miles (65 km) altitude. The chamber contains sensors to measure characteristics of projectile wakes, or targets to assess the lethality of direct impact.

Beginning in 1961, Delco Electronics (a subsidiary of General Motors) operated a laboratory in Santa Barbara, California, conducting advanced research on missile defense. As a part of this, Delco built a hypervelocity light-gun facility to make measurements concerning reentry physics and radar/optical signatures. In 1985, General Motors acquired Hughes Aircraft and merged it with Delco, forming Hughes Electronics; in 1988, the decision was made by Hughes to discontinue contracted work on missile defense, making surplus the hypervelocity gun equipment. A major sponsor of the Delco research had been the Key Technologies Division of the U.S. Army Strategic Defense Command in Huntsville. The Division Chief, Larry C. Atha, convinced Hughes to donate the equipment to the University of Alabama in Huntsville (UAH).

A contract was made with UAH to provide a building, reassemble the equipment, and continue with the missile defense research. The building, 6230 on Anderson Road in Redstone Arsenal, was constructed and Gerald R. Guinn, then Director of the Johnson Research Center at UAH, was named the Principal Investigator. Disassembly and transportation of the equipment from Santa Barbara to Redstone Arsenal was completed in 1990. Richard A. Hayami, who headed the aerophysics research work at Delco, was brought to Huntsville as a consultant. In 1991, the largest of the light-gas guns was made operational.

Over the next several years, all three of the light-gas guns were used in a wide variety of hypervelocity experiments for various Army organizations; up to 40 employees and students of UAH were involved. Experiments in hypervelocity impact included the development and evaluation of the lethality of kinetic energy weapons for strategic and tactical systems and kill assessment. Hypersonic flight experiments included studies of signature and hypersonic flow related to re-entry and the effects of high-speed flight on interceptor optical and radar sensors.

The building and its experimental equipment were purchased by the U.S. Army in late 1996; the facility, however, continued to be operated by UAH. As of 2013, this was the Test Range of the Aerospace Physics Center, a unit of UAHuntsville's Research Institute directed by Steven L. Messervy. Mark C. Zwiener was the Range Operations Manager, and Ross J. Cortez was the Lead Research Scientist.

X-Ray / Large Optics Test Facility

The testing of advanced X-ray detectors for space applications required a highly collimated X-ray beam with an intensity comparable to that of a source in deep space. In the mid-1970s, such a facility – the X-Ray Calibration Facility (XRCF) – was built at MSFC as Building 4708. It initially contained a stainless steel guide tube 305-m (1000-ft) long and 0.91-m (3.0-ft) diameter. On one end was an X-ray source and on the other end a vacuum chamber for instruments under test. The optically clean, thermally controlled instrument chamber was 6.2 m (20.5-ft) in diameter and 6.2-m (20.1-ft) long. Jack C. Reily, Jr. was the initial Facility Manager.

Large Optics & X-Ray Test Facility

The facility was first used for testing components of the High Energy Astronomy Observatory in 1977, and then used over the next 12 years for testing and calibrating X-ray detector payloads for sounding rockets, the grating spectrometer for the Geosynchronous Earth Orbiting Satellite, and the Soft X-ray Telescope for the Japanese Solar-A satellite. The instrument chamber was also used as a vacuum chamber in baking out the focal-plane structure in the Hubble Space Telescope.

Starting in 1989, the facility underwent a major upgrading in size and capability to perform testing and calibration of mirrors and scientific instruments for the Advanced X-ray Astrophysics Facility (AXAF – later the Chandra X-ray Observatory). The guide tube had two sections added: one 91-m (300-ft) long and 1.5 m (5.0-ft) diameter, and the second 122-m (400-ft) long and 1.2-m (4.0-ft) diameter. The total length was then 518 m (1,700 ft) – about one-third of a mile – with the large diameter where it joins the instrument chamber. The chamber, made of stainless steel, was increased to 22.9-m (75.2-ft) long by 7.3-m (24.0-ft) diameter; it was built in Pittsburgh, Pennsylvania, and transported by barge to Huntsville. James H. Newton and William E. Dickson were responsible for the improvements.

The source building has three X-ray generators: The Henke Electron Impact Point Source (EIPS) produces X-rays by focusing an electron beam onto a metal target. The EIPS operates over the energy range of 0.09 to 10 keV and projects the source as a spot 0.5 mm in diameter. The Penning Gas Discharge Source uses ionized argon gas interacting with aluminum cathodes to produce radiation at 95 and 500 eV. The Rotating Anode Source is an 18-kW Rigaku electron impact X-ray generator and operates between 6 to 40 kV and 10 to 450 mA. Three different anode materials and three different cathode configurations are available. Included with the X-ray sources is a step-driven rotational filter that can be used to limit the bandwidth (the maximum and minimum wavelengths) of the source.

All of these improvements were completed by 1991, and first used in the verification of the test article for the AXAF in September. Other improvements were then made, included new cryogenic pumps for the guide tube and instrument chamber (a pressure of 10^{-6} Torr can be achieved), and a pumped liquid nitrogen system for cryogenic shrouds in the chamber. These improvements were completed in early 1994. Charles L. Wyman, the Chief Engineer for AXAF, was a major user of the facility

In 1999, the instrument chamber was upgraded for testing in the optical spectrum without compromising the existing x-ray testing capability. Central to this was a 1- kW helium refrigerator and cryogenic shroud for testing detectors and mirrors operating at temperatures down to 19.8 K (-414 degrees F). Other new equipment included several wave-front measuring devices mounted to a 5 Degree of Freedom (5DoF) stage and a 5DoF optical mount.

Testing in Instrument Chamber

Mirrors up to 4.0-m (13-ft) diameter can be tested. The facility has two interferometers – instruments for optically measuring structural distortions that occur during cryogenic testing of mirrors. These instruments can detect thermal distortions as small as a few nanometers (billionths of a meter). The overall facility became a part of the MSFC consolidated optics capability under James W. Bilbro, MSFC's Chief Technologist.

In the early 20th century, the facility – the only laboratory in the world for testing large telescope mirrors in a space-simulated environment – has been used in developing a number of space telescope systems. The facility is presently finding major use in the development of the James Webb Space Telescope, scheduled for launch in 2018. As of 2013, Jeffrey R. Kegley was the Facility Manager.

Alabama Supercomputer Center

The Alabama Supercomputer Center (ASC) was conceived by Governor George C. Wallace in 1985, following his visit to Japan to see how that country was using this newest computer generation. An organization called the Alabama Supercomputer Authority (ASA) was formed as an Alabama public service corporation to oversee the final planning, budget formation, and operation of the supercomputer facility and associated statewide Alabama Research and Education Network (AREN). Prior to this time, all supercomputer centers were federally funded, thus making the ASC the first state-owned and operated supercomputer center in the Nation. James F. Woodward, Senior Vice President at UAB, was instrumental in getting the project started; Ray B. Toland served as the first CEO during 1987-88.

Operations at the ASC were through a professional services and facilities management contractor. Boeing Computer Services was awarded the first such contract in 1987; Melvin R. Scott was the Program Manager. The ASC and the

Alabama Supercomputer Facility

associated AREN went into preliminary operation in February 1988. The ASC is in a 24,000-ft^2 building at 686 Discovery Drive in CRP West.

In 1989, Ben B. Barnes was then hired as the CEO of the ASA and D. Wayne Whitmore as the COO. The initial staff was kept small and remains so to this day. Funding for the Authority comes from the Alabama Education Trust Fund, fees for services, and federal contracts and grants.

The services contract was rebid and won by Nichols Research Corporation (NRC) in 1993; David R. Ivey was the Program Manager. Late in 1999, NRC was merged into Computer Sciences Corporation (CSC); however, the ASC center staff remained the same.

The first supercomputer at the ASC, a Cray X-MP/24, went into operation in early 1988; this system had two central processing units (CPUs) and eight vector processing units, and had 32 Mbytes of memory. Costing around $6 million, it had less capability than most laptop computers of today. A few years later, the computer was upgraded to an X-MP/216, increasing the memory to 128 MB. A second supercomputer, an nCUBE 2 Model 10, was put in service in 1991; this was a massively parallel computer (by the standards of its day) that had 128 CPUs arranged in a hypercube topology.

In 1994 the Cray X-MP was replaced with a Cray C94A/264 system; this had two processors and eight vector units, and 512 MB memory. In 1999, the Cray C90A was replaced by a Cray SV1; this had sixteen CPUs, 32 vector processing units, and 16 GB of memory. The Cray SV1 was the last of the big-unit machines in the ASC; subsequent systems were clusters that can be incrementally expanded as needed.

Initially, the AREN was simply a mechanism for researchers at the Data Systems Management Division of the State of Alabama, the National Fertilizer Development Laboratory in Muscle Shoals, and a few select universities to access the supercomputer. This was a bridged, ethernet network, primarily consisting of 56 kbps circuits. The first Internet connection was through a T1 line (1.5 Mbps) from Birmingham to Atlanta; this was the first Internet Protocol (IP) network in Alabama.

As the ASC and its clients grew, so did the AREN. As of 2013, the AREN's Internet access had increased from 1.5 Mbps to 24 Bbps. The statewide backbone consisted of 11 network nodes interconnected with multiple 10 Gbps Ethernet circuits; included are three circuits each to Atlanta and Dallas. The AREN now provides Internet access and computer services to K-12 schools, colleges and universities, public libraries, and select other operations.

As of 2013, John M. Wade was the CEO of the Alabama Supercomputer Authority headquartered in Montgomery. In Huntsville, K. Wayne McVay was the ASC Facilities manager. Computer Sciences Corporation (CSC) remained the services contractor, with David R. Ivey still the CSC Program Manager with Paul H. Dugan the Deputy PM. The high-performance computer hardware included a Silicon Graphics Inc. Ultraviolet (SGI UV), a SGI Altix Cluster , and a Dense Memory Cluster (DMC).

The SGI UV has 268 CPU cores, 4160 GB of shared memory, and 19 Tbytes in the Panasas storage cluster. The computer system is configured with 256 Xeon E5-4640 CPU cores operating at 2.4 GHz and 4 TB of shared memory. The front-end node is configured with 12 Xeon E5-2667 CPU cores operating at 2.9 GHz and 64 GB of memory. This gives the entire system a floating point performance of 5194 GFLOPS.

The SGI Altix Cluster has 162 CPU cores, 1340 GB of shared memory, and 19 Tbytes in the Panasas file system. Each CPU is a 64-bit Intel Itanium 2 processor. The system consists of a SGI Altix 350 front end node with 1.4 GHz processors and Altix 450 nodes with dual core 1.6-GHz and 9.67-GHz processors. This gives the entire system a floating-point performance of 1035 GFLOPS. Sets of from 6 to 72 CPUs are grouped together into shared memory nodes.

The DMC has 2216 CPU cores and 16 Tbytes of distributed memory. Each compute node has a local disk (up to 1.9 Tbytes of which are accessible as /tmp). Also attached to the DMC is a high-performance Panasas storage cluster, which has 17 Tbytes of high-performance storage. The processing capacity of the DMC cluster is as follows: Conventional processing capacity - 29.52 TFLOPS; Single precision GPU capacity - 74.92 TFLOPS; Double precision GPU capacity - 23.53 TFLOPS.

Advanced Research Center

In December 1971, the Advanced Ballistic Missile Defense Agency awarded the System Development Corporation (SDC) a contract to build and operate what was then the most advanced data processing center dedicated to Ballistic Missile Defense in America. (The origin of this activity was described in the preceding chapter.) Called the Advanced Research Center (ARC), the initial primary computer was a CDC 7600 processor, a Seymour Cray-designed supercomputer. A large facility at 4800 Bradford Drive in the Cummings Research Park was built for the ARC. With participation by TRW and other subcontractors, the ARC also had extensive capabilities in software development. Throughout the 1970s and 1980s, the ARC was an important activity in BMD, as well as other defense programs.

The operation of the ARC was rebid in 1987, and the contract was won by COLSA Corporation; COLSA continues today as the ARC Prime Contractor. A new ARC facility, located at 5724 Odyssey Drive adjacent to COLSA's headquarters in CRP-West, contains 95,000 ft^2 of space, including 16,000 ft^2 of raised floors in three computer rooms.

In recent years, the ARC is one of the U.S. Army Space and Missile Defense Command's (SMDC) major research and development facilities. Organizationally, it is a facility in the SMDC Future Warfare Center's Decision Support Directorate, Information and Computational Engineering Division.

COLSA plans, procures, integrates, and maintains high-performance hardware and software and peripherals, providing simulation and scientific support for hundreds of scientists, engineers, and analysts from more than 65 different companies that are working on over 100 projects/programs in the ARC. As of 2013, John D. Welt serves as the COLSA VP / Director of the ARC.

For security reasons, the specific computer and network systems used at the ARC are not publicly disclosed. However, in the early 21st century the overall capabilities have been greatly improved and are described as follows:

High Performance Computing – The ARC is a High Performance Computing Modernization Program (HPCMP) Distributed Center partner. HPC hardware available to the ARC supports serial processing, massively parallel processing, and clustered computational resources with more than 2,000 high-performance computing processors and more than 1,000 scientific visualization graphic workstations.

Network Engineering and Communications – The ARC facility is interconnected through extensive network and communication architectures to numerous local and remote facilities. There are more than 100 local and wide - area networks, including high-speed connectivity via the Defense Research and Engineering Network (DREN) and the Secret DREN (SDREN).

DIVERSIFIED INDUSTRIES

In the period between 1970 and 1999, the number of high-technology industries in Greater Huntsville greatly expanded. Many of these were involved in government-sponsored research and development efforts supporting local Army and NASA programs. There was also a significant number in the commercial sector, particularly in electronics development and manufacturing. This section contains brief descriptions of the larger of these companies – mainly those having 100 or more employees – but a few smaller ones have been included.

Here it is noted that size does not necessarily indicate technological success; private operations with only a few employees sometimes have great accomplishments. This is especially the case for "start ups." In Greater Huntsville there are also many small offices – sometimes called field offices – representing outside firms seeking an opportunity in the area

Companies are listed in two main categories: those existing in Huntsville at the start of this period and already described in a prior chapter (herein given an Updating), and those newly opened in Huntsville during this period. Together with the name, their opening and closing (if applicable) years are designated.

Two of Greater Huntsville's oldest technical firms were lost in this period: Rohm and Haas in 1970, and Thiokol in 1996. Both had been involved in solid-propulsion research and engine development at Redstone Arsenal in government-furnished, contractor-operated facilities since 1949. In the early 1960s, Thiokol, with some 2,000 employees, was the largest firm in the area.

Updated Industries

High-technology industries in Huntsville primarily had their start during the 1950s and 1960s; a number of these were described in the preceding chapter. Several continued to evolve and have importance in this, the Diversification Era, and are included here with updating.

Chrysler Pentastar and Acuistar - Huntsville (1952-Updated) – As described in the previous chapter, Chrysler Corporation was an early entry into Huntsville's high-technology activities, with contract work on the REDSTONE missile and other DoD and NASA programs. During the 1960s, Chrysler built facilities at 102 Wynn Drive totaling 801,000 ft^2 in the CRP.

When NASA and defense work declined at the close of the 1960s, the Huntsville Operations of Chrysler diversified. Under Arthur E. Douyard, activities became as varied as manufacturing TOW missiles, special toilets for sea vessels, and closed-circuit TV systems. Automotive electronics, however, soon became the primary product; in 1977, a 225,000-ft^2 plant was built at 103 Wynn Drive for manufacturing this product. John E. Call was the Engineering Manager in the 1970s.

As diversification continued in the 1980s, a portion of the Chrysler Huntsville operations at 102 Wynn Drive was moved to 110 Wynn Drive and named Military/Public Electronic Systems under W. Frank Henley and Jewel Toney. In 1982, this division became Pentastar Electronics. Inc. (PEI), a wholly owned subsidiary of Chrysler. With some 450 employees, major activities were in computerized rail-traffic control and electronics for Army tanks. In 1997, the PEI segment was acquired and eventually became DRS Technologies, operating at the 110 Wynn Drive facilities.

The major activity at the original 102 Wynn Drive Facility became the Chrysler Electronics Division, primarily producing car radios and engine control computers, supporting the corporate automotive business. John L. Webster was Chief Engineer of this automotive electronics activity.

To handle the large production volumes required to support the automotive electronics business, Chrysler formed Acustar, Inc. as a subsidiary. In 1988, Chrysler Acustar created a huge complex with three buildings at 100 Electronics Boulevard, adjacent to the Huntsville International Airport and the Intermodal Transportation Center. (The original facility at 102 Wynn Drive was vacated and now houses the Huntsville unit of Calhoun Community College, but the facility at 103 Wynn Drive remains unused.)

Chrysler's Chairman, Lido Anthony ("Lee") Iacocca, came to Huntsville to dedicate the Acustar plant; this was Chrysler's first new plant in 20 years, and

Iacocca predicted that it would be "center stage in the automotive industry." Michael H. King was the Plant Manager, Claude E. Reese the Operations Manager, and Gary M. Marchenia the Chief Engineer.

The Chrysler Huntsville complex had R&D, administration, and manufacturing buildings with space totaling about 700,000 ft^2. By 1996, the plant had almost 3,000 employees, with some 200 in new product research and design; they produced 70 percent of all electronics found on Chrysler vehicles, including 7,000 radios per day.

Teledyne Brown Engineering - Huntsville (1953-Updated) – The first locally owned high-technology firm in Huntsville was founded in 1953, and gained prominence as Brown Engineering Company (BECO). Milton K. Cummings became the BECO President in 1958, Joseph C. Moquin was made EVP in 1959, and Raymond C. Watson, Jr. (author of this book) was hired in 1960, opening Huntsville's first contractor-owned research laboratories. In 1962, Cummings and Moquin started what would later be Cummings Research Park. Cummings retired in 1966, and Moquin became President.

BECO was acquired by Teledyne Corporation – the Nation's first highly successful conglomerate – in 1967, becoming Teledyne Brown Engineering (TBE). Personnel peaked at near 4,000 employees. During the Apollo Program, BECO / TBE was the largest support contractor, providing some 20,000,000 manhours of services. The Company was also the first local contractor involved in the Army's intelligence and missile defense activities.

Following conclusion of the Saturn-Apollo Program, Teledyne Brown Engineering experienced a severe drawdown – reaching a low of near 1000 employees. In the early 1970s, NASA's major activity was Skylab, America's first space station. With this, TBE began a major work segment in space payload integration. Following Skylab, TBE provided extensive support on the Apollo-Soyuz Test Project; in this, the Company made its first venture into building flight-qualified experimental equipment.

When Bell Telephone Laboratories (BTL) withdrew from defense contracting, TBE recruited a number of their senior engineers and scientists; these included R. Stephen McCarter, previously head of BTL's missile-defense radar unit. In 1971, McCarter led TBE in winning the highly sought effort as the System Engineering and Technical Assistance Contractor (SETAC) with the Army's newly formed Ballistic Missile Defense Command (BMDC). With success in several re-competes, this continued for many years.

In the mid-1970s, TBE had a significant role in technology upgrading efforts of the BMDC. For this, there were major activities in high-performance propulsion, synthetic-aperture radar, infrared optics, data processing, discrimination techniques, warheads and nuclear effects, and threat analysis.

Throughout the 1970s, TBE was involved in high-volume manufacturing. Contracts were received to produce 25,000 low-cost Ballistic Aerial Target Systems (BATS) at a facility in nearby Madison. For the U.S. Postal Service, the Company fabricated over 3,000 large, stainless-steel, semi-automatic bulk-mail transfer systems. At a leased facility in Decatur, TBE built large, structural steel assemblies and components for the nuclear power industry.

TBE had many contract efforts supporting the war in Vietnam. Products included television and night-vision devices for the Army and Air Force, command modules for the Navy's Swift Boats, and a large data-processing system for operation in Saigon by the CIA.

The Company entered the 1980s technically strong and an industrial leader. Executives were Joseph C. Moquin, President; Robert A. Rieth, VP Systems Division; Harold J. Harvey, then Robert A. K. Mitchell, VP Space Programs Division; William A. Davis, VP Space Defense; Henry (Bud) Burlinghof, VP Operations; and Charles E. Grainger, VP Administration.

In 1985, Rieth replaced Moquin as President. On the executive staff were Donald H. Spencer, Legal Counsel; William A. Meyer, Product Assurance; and Joe E. Sparks, Continuous Process Improvement. Contract Administration was under Eugene R. Andrzejewski. Growth in defense activities led to the formation of the Systems Analysis and Technology Division (primarily software-related activities) under VP Robert E. Alger, and Strategic Systems Division under VP John G. Jones. John R. Lee joined TBE as VP Finance in 1987. Employment returned to near 2,500, a strong recovery from the 1970s.

The first Space Shuttle was launched in April 1981, starting a two-decade-long effort for TBE in a Payload Mission Integration Contract (PMIC). In this, TBE designed, fabricated, and integrated most of the Shuttle's scientific payloads; a 4,000-ft^2 Class-100,000 clean room was built to support this effort.

President Ronald Reagan initiated the Strategic Defense Initiative (SDI) in March 1983. In an effort led by Walter M. Langley, TBE became one of the firms selected as a Center of Excellence in planning the SDI. By 1987, TBE had about 550 personnel engaged in the SDI program; technical study leaders included Douglas J. Bryant, Nicholas A. Passino, John D. Aitken, Samuel L. Grigsby, Harry J. Butcher, and Fred O. Leopard.

From 1982 through 1986, TBE conducted the System Technology and Materiel Program (STAMP) under contracts from MICOM and the Armament Materiel Readiness Command (AMRCOM) at Rock Island Arsenal. With John E. White as the Program Director, some 500 employees were involved with service-life extension efforts for Egypt on Soviet-built air-defense systems.

Henry F. McGill was the PM on the effort for MICOM, while the PM for AMRCOM efforts was Kenneth A. Ingram. Charles D. McWhorter was the Engineering Director and Edwin H. Seigler was the Program Administrator.

In the 1980s, TBE had significant activities in software verification and validation (V&V). Efforts at 11 agencies and commands were obtained; at its peak TBE was the largest V&V contractor serving the DoD. Also in the software area, TBE initiated an air defense modeling package that eventually evolved to be known as Extended Air Defense Simulation (EADSIM). Successfully applied in Operations Desert Shield and Desert Storm, EADSIM soon became one of the most widely used simulations in the world.

In 1984, the Company was awarded a large contract from the U.S. Army to develop and supply optical camouflage screens. A 210,000-ft^2 facility in Clark County, Alabama, was set up for the production. As the work expanded and radar screens were added, Earl S. Wallace, Jr. was brought in as the VP and General Manager.

In late 1987, a team composed of Teledyne Continental Motors, Teledyne Brown Engineering, and Teledyne Electronics was awarded a major contract to design the Army's future fighting vehicles. Called the Armored Systems Modernization, this activity continued for several years. TBE was responsible for the vehicle chassis, protective armor, and similar elements, as well as the systems analysis and simulation laboratory. Robert L. Stagg was the TBE Program Manager.

At the start of 1990, TBE employment was over 3,100, and the campus in the CRP had 10 buildings. James F. McGovern – earlier Undersecretary / Acting Secretary of the U.S. Air Force – became President in July 1991. McGovern established the position of Chief Engineer / Chief Scientist (CE/CS) in the President's

TBE Campus in 1990s

Office, consolidating overall responsibility for the diverse technical activities within TBE. Raymond C. Watson, Jr., (this book's author) who had left TBE in 1970, but had been continuously involved as a consultant, was appointed to this position. Other changes in this period included Eloy J. Torrez, VP Operations; Robert R. Scudamore, VP Strategic Systems; H. Gene Timmons, VP Advanced Programs; and former scientist/astronaut Owen K. Garriott, VP Space Programs.

The concept of "Core Competencies" was adopted for functional units. Most of the functional personnel were assigned to one of the core competencies, under the cognizance of a technical director. TBE's core competencies and associated technical directors were Systems Engineering, William Goldberg; System Integration, Charles R. Vedane; Software Engineering, Gene E. Sievert; Analysis & Simulation, John L. Regner; Optical Engineering (later Sensor Engineering), John R. Yanoski; and Hardware Engineering (later Hardware Development), Clyde L. Ware.

In 1993, Teledyne made a major consolidation of companies; in this TBE immediately added eight operations scattered across the nation. A new division, Commercial Operations, was formed under VP O. Charles Williams. McGovern left TBE in 1997, and Richard A. Holloway was named President. An annual award, the Harry J. Watson Engineer/Scientist of the Year, was initiated

Teledyne and Allegheny Ludlum, a steel and specialty metal firm, were merged in August 1996, forming Allegheny Teledyne, Inc. (ATI), with headquarters in Pittsburgh, Pennsylvania. In this, many of TBE's operations not located in Huntsville were transferred elsewhere. Incompatibilities between elements from the two former companies were significant, and, in November 1999, ATI became three new, separately traded firms. One of these was Teledyne Technologies, initially composed of 19 companies all dating (in some form) from the original Teledyne, Inc. With this, TBE returned to being a Teledyne wholly owned subsidiary. Raymond Watson, the CE/CS, served as Acting President during the transition.

SCI, Inc. - (1961-2001 Updated) – Space Craft Incorporated was founded by Olen B. King and several investors in 1961. From 1962, A. Eugene Sapp was King's closest associate. The initial plan was to build spacecraft for private firms, but after a few years it was realized that a new approach to business was needed; it then turned to general manufacturing of electronic assemblies and changed its name to SCI, Inc.

SCI obtained its first contract with IBM in 1976, providing subassemblies for desktop terminals. This opened the company to manufacturing services for original-equipment manufacturers – offering manufacturing services to original equipment manufacturers at lower cost and higher quality than OEMs could achieve themselves. Before the end of the decade, this was about half of SCI's activities. While still supplying communications and test equipment for the Navy and Air Force, the company was also making diverse products for private customers. SCI soon became Huntsville's second-largest private-sector employer.

SCI pioneered contract manufacturing – building full products to another companies design. Some of its most significant contracts involved building personal computers for IBM, computerized control systems for oil pipelines in

Saudi Arabia, and an automated toll collection system for the New Jersey Turnpike.

In the 1980s, SCI pioneered the surface-mount assembly process, a method that used smaller components and mounting devices to give space savings; this

soon became the design standard for the electronics industry. Recognizing the cost reduction by surface-mount technology, IBM fielded that work out to subcontractor SCI because of its high quality and low cost – the company had recorded only one failure per 50,000 assemblies, and its labor costs were about half of IBM's.

Olin King

During the 1980s, the union tried unsuccessfully to organize SCI's workers. By the end of this decade, SCI was the world's largest contract manufacturer. It was on Fortune magazine's list of the United States' 500 largest companies.

At its height during the 1990s, SCI operated 49 facilities in 19 countries and employed more than 35,000

people. The headquarters – topped with national flags – was at 2101 West Clinton Avenue, and there were plants at several locations in the Greater Huntsville area.

Eugene Sapp

SCI Headquarters

SCI was Huntsville's largest private employer with more than 4,300 personnel, and was the largest company in Alabama. Olin King retired in 1999, and longtime partner Eugene Sapp succeeded him. SCI was acquired by Sanmina in 2001, forming Sanmina-SCI Corporation; Sapp was Chairman and CEO. (See Sanmina-SCI Corporation in next chapter.)

Boeing - Huntsville (1962-Updated) – When NASA's Apollo program ended in the early 1970s, employment at Boeing Huntsville dropped to fewer than 100 people; facilities, however, remained on the campus at 220 Wynn Drive in the Cummings Research Park. Boeing Huntsville remained small until the 1980s, when work on the International Space Station began. By 1985, Boeing employment was about 1,500, leading to the start of a large Boeing-owned complex adjacent to the Huntsville International Airport. In December 1987, Boeing was selected to design and build the living quarters, a laboratory, and logistics modules for the Space Station. The hardware was designed in Company facilities and primarily built by Boeing at MSFC laboratories

In the late 1980s, Boeing divided its Huntsville operations into two primary facilities: central management and programs went to 950 Explorer Boulevard in the CRP West, and most engineering and local manufacturing to 499 Boeing Boulevard in the Jetplex Industrial Park at the Huntsville International Airport. In 1988, Boeing was awarded a large contract from MSFC for computing and communication operations; Huntsville employment grew to more than 3,800 in the early 1990s. Variations in government contracting during the 1990s, however, resulted in an employment decrease to about 1,900 by the end of the decade.

Lockheed / Lockheed Martin - Huntsville (1963-Updated) – Like many other contractor operations in Huntsville, the completion of the Apollo exploration of the Moon brought reductions to Lockheed operations in Huntsville. There was some support to the Skylab and associated ATM, but NASA activities did not return until Lockheed Space Systems Company of Denver was awarded the prime contract for the Space Shuttle External Tank (ET) in 1973. Although the actual fabrication of ETs was at the NASA-owned / Lockheed-operated Michoud Assembly Facility in New Orleans, the design and management required the establishment of a new facility in Huntsville at 4000 South Memorial Parkway. This 124,000-ft^2 facility in two main buildings was earlier occupied by General Electric. Over the years, Lockheed and MSFC worked to reduce the weight of the ET to increase overall efficiency – for each pound of weight reduction, the cargo-carrying capability of the shuttle spacecraft was increased almost one pound.

On the defense side, work for Army agencies at Redstone Arsenal continued at a relatively low level for most of the 1970s. In 1979, activities improved with the start of the Homing Overlay Experiment (HOE), conducted by Lockheed Huntsville for the Ballistic Missile Defense Agency. In 1984, a HOE showed the feasibility of destroying an incoming ballistic missile by colliding with it – the "hit-to-kill" concept. This successful HOE was highly important in the ballistic missile defense evolution, leading to elimination of nuclear warheads on interceptors.

Robert A. Drolet (BG USA Ret.) became the Director of Huntsville Operations in 1992; there were about 240 employees at that time. In March 1995, Lockheed Corporation merged with Martin Marietta, forming Lockheed Martin. Headquartered in Bethesda, Maryland, this was immediately one of the largest defense, aerospace, and technology companies in the world. After Martin Marietta was formed, the Missiles and Space unit continued in Huntsville with little change.

289

Northrop / Northrop Grumman - Huntsville (1963-Updated) – In 1963, Northrop Services, Inc. (NSI), a subsidiary of Northrop Corporation, opened a facility in Huntsville at 6025 Technology Drive. Initial work at NSI-Huntsville was performed for the parent Northrop; gradually, however, contract studies were obtained from NASA at MSFC and the Army at MICOM. Patrick R. Odom became the NSI-Huntsville manager in 1974.

MSFC awarded the firm a significant contract for supporting their newly formed Systems Analysis and Integration Laboratory (SAIL). Employment increased to over 250; this included many highly qualified engineers, physicists, and software specialists. A second building was added to the Technology Drive facilities, but contract participants often worked at MSFC facilities.

NASA's budget drastically reduced in the early 1970s, reaching its lowest point in 1975. The SAIL contract with NSI-Huntsville ended in 1978, and the Huntsville operation was forced to release most of its employees. NSI was dissolved as a subsidiary in about 1980, but a Northrop Field Office at 600 Boulevard South was retained. The Aerospace Systems and Information Systems sectors of Northrop opened Huntsville offices during the 1980s, but had only modest business success; there were usually less than 100 employees.

In 1994, Northrop Corporation acquired Grumman Aerospace, forming Northrop Grumman Corporation. In the following years, Northrop Grumman initiated several new activities in Huntsville.

M&S Computing / Intergraph (1969-Updated) – Led by James W. Meadlock, M&S Computing was founded in 1969. Their initial work was under government contracts for digital components. In 1974, they had their first commercial contract – with Nashville/Davidson County for geographical mapping application, using a revised version of their circuit graphics software.

By 1977, M&S Computing had about 200 employees, and opened their own facility on Highway 20 toward Decatur. In 1980, the firm released the first graphics terminal using raster technology; the dual 1280 x 1024 pixel screens established the industry standard for high-resolution displays – the same standard as today.

Also in 1980, the firm was renamed Intergraph (Interactive graphics) Corporation, and the next year became a publicly owned company. In a short time, the number of employees soon grew 500 percent, and square footage increased 500 percent with a new campus near the original facility. In 1982, Intergraph launched its first 32-bit standalone workstation. By the end of the 1970s, Intergraph had become the world's largest vendor of computer graphics systems.

After buying the rights from Fairchild Semiconductor, Intergraph started producing the C300 Clipper microprocessor in 1988, followed by the C400 in 1990. Also in that year, Intergraph became Huntsville's second Fortune 500 firm. In 1994, Intergraph introduced the industry's first Pentium-based workstations and first multiprocessor workstations, all based on Windows NT operating system. Near the end of the decade, Intergraph offered the first data and document management system specifically for the plant design industry, and also became the photogrammetry industry leader. Their facilities were at 170 Graphics Drive in Lowe Industrial Park.

New Industries

As previously noted, many new firms were started in Greater Huntsville during the last three decades of the 1900s – the Diversification Era. The following are mainly those that grew to have a hundred or more employees, but some smaller firms making special contributions to the technological evolution are also included. Certainly, some firms have been left out; these are mainly those that did not survive their formative years, and also some that are relatively undocumented. If no closing date is shown, the firm continued into the next century; for some of these there were significant changes and they are shown in the next chapter as Updated.

Hudson Metals, Inc. (1970-1982) – This foundry was formed in Huntsville by James R. Hudson and two of his sons, James R. Hudson, Jr., and Gary E. Hudson. The elder Hudson was experienced in foundry operations, being a highly skilled pattern maker and originally coming to Huntsville to work at John Blue Company's foundry. In 1970, Hudson Metals started with offices and a small foundry at 706 Mitchell Drive, NW. With Jim Hudson, Jr., as President, the firm provided gray-iron and aluminum custom castings for commercial customers throughout the Southeast. It quickly outgrew the initial facility and opened a much larger foundry at 500 Green Cove Road in SE Huntsville, close to the railroad spur to the Tennessee River. At maximum, there were over 75 employees. Although the firm was profitable, the owners had other objectives; they closed the foundry in 1982.

D.P. Associates (1970) – Recognizing the opportunity for a minority-owned data processing firm in the rapidly growing high-technology environment of Huntsville, Russell G. Brown and Nathaniel E. Griffin, Sr. co-founded D.P. Associates, Inc. (DPA) in 1970. Initially located in a small office on South Memorial Parkway, the firm soon qualified as an 8(a) organization and received a contract from the General Services Administration to operate a Federal Data Processing Center in Huntsville. By 1975, DPA had more than 100 employees and was located on Sparkman Drive; three years later and with 150 employees, it graduated from the 8(a) status and relocated to a facility on Old Madison Pike.

With Russell Brown as President, DPA soon won a support effort at the NASA/Marshall Space Flight Center Shuttle Project, providing cost modeling, professional analysis, software development, and configuration management. This contract lasted for 22 years, firmly establishing DPA as a respected and competent software firm. Corporate offices were at 4900 Century Street, NW. By 1990, DPA had about 450 employees, with an operation at the Stennis Space Center in Mississippi, and other branch offices in several states.

Russell Brown died in 2010, and his wife, Shirley C. Brown, continues the operation of D.P. Associates as the President and CEO.

Grumman Aerospace - Huntsville (1970-1994) – The Grumman Aircraft Engineering Corporation was founded by Leroy R. Grumman and several partners on Long Island, New York, in 1929; after several moves, it settled in Bethpage, New York, in 1937. During WWII, Grumman produced for the U.S. Navy the largest single-engine aircraft of that time: the TBF Avenger torpedo bomber and the F7F Tigercat and F8F Bearcat fighter aircraft.

Following WWII, Grumman continued with developing and producing naval aircraft. With the advent of NASA, Northrop shifted priorities and began a Space Steering Group. This led to the design and production for NASA of the Apollo program's Lunar Excursion Module (LEM) – later called Apollo Lunar Module (LM) – that landed astronauts on the Moon in 1969. During that same year, the company was reformed as Grumman Aerospace Corporation. The firm received the LEM contract in November 1962, and built 13 lunar modules. Primarily for coordination with MSFC, Grumman opened a Field Office in Huntsville in 1965.

In May 1970, MSFC awarded contracts to Grumman and to TRW for the Phase B design of two High Energy Astronomy Observatory (HEAO) satellites; their designs were completed in April 1971. This was coordinated by the local Field Office and is taken as the start of Grumman operations in Huntsville.

The Huntsville Operations eventually settled at 620 Discovery Drive, CRP; the Director was Leo P. Junen. Work included the Army's automated test-set development and production program under Stephen J. Pleva and support to MSFC's Space Station development under Dan R. Blenis. The local activities reached about 100 employees.

In 1994, Northrop Corporation acquired Grumman Aerospace, forming the Northrop Grumman Corporation. Most of the former Grumman activities in

Huntsville remained at 620 Discovery Drive, forming the Northrop Grumman Mission Systems operation.

New Technology (1971) – This research firm was formed in Huntsville by Chang-Sun (C. S.) Chang and his wife Alice S. Chang in 1971; the firm's name was later changed to New Technology. The Changs, both originally from China, became U.S. citizens before starting the company. Through the accomplishments of C. S. Chang (he is broadly published in prestigious journals), the firm became well known for work in analytical mechanics, particularly modeling and simulation.

In 1973, New Technology received its first support contract from the Army Missile Command for operations and maintenance of the Dragon Missile Flight Test Range. The firm soon expanded into a broad variety of activities including project management services; technical writing/editing and library management; data center operation and maintenance services; and computer operations, network management, and hardware maintenance. Today, the firm still operates, but mainly in consulting; C. S. Chang serves as the President, and Eugene J. M. Chang is the COO.

System Development Corporation - Huntsville (1971-1980) – Based in Santa Monica, California, System Development Corporation (SDC) was considered by many as the world's first computer software company. It was started by the non-profit RAND Corporation in 1955, serving as the systems engineering group for the Automated Air-Defense Ground System for the U.S. Air Force. Two years later, RAND spun off the group as a not-for-profit organization that provided expertise for the United States military in the design, integration, and testing of large, complex, computer-controlled systems. SDC became a for-profit corporation in 1969, and began offering its services to commercial as well as military customers.

SDC opened a marketing office in Huntsville in the late 1960s, and soon won a contract to design a computer-based management information system for the Sentinel Program, followed by a contract for Task and Skills Analysis for the successor Safeguard Program. Although these contracts were performed in Santa Monica, they gave SDC an opening to Huntsville's BMD activities.

A major barrier to BMD development was the means to detect the hundreds of objects in the reentry cloud, discriminate nuclear-armed targets from decoys, and guide the interceptors to the targets – all in a half-minute or less. For this, the data processor needed to handle near a million instructions per second. Bell Telephone Laboratories (BTL) proposed that this be accomplished by many processors working in parallel. To examine this, SDC received a subcontract from BTL to build a feasibility model called the Parallel Element Processing Ensemble (PEPE). With the hardware designed by Honeywell, a prototype PEPE showed the feasibility. In September 1971, as BTL was withdrawing from defense work, SDC was selected by ABMDA to implement the PEPE system in Huntsville. Burroughs won the contract for building the PEPE, and eventually delivered the system to ABMDA.

In December 1971, ABMDA awarded SDC a contract to develop and operate one of the most advanced data processing centers in America – the Advanced Research Center (ARC). The primary computer was the CDC 7600 processor, a Seymour Cray-designed supercomputer. For the ARC, SDC had constructed a large, two-story facility at 4800 Bradford Drive (the northeast corner of Bradford and Wynn). Gerald J. Hansen was the SDC Huntsville manager. Throughout the 1970s, the ARC was updated by SDC and used by Government agencies as well as many Huntsville contractors.

In the early 1980s, SDC was sold to Burroughs Corporation, but continued operations as SDC - a Burroughs Company. In 1986, Burroughs merged with the Sperry Corporation to form Unisys, and SDC lost its identity when it was folded into Unisys Defense Systems. The ARC was rebid in 1987, and the operating contract was won by COLSA.

Dixie MetalCraft (1972) – By the 1970s, Huntsville had a great capability in engineering space and defense hardware, but there were still only a few machine shops for supporting development activities. Dixie MetalCraft Corporation was opened by Leo E. Frassrand in Hazel Green (a few miles north of Huntsville) as a locally owned firm providing such services. Through the years, Dixie MetalCraft has expanded in capability and is now a full-service metal fabrication plant with some 80 employees. Operated by Gregory F. Richard, and in a 66,000-ft^2 facility located at 13887 Highway 231 North, it serves the aerospace, defense, and commercial fabrication industries.

BDM International - Huntsville (1972-1997) – Braddock, Dunn & McDonald, Inc., was founded by three Fordham University professors in 1959, and soon began operations in El Paso, Texas, providing expertise in missile development and testing. In 1962, Earle C. Williams was hired to lead the company and in 1970, moved the headquarters to McLean, Virginia. The corporate name was shortened to BDM in the mid-1970s, and then later it became BDM International (BDMI).

A Huntsville operation under John D. Aitken was opened in 1972. From its start, BDMI Huntsville was involved in missile defense, and had particular involvement in the analysis of directed energy weapons in the Strategic Defense

Initiative during the 1980s. In 1988, Ford Aerospace bought BDMI but continued its operation as a subsidiary. By 1992, BDMI had about 160 employees and was located at 950 Explorer Boulevard CRP West with Charles Pollock serving as General Manager

Among other activities, BDMI-Huntsville engaged in work for the Missile Intelligence Agency. In December 1994, a giant Russian transport plane made a mysterious landing at the Huntsville International Airport. The national news media soon disclosed that the plane carried the major elements of a Soviet air defense system that had been purchased by BDMI from Belarus, a country selling advanced military equipment left in their possession when the USSR collapsed.

In 1997, the BDMI subsidiary of Ford Aerospace was purchased by TRW and the name BDM was discontinued; the Huntsville operations were integrated with other TRW activities. In turn, TRW was acquired by Northrop Grumman in 2002, and the name TRW also disappeared.

Onan Corporation - Huntsville (1972-1997) – The Onan Company was founded by David W. Onan at Fridley, Minnesota, in 1920. From its start, Onan was a manufacturer of power-generation equipment using engines from other manufacturers. During WWII, Onan supplied 85 percent of the portable power-generators for the American armed forces. It became Onan Corporation in 1946, and in 1950, started building gasoline engines for its mobile generator sets; it quickly became a leader in the small genset field with a two cylinder, opposed piston, 1800-rpm engine. As the recreational vehicle (RV) market began, Onan gained over 90 percent share of the associated power generators.

Onan opened a 650,000-ft^2 production facility in Huntsville, Alabama in 1972. Located at 646 James Record Road near the Huntsville Municipal Airport, the facility initially concentrated on smaller generator sets for recreational vehicles, with most popular being their 2.5-kW units. Eventually, the Huntsville plant was producing generators with outputs up to 12.5 kW. Applications included emergency sets for homes and small businesses; for these, the plant also built automatic switching equipment.

Onan Corporation planned for new lines of diesel engines that would be manufactured at the Huntsville plant. Development started in 1977, and the production was launched in the early 1980s. Employment at the plant peaked at about 600; W. Wesley Woods was the Manufacturing Manager and Michael G. Rekoff led the engineering activities.

In 1987, Cummins Engine Company (now Cummins Inc.), the world's largest designer and manufacturer of diesel engines, acquired Onan and announced that the Huntsville plant would be phased out. The plant, however, actually continued manufacturing the new-type diesel engines until it was finally closed in 1998; Wesley Woods was the Operations Manager in this period. The Huntsville facility was later purchased by Navistar Diesel of Alabama LLC, and reopened in 2002.

General Products Partners, Inc. (1972) – Located at 739 Wall Road in Brownsboro, a few miles east of Huntsville on Highway 72, General Product, Inc. was established by Riley N. Hunter in 1972. Initially a sheet-metal shop, it soon grew into a full-capability facility for precision machining, fabrication, welding, assembly, finishing, and testing of metal components and subassemblies, primarily serving defense and aerospace prime contractors. Through the years, General Products had several national owners, but eventually was returned to local ownership. As of 2013, L. Mitchell Tucker was the GM with some 125 employees.

TRW - Huntsville (1972-2002) – Dating from the early 1900s, Thompson Products was a firm manufacturing automotive and aviation parts in Cleveland. In 1953, renowned microwave researcher Simon Ramo and his close associate, Dean E. Wooldridge, formed Ramo & Wooldridge Corporation as a subsidiary of Thompson Products. Ramo & Wooldridge soon won a contract to oversee the U.S. Air Force ICBM development program; from this, Simon Ramo is often called the father of the ICBM. The firms merged in 1958, becoming Thompson, Ramo & Wooldridge (TRW) Corporation, later officially shortened to TRW, Inc. Starting in 1961, TRW had a corporate marketing office in Huntsville, headed for many years by Steven Harris.

In early 1972, the TRW Los Angeles Spacecraft Division won a contract to build three High Energy Astronomy Observatory (HEAO) spacecraft for MSFC. James W. McCurry was sent to Huntsville to head a HEAO support group; this was located in the Clinton Building just off the Memorial Parkway. However, NASA's funding for HEAO was soon spread, leading to a several-year delay in first launch; consequently, TRW Huntsville's first office closed in April 1973.

Also in early 1972, the TRW Los Angeles Software Division sent James E. Long to open an office in Huntsville. This office, initially in a converted store on Highway 20 (Governors Drive) about half way to the airport, opened during the summer; its primary purpose was getting TRW involved with the local ballistic missile research work. In a short time, a contract with ABMDA to build a digital simulator of a BMD processing system was started by Lawrence R. Marker.

The early work eventually led to subcontracts from System Development Corporation (SDC) in establishing the Army's first Advanced Research Center

(ARC), a major computing activity located on the northeast corner of Bradford and Wynn Drives. For this, TRW moved to 213 Wynn Drive, a building near the ARC, and remained there for many years. In the early 1990s, this facility had about 350 employees; John E. Irby was the Huntsville Operations Manager.

During the 1980s and 1990s, TRW Space and Electronics Group (Redondo Beach, California) contracted to assemble and test the Advanced X-ray Astrophysics Facility (later called the Chandra X-ray Observatory). This was a major project of MSFC, and TRW Huntsville provided the local support.

In 2002, TRW was acquired by Northrop Grumman and the Huntsville activities were merged; today it is part of the Northrop Grumman Aerospace Systems and Information Systems sectors.

Science Applications Incorporated - Huntsville (1973) – A consulting firm named Science Applications Incorporated (SAI) was founded in the La Jolla suburb of San Diego, California, by nuclear physicist J. Robert Brewster in February 1969. The next year, branches were established in Los Angeles to support the U.S. Air Force's Space and Missile Systems Organization, and in Northern Virginia to pursue opportunities with the Washington, D.C. area.

Recognizing the business potential at Redstone Arsenal, a marketing office under Philip Miller was opened in 1972. Led by the Virginia Branch, a proposal was made to the U.S. Army for providing the independent software verification and validation (IV&V) on the Safeguard Anti-Ballistic Missile System. To the surprise of the defense community, the Safeguard IV&V contract was awarded to SAI. Hiring of a staff in Huntsville began in 1973; N. Ricky Byrn was the initial Technical Director. In a short time, the Huntsville Branch occupied the third floor of the Clinton Building at 2101 West Clinton Avenue. By the end of the decade, SAI Huntsville had about 160 employees.

Going into the 1980s, SAI Corporate had about 3,600 employees at some 50 locations, with 37 percent of the employees being the shareholders. International contracts had become significant, and the Company reformed as Science Applications International Corporation (SAIC) in 1982.

By the mid-1980s, SAIC Huntsville had its offices at 5725 Odyssey Drive in CRP West, and a staff of some 300 employees. With William H. Gurley as Senior VP and General Manager, SAIC Huntsville – called the Systems & Technology Solutions Business Unit – had considerable growth in the 1990s. It provided a diverse array of capabilities supporting the DoD, NASA, and other federal and commercial customers across the country. Core capabilities included software and hardware engineering, prototype development and testing, engineering tradeoff analysis, management services, modeling and simulation, system engineering and integration, education and training, cost analysis, technology assessment, and unmanned systems development.

Dynetics, Inc. (1974) – Headquartered in Huntsville, Alabama, Dynetics was founded by Herschel A. Matheny and Stephen M. Gilbert in 1974. All of the initial employees had previously been with Teledyne Brown Engineering; some, including Gilbert, had earlier served as engineers at Bell Telephone Laboratories. Beginning activities of Dynetics centered on radar analysis and signal processing.

After starting in a small office suite on Governors Drive, Dynetics established facilities at 110 Wynn Drive, and then at 306 Wynn Drive, Cummings Research Park. Activities expanded into other areas of missile defense and intelligence, including infrared and electro-optical sensors, missile system analysis, software development, modeling and simulation, and foreign materiel exploration. In addition to Stephen Gilbert, key technical personnel included Marc J. Bendickson, Thomas A. Baumbach, Mervin C. Budge, and R. Duane Hays. Staff advisors included Jack D. Wilson and Rankin A. Clinton, Jr., both prior Directors of the Missile Intelligence Agency.

In 1982, Dynetics was the first firm to open a facility in the Western Addition to Cummings Research Park, establishing a campus at 1002 Enterprise Way. Dynetics became fully employee-owned in 1989.

BASF Catalysts - Huntsville (1974) – BASF SE, headquartered in Ludwigshafen, Germany, is the world's leading chemical company; BASF's Catalysts, a global division with central offices in Iselin, New Jersey, is the world's leading supplier of environmental and process catalysts. In 1974, the division opened the Mobile Emissions Catalysts facility in Huntsville; its sole product is catalysts for automotive catalytic converters. In use, these catalytic converters reduce by over 99 percent the pollutants otherwise produced by automobile gasoline engines; the Huntsville facility has produced well over 300 million catalysts for automotice converters. The Huntsville facility is located at 9800 Kellner Road in Lowe Industrial Park. As of the start of the 2000s, John Brock was the Site Manager; there were about 500 employees.

Schwarze Industries, Inc. - Huntsville (1974) – Schwarze Industries designs, manufactures, maintains, and sells one of the most comprehensive lines of power sweeping equipment in the world. Their products include specialized models for parking area cleaning, industrial sweeping, airport runway sweeping, street sweeping, milling cleanup, and storm -water runoff management; they also have a line of velocity-filled pothole patchers. The firm was founded by

Mark Robert Schwarze in 1974; its main facilities are at 1055 Jordan Road in Chase Industrial Park. Schwarze Industries also has an assembly facility in Brisbane, Queensland, Australia, that opened in 1996.

In 2000, Schwarze Industries was acquired by the Alamo Group, headquartered in Seguin, Texas. Founded in 1969, the Alamo Group has 18 principal production facilities worldwide, mainly specializing in right-of-way and agricultural maintenance equipment. Schwarze Industries, Inc. continues operating in Huntsville as a division of the Alamo Group. As of 2013, Howard W. May was the Schwarze President, Gregory A. Heyer the VP Marketing & Product Management, Kevin P. Lozen the Huntsville Plant Manager, and Dale D. Glubrecht the Design Engineering Manager; there are some 150 employees in Huntsville.

Excellance, Inc. (1975) – After serving as the technical leader for aerospace firms Brown Engineering and SPACO, George F. Epps applied his engineering capabilities in a totally different area – customized emergency vehicles. In 1975, Zeb D. Overstreet – who had earlier built similar vehicles – joined with Doris Roberts Humphrey and George Epps in founding Excellance, Inc., in Madison, Alabama.

Excellance is one of the oldest and most respected manufacturers of emergency vehicles in the world. It was established with the mission of designing, developing, and manufacturing high-quality, all-aluminum modular ambulances meeting Federal Speculation KKK-A-1822. Operating in a 34,000-ft^2 facility at 435 Lanier Road in the Madison Industrial Park, Excellance builds two general types of modular ambulances; both are produced by attaching a welded, all-aluminum, box-type patient compartment to a standard rear-wheel cab chassis.

All vehicles are custom-built to customer's specifications. Their customer base includes the royal family of Saudi Arabia – supplied with some of the largest and most impressive emergency vehicles ever manufactured. Multiple vehicles have also been shipped to Kuwait and Ukraine, of note because of designs to meet harsh environmental conditions. The company built the Major Emergency Response Vehicle (MERV) for New York City in 1992; capable of transporting up to twelve patients, MERV is the largest known ambulance in the world. They also fill orders from the military for special-purpose, one-of-a-kind vehicles for highly rugged applications.

As of 2013, Excellance had some 100 employees, and produces about 110 units per year. It remains privately held and operated by three generations; George Epps, a Registered Professional Engineer, is the founder and consultant, Charles T. Epps is President and CEO, and A. Benjamin Epps is the COO.

Coyne / Taylor-Whorton / Norris Cylinder - Huntsville (1975) – In 1975, Coyne Cylinder Company was opened in Huntsville by Thermodyne Industries of Saint Louis, Missouri, to manufacture compressed gas cylinders. The plant was located at 531 Green Cove Road near a railroad spur in the southern part of the city; Gerald Nelson was VP / General Manager, and Michael C. Camp was VP for Manufacturing. By the 1990s, there were about 350 employees. Taylor-Whorton Cylinders of Harrisburg, Pennsylvania, acquired Coyne in 1996; then it was sold to Norris Cylinder of Longview, Texas, in 2010. Now managed by Michael C. Camp, Jr., it continues producing gas cylinders at the original facility; it is recognized for setting the standard in high-pressure and acetylene containers meeting ASME requirements.

ADS Environmental Services / LLC – Huntsville (1975) – Peter D. Petroff, an engineer from Bulgaria and the Deputy Director of Research at NASA MSFC, developed an ultrasonic instrument for measuring the velocity of particulates in flowing fluids. In 1975, Petroff and his three sons founded American Digital Systems (later changed to ADS Environmental Services); facilities were at 5025 Bradford Drive in CRP. The instrument was quickly accepted by cities for mapping waste-water flow. In 1989, controlling interest of ADS was sold to the Swedish firm Axel Johnson, Inc., but the operation continued in Huntsville with Ralph G. Petroff as President / CEO and Mark P. Petroff the EVP. ADS established offices in four other countries, becoming the largest sewer diagnostics firm in the world. Over the next two decades, ADS developed a wide variety of fluid instrumentation systems, including software for data analysis and situation predictions. In January 2008, ADS was acquired by IDEX Corporation, a global fluids and metering firm headquartered in Lake Forest, Illinois, and the name was changed to ADS LLC. As of 2013, the Huntsville facility was located at 1300 Meridian Street (part of the old Huntsville Industrial Complex), and there were 20 field offices worldwide; Hal R. Kimbrough was the General Manager.

T-H Marine Supply, Inc. (1975) – 200 Finney Drive. A family owned and operated activity, this firm manufactures a huge variety of products for the recreational marine industry. Their boat products are found on virtually every recreational boat built in the United States. As of 2013, Jeffrey Huntley was the CEO/president and there were about 100 employees.

Martin Technologies (1976-2009 – Anthony M. (Tony) Martin, a native of London, England, began his industrial career as an apprentice toolmaker at age

14. In 1954, Tony and his brother, Peter, founded Martin Brothers (Toolmakers) Ltd, but eventually decided to immigrate to America. Selecting Huntsville as a new location, Martin Brothers Toolmakers, Inc. was opened at 301 Nick Fitcheard Road in 1976. This manufacturer of metal and powder-metal components later changed its name to Martin Technologies, Inc. and opened a facility in CRP at 4906 Research Drive. A family-owned firm, Tony Martin was the Chairman, Kim Martin the President, and Neal Martin the VP. The firm had a wide range of engineering and fabrication capabilities, providing total manufacturing solutions from design through assembly. Parts were fabricated from iron, copper, stainless, nickel, or titanium solids or powders. In early 2009, Martin Technologies, Inc. filed a voluntary petition for reorganization under Chapter 11 Bankruptcy and the firm closed.

Nichols Research Corporation (1976-2000) – Nichols Research Corporation (NRC) was founded in Huntsville by former McDonnell Douglas employees Roy J. Nichols and H. Chris Horgen in 1976. Their focus was on optical technology and sensor systems for the military and for the space program. In the 1980s, President Ronald Reagan's Strategic Defense Initiative

(SDI) called for the deployment of a space-based surveillance and weapons system; NRC's optical sensor expertise placed it in a prime position to take part in the SDI Program. By 1986, SDI contracts and subcontracts with other SDI contractors constituted 86 percent of annual sales; Nichols went public that year.

Roy Nichols

Chris Horgen

NRC was also gaining new customers, with contracts from White Sands Missile Range in New Mexico and Eglin Air Force Base in Florida. In addition, the company was profiting from the Small Business Innovative Research (SBIR) program, winning 15 contract awards in 1987. By 1988, NRC had grown to 500 employees, and opened a 40,000-ft^2 headquarters building at 4080 South Memorial Parkway.

The company started business acquisitions and gained more new customers, particularly the Army Missile Command (MICOM). MICOM work was on the Fiber Optic Guided Missile Program; Foreign Missile Subsystems and Technology Analysis; Guidance and Control Support; and a contract under the Combined Allied Defense Experiment. At this time, Chris Horgen was CEO, Lawrence R. Greenwood the President, and Roy Nichols the CTO.

In July 1994, NRC acquired Communications and Systems Specialists, a company based in Maryland that specialized in computer simulations and I/T services for NASA and the intelligence agencies. This was followed with the acquisition of Computer Services Corporation, based in Birmingham, Alabama, in 1995, adding that company's base of health-care information services. Soon after this, NRC acquired another Alabama firm, Conway Computer Group, which provided software and I/T services to the insurance industry.

These acquisitions helped the company secure prime commercial contracts, including a large contract with Federal Express for multimedia training services, and another with the Centers for Disease Control in Atlanta for computer services. Its I/T capacity increased to 35 percent of the company's revenues.

NRC continued to expand through acquisitions. In June 1996, the company purchased Advanced Marine Enterprises, Inc., a maker of advanced simulation and virtual reality technology for naval and marine applications. NRC also acquired TXEN, Inc., a database management provider for the health care, insurance, and third-party administrator markets in June 1997. NRC also teamed with Medifinancial Solutions, Inc. of New Jersey to form Healthshares LLC, a joint venture engaged in providing integrated information systems and support services to the health care industry.

By 1999, NRC had 3,000 employees at 30 locations nationwide. The federal market accounted for about 80 percent of the Company's revenue, and health care accounted for 10 percent. It was a prime target for acquisition. In July 2000, Computer Sciences Corporation (CSC) announced the acquisition of Nichols Research Corporation. Existing NRC contracts were novated to CSC, and work continued for several years in the Memorial Parkway facility under CSC's name.

Solar Unlimited (1976-2003) – While working on a large solar power system that MSFC was developing for the Department of Energy, Donald R. Bowden recognized the opportunity of using this technology for water and space heating systems in residences and small businesses. Solar Unlimited Incorporated (SUI) was formed in 1976; facilities were in the Huntsville Industrial Complex (HIC) – the former Lincoln Mill at Oakwood and Meridian Streets. With Lewis P. White, Jr. as Chief Engineer, SUI developed an assortment of solar-based equipment; six patents were eventually obtained. Their primary product was an integrated solar energy system – solar collector, thermal storage and distribution units, and controls – trade named Suncatcher and sold as an installation. In 1980, a huge fire destroyed most of the HIC buildings, and SUI restarted at Traylor Island off Memorial Parkway.

Techni-Core (1978) – As the space-related activities were continuing to decrease in Huntsville and missile-defense was on the rise, Mildred June (Dawn)

Reynolds founded a firm to meet personnel needs in this transition. Techni-Core Professionals, Inc. opened in February 1978, primarily as a temporary contract personnel service for aerospace engineers and other professionals. Using her contacts with highly qualified personnel, Dawn Reynolds enlarged Techni-Core's activities to include subcontracts for supporting technical projects, computer-based documentation services, and audio/video services. Doing business as Techni-Core Engineering, offices were at 3319 Memorial Parkway, South. Initial work was primarily in supporting Boeing in the development of the Main Laboratory Module of the International Space Station. In 2010, the firm was officially renamed Techni-Core Corporation, and facilities were at 4681 Research Park Boulevard; there were some 75 employees. A subsidiary firm, Techni-Core Network Services, was formed with Christina D. Reynolds as the CEO.

Delta Research (1978) – Louis B. Weiner and Michael P. Radke founded Delta Research, Incorporated, in 1978; shortly thereafter, the firm occupied facilities at 315 Wynn Drive in CRP. With Radke as Chairman and Weiner as President, this employee-owned firm provided research, technology development, engineering analysis, and test and evaluation of defense systems.

In 1988, Delta Research received a contract from the U.S. Missile Defense Agency to perform extensive support in developing the Arrow air defense system, a joint program of Israel and the United States. In 2006, the Jerusalem Award for Cooperative Research and Development was awarded to Weiner and Delta Research by the Ministry of Defense of the State of Israel, for contribution to the defense of Israel.

In 2000, Delta Research established a video studio to create digital productions for government entities and defense contractors; directed by Eric R. Corbett, this has become a leading animation group for such services. The facilities of Delta Research are now located at 996 Discovery Drive CRP West.

SPARTA, Inc. - Huntsville (1979) – SPARTA, an acronym for Systems Planning Analysis Research and Technology Associates, was formed as a California corporation by Wayne R. Winton and Robert J. Vickery, Jr., in 1979. Winton had his office in Huntsville, while Vickery's office was in Laguna Hills, California. Previously, Winton had been an engineer with MICOM and then with SAIC.

From the start, the firm was an employee-owned company; as it grew, employees owned about 98 percent of the stock, including the officers and directors as a group with 13 percent. Winton was the Board Chairman / CEO.

SPARTA described itself as an engineering company serving U.S. defense and intelligence agencies. Its first Huntsville facilities were in the Cummings Research Park at 4901 Corporate Drive, and much of the early work was for the Missile Intelligence Agency.

As SPARTA matured, it operated with three major business units: Missile Defense, Mission Systems, and Hardware Systems. In Huntsville, Carl T. Case served as President of the Missile Defense Sector. Robert C. Supucha was President of the Mission Systems Sector. In 1991, R. Stephen McCarter joined SPARTA and became President of the Hardware Systems Sector; earlier, McCarter had headed radar research at Bell Telephone Laboratories. By the end of the century, SPARTA had grown to over 1,000 full-time equivalent employees nationwide at several locations, the largest being at Huntsville and Arlington, Virginia.

Motorola - Huntsville (1979-1998) – In 1928, Paul V. and Joseph E. Galvin started Galvin Manufacturing Company in Chicago. The company's first product was a device that enabled battery-powered radios to be operated on household electricity. To survive in the Great Depression, they designed a low-cost automobile radio using the trade name Motorola. In the late 1930s, Galvin developed for the Army a light-weight two-way AM radio. In 1940, the set was improved to become the 5.0 lb (2.2 kg) Handie-Talkie SCR-536; during the war, this became an icon, with 130,000 sets manufactured. Following the war, Gavin Manufacturing entered the television field, and by the 1950's the firm – now named Motorola – became a giant in electronics. In 1974, they introduced the MC6800 microprocessor, and this was soon central to automotive applications.

Recognizing that data communication systems were the wave of the future, Motorola acquired Huntsville-based Universal Data System (UDS) in 1979. Mark C. Smith, founder of UDS, became a Motorola Vice President, and the Huntsville operation continued as Motorola UDS, making data-communication modems and data communication test equipment – mainly a network analyzer for verifying data integrity in communication systems.

A lawsuit brought by Carter Electronics of Dallas against AT&T dragged through the courts for a decade. This finally resulted in the FCC ruling that AT&T could not prohibit the attachment of Carterphones (a device incorporating an acoustic coupler) to the Bell Telephone lines. Coupled with this was a 1984-ruling by the U.S. Supreme Court that AT&T was a monopoly and could not control related electronic coupling devices.

Motorola saw the huge market potential resulting from these rulings, and had Mark Smith and his engineering staff initiate new products. Smith, however, saw the Carterphone rulings as an opportunity to start his own firm, and left Motorola to began ADTRAN at the end of 1985.

Although almost all of the Motorola engineering followed Mark Smith, the staff was slowly rebuilt and eventually came up with a number of new products. Among these were the SURFRSeries 56K modems that include data communications and fax software. As cable television became popular, Motorola introduced the CyberSURFER cable modem. Another major product was modems to successfully operate over analog cellular communications systems. This brought in a contract with United Parcel Services for 60,000 combined modem and cellular phone units, one for every UPS vehicle.

By the mid-1990s, modems *per se* were being replaced by simpler and higher-speed technologies. In early 1998, Motorola agreed to sell the Huntsville operations to an international investor group. The sale, however, did not go through, and in May it was announced that the Huntsville plant was being closed; the facility had 600 employees at that time.

Dunlop Tire - Huntsville (1979-2003) – The pneumatic tyre (tire) was patented by veterinary surgeon John Boyd Dunlop in 1888; commercial production began in Belfast, Ireland, in 1890. The American Dunlop Tire Corporation was established in 1893, with a factory in Buffalo, New York.

Greater Huntsville gained a diverse manufacturing activity when Dunlop Tire opened a local manufacturing plant in 1970. Located at 248 Dunlop Boulevard just east of the Huntsville Airport, the facility eventually reached 1.2-million-ft^2 in closed areas and, at a peak, had about 1,500 employees; it had a capacity of making 27,000 tires daily. D. L. Gover was the initial Plant Manager. Tires, particularly those for motorcycles, were also developed at this facility. There was a large proving ground for testing automobile tires, and a 100-acre dirt track with a water sprinkler system for developing tires for motorcycles.

In June 1994, union workers at the Huntsville plant went on a 10-week strike, and, even after it was settled, working relationships continued to decline. The Goodyear Tire and Rubber Company acquired 75 percent interest in Dunlop in 1999, and in October 2003, Goodyear announced that the plant was closing. Philip VanHoutan was the last Plant Manager.

SRS Technologies - Huntsville (1979-2008) – Mohindar S. Sandhu founded SRS Technologies, Inc., at Newport Beach, California, in 1970, and then served as the firm's CEO. An operation was opened in Huntsville, Alabama, in 1979; this was later designated as the SRS Systems Technology Group. SRS in Huntsville eventually occupied facilities at 500 Discovery Drive, CRP West. C. Wendell Mead, a nationally recognized expert in missile defense, was the initial

General Manager, and served to 1990. At that time, Harold L. Pastrick was named Corporate Vice President and General Manager of the Systems Technology Group – later Systems Solutions Division.

SRS Huntsville specialized in engineering and scientific services; systems simulation and analysis; computer systems and network integration; software development; and project management and technical support. Technical leaders included Paul A. Gierow and Joe L. Thurman. At maximum, SRS Huntsville had near 100 employees.

SRS Technologies was acquired by ManTech Corporation of Fairfax, Virginia, in 2007, and the Huntsville operation was merged with that of ManTech. This soon became ManTech Nexolve Corporation and the name SRS was no longer an identity.

CAS, Inc. (1979) – Fredric H. Clark, William H. Stender, Jr., and a third unnamed partner founded CAS, Inc. in 1979. The primary founders had extensive civil-service experience in missile-based air- and theater-defense system development, and quickly built CAS into a well-respected defense contracting firm. With Fred Clark as the President and Bill Stender as the CEO, CAS provided system engineering and analysis support for theater missile defense, air defense, aviation, and land combat missile systems, as well as related surveillance, battle management, command, control, communications, and computer systems. With facilities at 650 Discovery Drive in CRP West, CAS had over 600 employees by the early1990s.

Well-known products from CAS included Pelorus Blue and Pelorus Gold, modular simulation applications designed to support real-time distributed exercises, and, Flotsam, a simulation and tactical data collection / analysis tool.

Control Dynamics / bd Systems - Huntsville (1980-2010) – Upon retiring from NASA MSFC, Sherman M. Seltzer and H. Eugene Worley, Jr. formed Control Dynamics, Inc. in 1980. Seltzer and Worley were already recognized experts in flight control and structural dynamics, respectively, and the firm soon became highly regarded in these fields; their facilities were at 600 Boulevard South. In 1986, the firm was acquired by Logicon, Inc., an information technology firm of Torrance, California, and was renamed Logicon Control Dynamics Company, Inc. Seltzer and Worley remained as technical leaders, and F. Brooks Moore, earlier Director of the Astrionics Laboratory of MSFC, was appointed President in 1990.

In 1994, bd Systems, Inc., a woman-owned, minority firm also headquartered in Torrance, acquired the Huntsville Control Dynamics operations from Logicon. Called Control Dynamics Division of bd Systems, R. Joseph Clayton eventually became the Operations Manager, and Brooks Moore the Senior Advisor. They were acquired by SAIC in 2006; with Scott M. Pollard as Director, they operated as an SAIC subsidiary until 2010, and then were absorbed into the SAIC Huntsville operations.

Applied Research (1980-1994) – Harold B. Jeffreys and Larry Z. Kennedy, then with SAIC, saw an opportunity for a strong, independent research operation in Huntsville and founded Applied Research Inc. in 1980. Within a short time, the firm obtained NASA and Army contracts and their exponential growth began. Throughout their initial decade, Applied Research developed a highly qualified staff of engineers, scientists, and analysts. Their expertise covered many topics from cost estimating to astrophysics; they published widely and obtained a number of patents. Their facilities were at 6700 Odyssey Drive, CRP West. In 1989, James B. Odom, previously Director of Science and Engineering at MSFC, joined Applied Research as President /CEO; other key officials at that time included Harold Jeffreys, Chairman, and James D. Ratliff, VP Operations. They soon reached some 270 employees and, with a good business backlog, were ripe for acquisition. Applied Research was bought by SAIC in 1994.

COLSA Corporation (1980) – Francisco J. (Frank) Collazo, a native of Puerto Rico, and his wife, Carmen, started COLSA Corporation in the garage of their Huntsville home during 1980. The two-employee company offered systems analysis and engineering in support of air-defense systems materiel development. COLSA grew rapidly, moving from the Collazo's garage to small offices on South Memorial Parkway, to Sparkman Drive, and to its current facility at 6728 Odyssey Drive in CRP West. In 1983, the SBA under the 8(a) program certified COLSA as a socially and economically disadvantaged company. Franciso Collazo was President and Clarence Tidwell was VP.

Early contracts that contributed to COLSA's growth included support services for the Royal Saudi Arabia Defense Force and system integration for the FOG-M missile. COLSA's specialties were primarily in information services, programmatic support, system engineering and integration, intelligence support, and test and evaluation. In 1987, COLSA became the Prime Contractor for the Advanced Research Center (ARC), a supercomputer facility supporting ballistic missile defense. There were about 400 local employees.

Frank Collazo

COLSA was selected as the National Small Business Prime Contractor of the Year in 1988. The Cogswell Award for Excellence in Security Operations was received by COLSA in 1997 and again later in 2008. At the close of the century, COLSA also had business activities in Washington DC, Florida, California, Alaska, and Hawaii. Clarence Tidwell was President, and Allen Sullivan and Anthony DiRienzo were Senior VPs.

BAMSI - Huntsville (1981-2000) – Hugh M. Brown founded Brown Association Management Services, Inc. as a minority-owned firm at Titusville, Florida in 1978; from the start, the firm did business as BAMSI, Inc. By 1981, the firm was a subcontractor at NASA MSFC, and then won the prime contract to provide management support services for MSFC in 1989. Extensively using disadvantaged subcontractors, BAMSI continued with work in Huntsville until about 2000. All of their work was performed in facilities of MSFC. Marvin H. Lindsey was the Program Manager.

Many in Greater Huntsville first heard of BAMSI through a 1998 news article about employee David Glover, a chemical systems supervisor, collecting human hair at barber and beauty shops for an experiment at MSFC in absorbing oil spills. This technique had been suggested to MSFC by Phillip A. McCrory, a Madison, Alabama, hairdresser, after watching on television an otter's fur being cleaned following the 1989 oil spill in Alaska's Prince William Sound.

Bastion Technologies - Huntsville (1982) – MSFC , Bldg. 4666; Corp. Headquarters, Houston, Texas. Bastion Technologies provides safety and mission assurance services to MSFC and other NASA centers. Randy Reed was the local Program Manager, with Coy Newton the Technical Director. As of 2013, there were about 110 local employees and an off-site office was at 700 Boulevard South.

Cybex Computer Products (1981-2001) – A group of Huntsville engineers, organized by Remigius G. Shatas, founded Cybex Computer Products Corporation in 1981. Stephen F. Thornton was brought into the company to develop a product line of peripheral equipment for the microprocessor industry. Initially, the products centered on KVM (keyboard, video, and mouse) equipment switching, consolidating PC and server control.

Over the next years, the company expanded its products into a line of scalable, platform-independent systems that gave administrators local and remote control of network hardware devices. Key officials included Stephen Thornton, President; Remigius Shatas, VP Systems; Robert R. Asprey, VP Engineering; and James R. Henderson, VP Manufacturing. Facilities were at 4912, Research Drive, CRP; there were also centers in Shannon, Ireland, and Steinhagen, Germany.

By the end of the century, Cybex had become one of the largest manufacturers of KVM equipment in the Nation. They had some 200 Huntsville-based employees and about 350 worldwide. In 2001, Cybex merged with its principal competitor, Apex, Inc. of Redmond, Washington, and formed a new company, Avocent Huntsville Corporation.

Goldstar / LG Electronics - Huntsville (1982) – Lucky-Goldstar, a conglomerate-like firm, was formed in South Korea by the Koo and Huh families in 1947, and quickly grew to become dominant in the national chemical manufacturing business. Goldstar Company, Ltd, was created by Lucky-Goldstar in 1958, and was Korea's first venture into the electronics industry. It rapidly expanded during the 1960s, branching into the manufacture of appliances.

In 1983, Goldstar Company opened a facility in Huntsville, becoming the first Korean manufacturer to have a plant in the United States. Located at 201 James Record Drive adjacent to the Huntsville International Airport, Goldstar Company averaged 380 employees producing 800,000 color television sets and 300,000 microwave ovens a year by the late 1980s. The company added parts distribution and a customer call center in 1987. Kevin Kim was the local manager.

By 1992, all production of television sets and microwave ovens had been transferred to Mexico.; employment soon dropped to about 150. However, major increases were soon made in the parts distribution, servicing, and consumer electronics support – this latter responding to near 5,000 calls daily in English, Spanish, and French. The name of the company changed to LG Electronics in 1997.

DESE Research (1982) – Wallace E. Kirkpatrick, a former U.S. Army civil-service executive, formed DESE Research, Inc., in 1982. This veteran-owned small-business firm is a knowledge-based organization conducting theoretical and analytical research in the fields of Defense, Energy, Space, and

Environment – thus the name DESE. The main offices are at 315 Wynn Drive in the Cummings Research Park, and there are activities at Washington, D.C., and Leavenworth, Kansas.

Earlier, Kirkpatrick had worked in the missile defense field for a half century and is an honoree in the Distinguished Civilian Wall of Fame. After forming

Wallace Kirknatrick DESE, he continued a strong personal involvement in national security. Cognizant of the rapidly changing global security environment, the proliferation of technology, and the evolving threats to the United States and its Allies, Kirkpatrick prepared a paper that received the Gnosis Award from the United States House of Representatives in 1983, the awarded being presented in person by the prominent scientist Edward Teller.

As DESE Research matured, Wally Kirkpatrick was the CEO, Stephen J. Kirkpatrick was President, and Frederick W. Chiverton the VP Operations (later President).   In addition to technical accomplishments, the firm became recognized for high ethical standards and leadership in community services. Although in his late 70s, Wally remains the DESE CEO, a highly respected community leader, and advisor to members of Congress.   In recent years Michael A. Kirkpatrick has served as DESE President.

Pratt & Whitney - Huntsville (1983) – Pratt & Whitney Aircraft Company began in 1925, building one of the most successful aircraft engines. In 1929, Pratt & Whitney, Boeing, and several other aircraft companies formed the conglomerate United Aircraft and Transport Corporation (later called United Technologies Corporation – UTC), with Pratt & Whitney as a business division headquartered in East Hartford, Connecticut, and plants in six other states. Pratt & Whitney opened an office in Huntsville in the early 1960s, when its RL-10 rocket engine was selected for use on the S-IV second stage of the Saturn I vehicle. As the Space Shuttle was undergoing final design in the early 1970s, the Pratt & Whitney RS-25 was developed for the main engine; this engine will also likely be used on MSFC's forthcoming Heavy Launch Vehicle (HLV). In 1983, Pratt & Whitney opened a formal operation – Government Engine & Space Propulsion – in Huntsville at 188 Sparkman Drive.

Vintage / Qualitest Pharmaceuticals (1983) – In 1983, William S. Propst founded Vintage Pharmaceuticals in Huntsville to produce generic prescription and over-the-counter drugs; it was located on Vintage Drive, just off Moores Mill Road, in Chase Industrial Park. William Propst was the CEO, and William S. Propst, Jr., was the President. In a short time, the plant had about 400 carefully screened employees. Its products included tablets, capsules, liquids, suspensions, suppositories, creams, and ointments. Generic drugs have the same active ingredients as the brand-name equivalent for which the original patents have expired.

As completed, the highly secure complex has three buildings: One is a 180,000-ft^2 facility for manufacturing and packaging liquids and other soft products. A second building is a 320,000-ft^2 manufacturing and packaging plant for solid-dose products. Both of these facilities have laboratories for research and development and for quality control. The third is a 280,000-ft^2 building housing offices and a distribution facility. The firm also has a 38,000-ft^2 facility in Charlotte, North Carolina; in 1986, this was made a separate firm, Qualitest Pharmaceuticals with headquarters in Huntsville.

Pace and Waite, Inc. (1984-2003) – This firm was formed by Robert E. Pace and Jack H. Waite at Huntsville in 1984. From the beginning, Jane G. Waite served as the CEO. With headquarters at 7501 South Memorial Parkway, this firm offered expert services in configuration management. They also provided high-end engineering support on subcontracts to several major corporations.

With Paul E. Dixon serving as the company President, Pace and Waite was awarded a five-year contract from NASA to provide center-wide document services at Marshall Space Flight Center starting in 1996. Activities included managing data requirements, checking engineering drawings, and operating the center's documents release desk. The firm had about 160 employees at that time. This contract ended in 2001, but was recompeted and Pace and Waite again won the contract. COLSA acquired Pace and Waite in 2003.

Bowden Industries (1984-2003) – Donald R. Bowden had formed Solar Unlimited Incorporated in 1976 (see earlier entry in this section). With federal tax credits for residential use of solar energy set to expire and subsequent decrease in SUI business expected, Bowden turned his efforts toward a water-based parts cleaning system that SUI had developed and patented. In 1984, Bowden Industries, Inc. was opened, devoted to developing and manufacturing cleaning systems for high-technology equipment; William L. Seale was the Engineering Manager. Bowden Technology's environmentally safe system was shown by independent tests to be superior to traditional cleaning systems using solvents. By the 1990s, the firm had near 100 employees and was located at 1004 Oster Drive. In 2003, the Liquid Turbo-Charger was sold to the Cleaning Technologies Group in Cincinnati, and Bowden Technologies was closed.

Huntsville and Madison County Railroad Authority (1984) – Although not an industry *per se*, the railroad that runs from the center of Huntsville south to the Tennessee River has played a role in Greater Huntsville's evolution. In the late 1880s, the Nashville, Chattanooga and St. Louis (NC&StL) Railway) opened a line between Nashville and Huntsville; a spur extending to Whitesburg on the Tennessee River was added in 1892. Through mergers, this came under the L&N Railroad and then the Seaboard Railroad. This spur was important for the munitions plants on the Huntsville and Redstone Arsenals during WWII, and later served commercial plants built along the river.

HMCRA Engine 8933

Seaboard decided to close this tiny line; to keep it alive, the Huntsville and Madison County Railroad Authority (HMCRA) – Alabama's first public rail authority – was formed in 1984. With offices at 119 Woodson Street, the HMCRA continues the operation of this 14-mile freight rail line as a vital service to a number of industries; as of 2013, Karen P. Monroe was the Manager.

Uwohali Incorporated (1985-2000) – Native American W. Diane Weston founded Uwohali Incorporated in 1985, the name coming from the Cherokee word for Eagle: *Uwo Hali*. With facilities at 300 Sparkman Drive in CRP, Uwohali offered engineering services in logistics, software development, and computer acquisition; Charles D. Stroud was VP Technical Operations. Initial customers were Marshall Space Flight Center and the U.S. Army Strategic Defense Command. In less than a decade, Uwohali had built a facility at 3317 Triana Boulevard, and had about 100 employees. Diane Weston co-founded the Women's Economic and Development Council in 1994, providing opportunities for women to make substantial contributions to the Greater Huntsville community. Unfortunately, Weston died in 2000, and Uwohali soon closed.

Phase IV Systems / Operations – Huntsville (1985) – Radar technology has played an important role in Huntsville's evolution since the area was first involved in missile activities. George T. O'Reilly, Michael D. Fahey, and Stanley E. Prevost, radar specialists previously with MICOM, formed Phase IV Systems, Inc. in 1985; they occupied facilities at 3405 Triana Boulevard. Within a few years, the company evolved into an employee-owned firm that specialized in the design, development, testing, and analysis of radar and other related electronic systems. Their facilities include 42,000 ft^2 in the main area, plus 9,000 ft^2 in a separate manufacturing subsidiary, PIVC LLC. Phase IV laboratories have equipment for the analysis and development of RF hardware up to 110 GHz; complementing this are high-speed digital and signal processing capabilities as well as a Class-10,000 clean room. PIVC fabricates custom RF components and sub-assemblies, and includes a well-equipped machine shop.

In 2004, Phase IV Systems was acquired by Technology Service Corporation (TSC). Headquartered in Silver Springs, Maryland, TSC provides engineering services in radar and surveillance systems to military, intelligence, and civil aviation customers; it was founded by Peter Swerling, an influential radar theoretician. The Huntsville activities continue as Phase IV Systems Operation, a subsidiary of TSC. As of 2013, Michael A. Pope was VP / Division Manager.

InfoPro Corporation (1985) – Native American Joann P. Longshore and her husband William R. Longshore, Jr., formed InfoPro Corporation (IPC) in 1985. Operating in the Parkway Office Center at 3313 South Memorial Parkway, InfoPro offered computer support services to commercial professional firms. In 1987, they changed their business objective to computer maintenance, and soon obtained a contract to serve as Zenith Data Systems first nationwide warranty service provider. Within two years, IPC also became an authorized service provider for Compaq, Hewlett Packard, and IBM, and also obtained service contracts from the Army Missile Command and the Army Space and Missile Defense Command.

In 1996, IPC expanded its maintenance activities to include electronic security and access control. At the turn of the century, IPC obtained 8(a) Small Business certification. At this time they further expanded their activities to provide unexploded ordnance (UEX) and ordnance and explosives (OE) services, including chemical weapon munitions (CWM) removal.

With facilities at 6705 Odyssey Drive CRP West, IPC is an industry leader in UXO/OE services, and now also provides security personnel, access control systems, and vulnerability and risk assessments to a variety of government and commercial customers. In 2013, with Joann Longshore as the CEO and William Longshore the President, InfoPro had more than 400 employees supporting customers throughout the United States.

Sigmatech, Inc. (1986) – Gurmej S. Sandhu, a native of India and holder of a doctorate in engineering from the University of Alabama in Huntsville, founded Sigmatech, Inc. in 1986. Sigmatech's initial focus was on providing systems engineering and technical assistance to the U.S. Army organizations on Redstone Arsenal. With offices at 690 Discovery Drive, CRP West, by the turn of the century Sigmatech was providing management, technical, and training services to a variety of Federal clients. Sandhu was President and CEO; Stephen G. Case was VP Operations.

As Sigmatech continued to grow, it moved to 4901 Corporate Drive in CRP. Offices were also established in Colorado Springs, Colorado; Arlington, Virginia; and Aberdeen, Maryland. At a celebration of its 25th year of business, Sandhu commented that starting and growing a business is similar to having a child. "It is the vision of every parent to have a child grow up and be successful. It is the same in founding a business. As a first-generation immigrant, I'm grateful to this nation for giving me this opportunity."

Activities now include systems and software engineering, test and evaluation, integrated logistics support; international programs/security assistance, and training and e-learning. As of 2013, there are about 200 employees and their key officers include Joseph L. Bergantz (MG U.S. Army, Retired), CEO and President, and Douglas J. Bryant, Senior VP Corporate Development.

Brown Precision, Inc. (1986) – Formed under the name of SEMCO, Inc. in 1964, this small machine shop was acquired by Roland E. Brown and reformed as Brown Precision, Inc. (BPI) in 1986. A CPA and former CFO of Intergraph Corporation, Roland Brown made large investments in BPI, building it into a comprehensive, precision machine shop. They specialize in medical and aerospace products.

Located at 90 Shields Road, just north of Chase Industrial Park in northeast Huntsville, BPI has added state-of-the-art equipment as it evolved. BPI facilities cover about 50,000 ft^2, and the staff is some 85. Capabilities include large and small 5-Axis CNC machining; conventional and 4-axis CNC machining; CNC waterjet machining; wire and plunge EDM machining; CNC large routing and turning (up to six-axis); sheet-metal fabrication; and mechanical assembly. An engineering staff provides design, reverse-engineering, and quality-control capabilities. As of 2013; Gregory Brown and Daniel Brown were co-CEO and Shannon Bell the VP Engineering; there were about 105 employees.

Madison Research Corporation (1986-2006) – Johnny Lee (John) Stallworth is a former American football wide receiver who played 14 seasons for the Pittsburgh Steelers. Stallworth played in three Pro Bowls and was the Steelers two-time MVP; he was inducted into the Pro Football Hall of Fame in August 2002. While playing professional football, Stallworth also completed an MBA degree at his alma mater, Alabama A&M University.

In 1986, Stallworth co-founded Madison Research Corporation (MRC) in Huntsville with his wife, Florastein C. (Flo) Stallworth; Jack E. White; and Sam M. Hazelrig, a retired Army engineer. With headquarters at 401 Wynn Drive in CRP, Madison Research specialized in engineering and software development, systems design, modeling, and

John Stallworth

analysis for the U.S. Army, U.S. Air Force, NASA, and other domestic and international government agencies.

By 2000, with Stallworth serving as President and CEO, the company had more than 650 employees in Huntsville and five regional offices; *Washington Technology Magazine* named MRC one of the 25 top minority-owned businesses. In Huntsville, a major activity was serving as the prime contractor for SMDC's Simulation Center that included a Cray X1 supercomputer system.

In October 2006, MRC was sold to Wireless Facilities, Inc. (WFI) of San Diego; Samuel N. Liberatore took over as President of the WFI Madison Research Division. Two years later, WFI was bought by Kratos Defense and Security, and this became the Kratos Huntsville operations.

Booz, Allen & Hamilton - Huntsville (1986) – Founded by Edwin G. Booz as Booz Analysis in 1914, James L. Allen joined the firm and its name became Booz Allen in 1936. Since 1940, when Booz Allen helped the Navy prepare for World War II, the firm has expanded to develop expertise in technology, engineering, and analytics, offering a full range of consulting services. Carl L. Hamilton was added as a partner, and Booz, Allen & Hamilton became a public company in 1970; the name Booz Allen, however, is still commonly used.

With headquarters in Tysons Corner, Virginia, Booz, Allen & Hamilton operates with offices spread across the Nation. In 1986, Booz Allen opened an office in Huntsville; James A. Baird was a Booz Allen Vice President managing this office, and facilities were at 1525 Perimeter Parkway. From its opening, the Huntsville operation has centered on defense activities.

The parent firm split into two parts in 2008; the part focusing on U.S. government contracting retained the name Booz, Allen & Hamilton, with the stock majority owned by private equity firm, The Carlyle Group. As of 2013, the Huntsville operation was at 6703 Odyssey Drive, had about 225 employees, and Rosalie W. Allen was the Principal and Office Lead.

ADTRAN (1986) – In 1984, the Supreme Court ruled that AT&T had been operating a monopoly in the telephone service and equipment markets. AT&T was then prohibited from manufacturing telecommunications equipment. Mark C. Smith (1940-2007), founder of Universal Data Systems in Huntsville and then Vice President of Motorola UDS, saw this as opening an opportunity to manufacture a wide range of high-speed digital products to widen the application and performance of the nation's hard-wire telephone system.

Mark Smith with John A. Jurenko, Lonnie S. McMillian, and several others left Motorola and formed Huntsville-based ADTRAN, Inc. (for "advanced transmission"), opening in January 1986. ADTRAN would develop and manufacture digital transmission telecommunications equipment – the technology of the future – for telephone switching and distribution in central offices, and their local loops connecting the central office with individual users, getting the maximum performance out of copper-wire telephone lines.

There were seven new regional Bell systems and more than 1,300 independent telephone companies throughout the Americas; these involved about 30-thousand central offices and some 160-million local loops in the U.S.

alone; ADTRAN was formed to pursue complete digital systems for these markets.

By 1991, ADTRAN had gained the leading share in central equipment and local-loop equipment; this was followed by what was called customer premises equipment. Later, they pursued a third market: original equipment manufacturers (OEM) – making its digital products with nameplates of other manufacturers.

Mark Smith and ADTRAN Campus

ADTRAN's first facility was at 4955 Corporate Drive in the CRP. An 81-acre plot was purchased in CRP-West, and an ADTRAN campus was started at 901 Explorer Boulevard in 1992. It had about 250 employees by 1994. Over the next several years, the campus grew to have over one million ft^2 in facilities for the corporate head-quarters, engineering, and manufacturing. By the turn of the century there were around 1,500 employees. During the major growth years, Mark Smith served as Chairman and CEO, and Howard A. Thraikill was President and COO.

Research Genetics (1987-2000) – The biotechnology firm, Research Genetics, Inc., was started in Huntsville by James R. (Jim) Hudson, Jr., in 1987. Hudson, with degrees in chemistry and physics, had previously been successful in operating a family-owned foundry, Hudson Metals, but had become interested in molecular biology. After pursuing graduate studies in this field, he entered a new professional career at age 45.

Starting in a small rented facility, Research Genetics was quickly successful. In a few years, it moved into a modern, five-story laboratory facility at 2130 South Memorial Parkway. Research Genetics' product lines included DNA microarrays and custom software for microarray data analysis; PCR primers that amplify selected genes; genetic markers used to locate disease genes; and custom-made DNA. D. Troy Moore was responsible for technology licensing, new product development, and whole genome resource distribution.

The Human Genome Project, coordinated by the U.S. Department of Energy and the National Institutes of Health, began in the early1990s. Research Genetics quickly became an integral partner, providing biochemical reagents to participating laboratories. During the latter half of the 1990s, Research Genetics became the world's leader in genetic-linkage products, offering microarrays of

30,000 different human genes – the world's largest collection of commercially available, sequence-validated clones. About 250 persons were employed.

In 2000, Research Genetics was sold to Invitrogen Corporation, a California-based provider of products and services for molecular biology and cell biology research. Invitrogen continued the operation in Huntsville for a while, but then merged it into their San Diego operations.

Quantum Research International Inc. (1987) – Founded by D. Frank Pitts at Huntsville in 1987, Quantum Research International (QRI) is located at 991 Discovery Drive in CRP West. The headquarters has 35,000 ft^2 of office space and conference rooms with video-conferencing capabilities. In addition, QRI has operating facilities in five other states; personnel are also located on-site with customers throughout the United States and abroad.

QRI's areas of expertise include support of systems acquisition, test, and evaluation; logistics and fielding support; cyber/ IT support; C4ISR support; weapon system analysis; space operations and control; and audio/visual technology applications; and foreign threats and weapons analysis. The firm has a major ongoing activity for the U.S. Army's Future Warfare Center; Kenneth H. Dryden is the Program Manager. QRI's Foreign Threat Systems Division, managed by James C. Bush, supplies equipment, spare parts, and manuals necessary for operating, maintaining, and repairing foreign threat systems.

As of 2013, QRI had over 300 employees. Frank Pitts continued as the President and CEO, and Deborah L. Fraley was Manager of Business Development.

Time Domain (1987) – The concept of ultra-wideband (UWB) technology in radar originated in the 1940s, but its application required generation of extremely short electrical pulses and signal processing far in excess of anything at that time. Additionally, the use of UWB transmission was restricted by the Federal Communication Commission (FCC) because of its potential interference characteristics. As digital processing evolved – particularly micro-circuit devices – UWB technology came back into practical consideration.

In June 1986, Larry W. Fullerton at Complexx Systems, a small research operation in Huntsville, filed a patent application for a Time Domain Radio Transmission System; this incorporated electromagnetic signals modulated by pulses of pico-second (10-12 sec) duration. In the same period, Fullerton designed silicon-germanium chip sets, called PulsON, for implementing transmitters and receivers.

In 1987, Time Domain, Inc. was founded in Huntsville by Fullerton and a financial backer. The initial patent was awarded 10 May 1988, and many other patent applications followed. In a facility at 4955 Corporate Drive, CTO Fullerton developed chip sets and made prototypes for wall-penetrating radars, noise-immune communications, and a number of other UWB systems. Rachel Reinhardt was VP Engineering. Time Domain received several R&D contracts from government agencies. Finished products and FCC approvals, however, were not forthcoming.

By the early 2000s, Time Domain was heavily in debt and had no revenue-producing products. FCC approval of UWB technology was made in 2002, but other firms had already prepared competing systems. Fullerton resigned in 2007. Pharos Capital Group, which owned about 75 percent of Time Domain, foreclosed on all of the company's assets in 2010, and a new Time Domain was formed. The new Time Domain is concentrating on government services and licensing two UBW radar applications patented and approved by the FCC.

Summa Technologies (1987-2007) – E. C. (Pony) Lee founded Summa Technologies, Inc., as a minority-owned small business in 1987. With offices and manufacturing facilities at 140 Sparkman Drive in Huntsville, the firm's products included precision-machined parts, robotic and mechanical systems, and other subsystems for defense, aerospace, and industrial customers. Pony Lee was the President and Gene Berryman was the Engineering VP. Within several years, there were about 250 employees. Manufacturing facilities were later added in Cullman, Alabama, and Lebanon, Kentucky.

In 2007, AAR Corporation purchased Summa; the Huntsville operations continued as AAR Aerostructures & Interiors and AAR Integrated Technologies.

Prozone International / Water Products (1987) – Ozone has been used for over a century by large water systems worldwide; traditionally, however, ozone systems have been complex and costly. In 1987, physicist Ronald L. Barnes founded Prozone International, Inc. to exploit his concepts for ozone-based water purification using aerospace technologies – these concepts originated from his work on water systems for the International Space Station. In laboratory facilities at 1019 Old Monrovia Road in Huntsville, Prozone developed new methods for generating ozone (O_3) using low voltage (the previous methods used a high-voltage discharge), and applied this in designing a variety of water-purification products.

By the turn of the century, the firm was advertising new purification products for use with residential pools and spas, as well as small air purifiers for homes and offices. These evolved to large commercial ozone-based systems used worldwide. The firm took the name Prozone Water Products, Inc. and relocated to several buildings around 2610 6th street. Ron Barnes has published many papers concerning ozone and its applications, and, to date, 34 patents have been awarded. Cherie Barnes Brook is the President and Ron Barnes the CTO.

CFD Research Corporation (1987) – Ashok K. Singhal and Sangeeta Singhal co-founded CFD Research Corporation (CFDRC) in 1987, occupying offices at 3325 Triana Boulevard in Huntsville. Early key personnel included Andrzej Przekwas, VP Research, who led in developing applications in biotechnology, military medicine, and soldier protection. Clifford E. Smith was VP Engineering, personally specialized in propulsion systems. Sami D. Habchi was an early addition, leading the firm's aeromechanics business including R&D and engineering services in support of NASA and DoD programs. Vincent J. Harrand joined the firm in 1991, heading CFDRC's commercial software development and its transitioning to U.S. Government agencies.

As activities expanded, CFDRC moved to 215 Wynn Drive, occupying the 5th floor in 1996. By 2000, the firm had about 120 employees, with a large number holding doctoral degrees. In 2008, the operations were organized into two divisions: Aerospace and Defense Solutions led by Sami D. Habchi, and Biomedical and Energy Technologies with Kapil Pant as the leader. At that time, the CFDRC biomedical laboratory became one of the original tenants in the newly opened HudsonAlpha Institute building.

Joseph M. Cosumano, Jr. (LTG US Army, Retired), joined CFDRC as President and COO in early 2013. In November of that year, CFDRC opened their new 29,000-ft² facility at 701 McMillian Way on the HudsonAlpha campus.

ERC (1988) – In 1977, Y.C.L. Susan Wu – a native of China and the first woman to earn a Ph.D. in Aeronautical Engineering from California Institute of Technology – and two other professors at the University of Tennessee Space Institute (UTSI) formed Engineering Research and Consulting, Inc. as a part-time activity. In 1988, Susan Wu left UTSI and reestablished the firm as ERC, Inc. with facilities in Tullahoma, Tennessee. For the first few years, ERC's growth was slow, but steady; in 1996, ERC moved to Huntsville to be closer to potential customers.

With offices in CRP at 4901 Corporate Drive, ERC grew very well; they provided engineering and management services, modeling and simulation, test and evaluation, information technology service, and operations and maintenance support to government agencies in Huntsville and also five other locations. As of 2013/14, the firm had about 600 employees nationwide; number locally undefined). Ernest Wu was the CEO, Kenneth C. Frame the President, and Susan Wu remained the Chairman

Quality Research (1988-2003) – Alabama Quality Research, Inc. was founded by Dusit Charern in 1988; a privately owned firm, the name was changed to Quality Research, Inc. (QRI) in 1990. Their initial office was downtown on West Side Square, but they soon moved to 150 West Park Loop. By 1994, QRI had less than 25 employees; however, after being designated as minority owned (8-A), contracts dramatically increased. Their primary capabilities were in modeling and simulation, information management systems, and enterprise IT solutions.

With Charern as the CEO and Gary S. Ryan as President, QRI relocated to 4901 Corporate Drive CRP and had over 200 employees by 1999. Its primary customers were the U.S. Army Corps of Engineers, Missile Defense Agency, U.S. Air Force, U.S. Army Training and Doctrine Command, and Space and Missile Defense Command. In 2000, QRI received the Region IV Small Business of the Year award from the Small Business Administration.

Since 1992, QRI had been a small-business protégé of SAIC under a formal DoD Mentor-Protégé agreement. By 2003, QRI had near 600 employees in Huntsville and 20 other locations throughout the United States. In February 2003, SAIC purchased QRI, merging the activities into SAIC's general operations.

Morgan Research (1988-2010) – Sharon D. Morgan and her husband Timothy D. Morgan founded Morgan Research Corporation shortly after they were married in 1988. Their first facility was at 2707 Artie Street, and it took five years for the firm to have 25 employees, but then exponential growth began. Morgan Research engaged in a variety of hardware and software development activities and engineering services, including test and evaluation, technology insertion, systems integration, and systems engineering. There were also technical data management, logistics support, training, and acquisition support services.

By 2003, there were over 200 employees in facilities at 4811 Bradford Drive. With Sharon Morgan as the CEO and Tim Morgan the President, Albert K. Killen was Senior VP Engineering and J. Michael Coward was VP Software.

In 2006, the Morgans sold the firm to Stanley, Inc., an information technology company based in Arlington, Virginia. Stanley continued the Morgan Research activities as a Huntsville Division with Albert Killen as the General Manager. In 2010, Stanley was acquired by the CGI Group, a leading provider of information technology and business processing services.

Advanced Optical Systems (1988) – Richard L. Hartman, the former Director of Research at the U.S. Army Missile Command, formed Advanced Optical Systems, Inc. (AOS) in Huntsville in 1988. With facilities at 6767 Old Madison Pike, AOS specializes in optical signal processing, including automatic target recognition and remote attitude measurement systems. Hartman serves as CEO, and Keith B. Farr is COO; both are also Principal Scientists.

AOS products with wide recognition include AIRprint, a non-contact fingerprint reader that finds hands and extracts fingerprints anywhere in a large volume such as at an airport. Another product is the Hermes, a proximity, location and ranging system giving drone cargo aircraft the ability to autonomously pick-up and deliver supplies. Both of these products have received distinguished Edison Awards, one of the highest accolades a company can receive in the name of innovation and business.

Tec-Masters (1988) – Marvin P. Carroll founded Tec-Masters, Inc., a minority owned, high-technology firm in Huntsville, in 1988. For much of the time since opening, their facilities have been at 1500 Perimeter Parkway. From inception, Tec-Masters provided engineering and logistical support for Army programs. Under contracts with NASA, support in advanced technology and space programs were added in 1991. In 1992, Tec-Masters was awarded a Systems Engineering and Technical Assistance (SETA) contract from the Army; since that time, subsequent SETA contracts were awarded in 1997 and 2003. Starting in 1994, Tec-Masters has provided software development and information systems support, as well as modeling and simulation, as components of its engineering services.

In recent years, support services have been expanded to Navy and Marine Corps customers. Capabilities in training led to offices being opened in Lawton, Oklahoma, and Orlando, Florida. Tec-Masters' eTeaching Institute is an on-line Internet service that provides live teacher-moderated electronic class rooms through Internet access. As of 2013, Marvin Carroll remained as President / CEO; Ralph Johnson was EVP, and John B. Broyles; VP Operations. There were about 150 employees.

Amtec Corporation (1988) – Amtec is a Huntsville-based, privately held company serving DoD, NASA, and commercial clients. The firm was founded by William P. (Bud) Albritton, Jr. in 1988, with headquarters and offices at 500 Wynn Drive and shops and laboratory facilities at 4808 Bradford Drive. Amtec provides a broad variety of services in engineering design and development, system integration and analysis, modeling and simulation, test and evaluation, and specialty / prototype manufacturing. They also offer ammunition and explosives safety, certification, and refresher training at locations across the nation as well as on line. Amtec is the Southeastern U.S. distributor of MiniTec Profile System, a line of aluminum extrusion framing products.

As of 2013, Amtec had about 375 employees, and the corporate offices had relocated to 654 Discovery Drive. Bud Albritton was the President; H. Marshall Arney, the General Manager; E. Earl Hughes, Chief of Operations; and Nathaniel G. Albritton, Director of Engineering & Technology Development.

Raytheon Company - Huntsville (1989) – In 1922, the American Appliance Company was established in Cambridge, Massachusetts An early product was a gas rectifier vacuum tube, quickly adopted by radio receiver manufacturers to eliminate batteries. The firm's name was changed to Raytheon Manufacturing Company in 1925, and by the 1930s it was one of the largest suppliers of radio tubes in the world. During WWII, it was the primary source of magnetron tubes, and also produced microwave radars. Following the war, it developed the microwave oven as a commercial product, but soon turned to missiles and their guidance systems. They gained widespread recognition with the LARK, SPARROW, and PATRIOT missiles. The name Raytheon Company was adopted in 1959.

Although Raytheon had long been a supplier to Army agencies at Redstone Arsenal, it did not establish a full operation in Huntsville until 1989. At that time, Benjamin R. Harrell opened a facility for Raytheon at 306 Wynn Drive in CRP; it soon had about 460 employees engaged in the design, analysis, development, fabrication, and supply of missile systems, systems engineering, and logistic support. Later, Raytheon Technical Services Company incorporated as an Alabama LLC, and opened a facility at 401 Jan Davis Drive in CRP West; Robert K. Wolf managed this branch.

Jacobs Sverdrup Technology - Huntsville (1989) – Founded in 1928, Sverdrup & Parcel became one of the foremost facility development firms in America. In 1950, they were awarded a contract from the U.S. Air Force to build the Arnold Engineering Development Center (AEDC) at Tullahoma, Tennessee. When completed, AEDC was privatized; Sverdrup and Parcel then established a separate subsidiary organization called ARO (Arnold Research Organization), Inc., to operate it. Sverdrup & Parcel became the Sverdrup Corporation in 1977. Sverdrup opened a Huntsville office in 1989 and was awarded a major engineering services contract for MSFC.

Jacobs Engineering, founded in 1947, was an international engineering, architecture, and construction firm. In 1999, Jacobs Engineering acquired Sverdrup Corporation and merged the two firms. They formed a division called

Jacobs Sverdrup Technology, Inc. with headquarters at Tullahoma, Tennessee; this took over the activities in Huntsville. For operational purposes, the Huntsville activities eventually became the Jacobs ESSSA (Engineering, Science, Systems, and Augmentation) Group, and the Jacobs ITSS (Information Technology and Support Services) Group. Most of the work is in government facilities, but both also have activities at 1500 and 1525 Perimeter Parkway.

As of 2013, Jacobs ESSSA has about 645 Huntsville employees; Randal W. Lycans is a Jacobs VP / ESSSA Group General Manager, and former astronaut N. Jan Davis is the ESSSA Group VP / Deputy General Manager. There are some 100 employees with Jacobs ITSS; Gerald Burnett is the ITSS Group VP / General Manager, and Michael Williams the Operations Manager.

Baron Services, Inc. (1990) – Baron Services is a weather technology company based in Huntsville. Founded in 1990, by Robert O. (Bob) Baron after 22 years of experience in broadcast meteorology, the company develops weather systems that aid in the detection and dissemination of weather information to customers in the broadcast, government, aviation, marine, and automotive fields. Baron's products and services include Doppler weather radar systems, weather-tracking software, personal weather-alerting services, and weather equipment installation and maintenance.

In 1992, Baron Services developed the OmniWxTrac system, the first street-level storm-tracking system for broadcast television. OmniWxTrac was followed by many additional weather tools, including the SAF-T-Net alert system giving warnings over individuals' cell phones. Baron Services, together with L-3 Communications, developed a dual-polarity radar system, providing both horizontal and vertical images; they were them selected to upgrade 171 U.S. National Weather Service, Department of Defense, and Federal Aviation Administration NEXRAD (Next Generation Radars) to dual-polarization.

With about 95 employees, Baron Services' facilities are located at 4930 Research Drive in CRP. Bob Baron is the President and CEO; he was recently named Entrepreneur of the Year by his alma mater, the University of Tennessee. Robert O. Baron, Jr., is the Chief Products Officer.

MEVATECH Corporation (1989-2003) – Nancy E. Archuleta, an accountant, founded MEVATECH in 1989. C. Thomas Houser, who had many years of experience at defense firms, was hired as President. Facilities were at 1525 Perimeter Parkway. MEVATECH's technical capabilities were in software engineering, systems design, development and integration, and technical support systems. In management services, MEVATECH provided a range of financial management solutions. By the early 2000s, MEVATECH had about 500 personnel in 14 locations across the U.S. A prime for acquisition, MEVATEC was bought by BAE Systems in 2003.

Computer Systems Technology, Inc. (1989-2003) – Computer Systems Technology, Inc. (CSTI) was a Woman- and Minority-Owned firm established in 1989. Bobby A. Bradley, a native of Huntsville, founded CSTI, then served as its President and CEO. With offices at 1525 Perimeter, CSTI centered on information technology solutions: IT product procurement, IT engineering, program management, and logistics services. Jay W. Newkirk was the Senior VP, responsible for business development. By 2003, CSTI had some 200 employees, and was acquired by SAIC.

Westinghouse Electric – Huntsville (1989-1997) – Although a representative of Westinghouse Electric Corporation had an office in Huntsville as early as 1960, a formal operation was not established until 1989. As a part of the Westinghouse Industries and Technology Group. a facility was opened at 917 Explorer Boulevard in CRP West; James R. Rambo was the Manager, and there were about 80 initial employees. Under a subcontract from Martin Marietta, the Huntsville operation engaged in the engineering and critical components assembly of the guidance system for the LONGBOW missile system. In 1997, Westinghouse Electric discontinued its Industries and Technology Group and the Huntsville operation was closed.

AEgis Technologies (1989) – William F. Waite and Steven S. Hill co-founded AEgis Technologies Group, Inc. in 1989. A privately held small business, the firm provides advanced technology and expert consulting services to industries throughout the world. From the beginning, Waite has served as Chairman and CTO, while Hill was the President and CEO. AEgis is best known for its work in modeling & simulation (M&S). Their M&S products and services include simulation software and training simulators; geospatial databases; 3D models; war fighter exercise support; verification, validation, and accreditation; and hardware-in-the-loop and man-in-the-loop simulation.

In the new century, AEgis has had considerable growth. They occupied a new complex at 410 Jan Davis Drive in CRP-West, and also opened offices in Washington DC; Albuquerque, New Mexico. For the 2012 Summer Olympics in England, AEgis was awarded a contract to develop a detailed, 3-D digital model of London.

As of 2014, AEgis reached over 130 professional employees. William F. Waite was Chairman and CTO, and Steven S. Hill, President and CEO. Technical executives included Lance W. Cooper, EVP Defense Services; David

E. King, EVP Technical Solutions; John R. Anderson, VP Huntsville Operations; and Mark E. McDaniel, VP Simulation Development.

Camber Corporation (1990) – Walter P. Batson, Jr. was the principal founder of Camber Corporation. Established in 1990, Camber's initial office was at 2905 West Corporate Boulevard in Huntsville. Batson served as its initial President, developing a business plan that resulted in early growth. Robert L. King was Operations Manager, and later became EVP. After starting with computer programming services, Camber spread into other activities, including modeling and simulation, engineering, training, and logistics. In 1992, an office was opened in Washington, DC, and soon became Camber's largest element. In 1999, Batson took the position of CEO, and B. Joseph Alexander, who had led the success in Washington, moved to Huntsville and was named President. Camber became an ESOP-owned company in 2002. With headquarters relocated to 635 Discovery Drive, the firm had a total of about 360 employees in 2014.

Chandler-May, Inc. (1990) – 133 West Park Loop. This company designs, develops, and manufacturers unmanned aerial systems and shelter-enclosed command and control systems for tactical applications. Chandler-May, and its facility in San Luis Obispo, California, has experience with K-MAX unmanned helicopters, Desert Hawk UAVs, and Persistent Threat Detection System aerostats. Jesse May was the President and Jay Chandler the EVP. In November 2012, Chandler-May had about 180 employees. It was acquired by Lockheed Martin, becoming the Unmanned Integrated Systems of LM. Their new products include the Fury UAS and the SharkFin that provides navigation control, video display, and payload control in one integrated, commercial software package.

AI Signal Research, Inc. (1990) – 2001 Nichols Drive. Jen-Yi Jong founded ASRI as a minority- and woman-owned business to provide vibration analysis tools and services. In 1993, the firm won its first competitive contract with NASA MSFC to develop vibration signal analysis capabilities for analyzing the Space Shuttle Main Engine. As of the end of the century, ASRI had about 200 employees.

Taos Industries (1991-2010) – From headquarters in Madison, Alabama, Taos Industries, Inc., provided commercial and military spare parts and systems from around the world to U.S. Federal and State government customers. It was formed by David B. Hogan in 1991, shortly after Hogan retired from a U.S.

Government career in foreign intelligence. Initially a family business, it competed very well with large firms such as BDM in acquiring Soviet weapon systems and spares for domestic intelligence agencies and friendly foreign militaries. Two sons held key positions – D. Craig Hogan as President and Steven P. Hogan as CFO; tragically, both sons were killed in an aircraft accident in January 2005.

John R. Hamilton, who had extensive military experience in foreign weapons, joined Taos as a minority partner and VP for Business Development in 1994, was named COO in 1996, and then became President in 2005. Under Hamilton's leadership, the company grew to a global business operating in over twenty countries. During 2004-2007, business greatly surged as Taos was one of the few government contractors experienced in buying the foreign equipment needed in Iraq. Taos was honored in 2003 as the Huntsville and Madison County Small Business of the Year, and Hamilton himself was selected by the Chamber of Commerce as Small Business Executive of the Year in 2005.

In October 2006, Taos was acquired by Agility Defense and Government Services (D&GS); this is an American subsidiary of Agility, a publicly traded global logistics company headquartered in Kuwait and controlled by the family of Tarek Abdulaziz Sultan Al-Essa. John Hamilton was first retained as the COO of Taos, and then made President and CEO. Although a wholly owned subsidiary of Agility D&GS, Taos had its own Board with Hamilton as Chairman and kept U.S. clearances for classified work.

The Board restructured Taos in September 2008, and Joseph M. Cosumano, Jr. (LTG U.S. Army, Retired) became President and CEO. Hamilton returned to the position of COO, but elected to retire a year later. In July 2009, Taos was awarded a very large contract to provide tailored services ranging from base operations, transportation, warehousing, and material management, mainly in support of U.S. Army's Afghanistan operations.

In the fall of 2009, the U.S. Government alleged that the Agility corporate office in Kuwait had overcharged on contracts; although Taos was not directly involved, it was financially crippled. In early 2010, the Government imposed a suspension on Agility, its D&GS Subsidiary, and 120 Agility-owned companies, including Taos Industries. It was necessary for Taos to close; Cosumano remained for some time, gradually terminating the operations. After long legal procedures, the suspension was eventually lifted and operations in Huntsville were resumed, but now conducted as a small branch office of Agility D&GS.

Analytical Services / ASRC Federal - Huntsville (1992-2007) – Irma Loya Tuder, a financial executive from El Paso, Texas, moved to Huntsville and founded Analytical Services, Inc. (ASI) in 1992. With initial offices at 555

Sparkman Drive, this woman- and minority-owned firm was soon very successful in providing technology, management, and organizational solutions to both government and commercial customers. By the turn of the century, ASI had over 120 employees and had moved to 689 Discovery Drive, CRP West. Irma Tuder continued as President and CEO, Jack D. Conway was Senior VP and Operations Director, and Murray D. Hillman was VP Science & Technology. In May 2007, ASI was sold to ASRC Federal Holding Company of Beltsville. Maryland.

Computing Technologies, Inc. – Huntsville (1992) – 4101 University Square; Corp. Headquarters, Fairfax, Virginia. A minority-owned, minority-operated firm, William Starks was the VP and there were about 115 employees.

Logistics & Environmental Solutions Corporation (1992) – Doing business as LESCO, Logistics & Environmental Solutions Corporation was started by Anita B. Williams as a woman- and minority-owned firm. Incorporated in June 1992, and certified 8(a) in January 1995, Anita Williams was CEO and Paul D. Smith was President. LESCO specialized in four core areas: base operations support, engineering services, professional services, and warehousing and distribution. Corporate offices were at 4845 University Square. Within a few years, LESCO had some 200 employees, most of whom are performing services at seven locations (mainly military bases) across the U.S. and overseas LESCO was acquired by Donald W. Miller in 2010.

MagneTek / Universal Lighting - Huntsville (1993) – MagneTek, Incorporated, headquartered in Los Angeles, California, was one of the largest producers in the United States of electrical power products. In 1993, MagneTek opened an advanced development center in Huntsville, renovating an existing 55,000-ft^2 facility at 1430 Wall Triana Highway. The Huntsville operation was involved in research, development, and manufacturing of products carrying the trade name Triad; their primary product was an electronic ballast used in fluorescent lights. In a short time, there were some 300 employees. Robert E. Bell was the Operations Director and R. Patrick Gill the Process Engineering Manager.

By 2000, the Huntsville facility had doubled in size, employees numbered over 500, and the operation was named the Manufacturer of the Year by the Alabama Development Office. Early in the new century, MagneTek, Inc. was involved in a series of mergers and acquisitions, and eventually became Universal Lighting Technologies headquartered in Nashville, Tennessee. Due to international competition and technical factors, the Huntsville operations greatly declined; as of 2013, the local facility was a service operation.

System Studies and Simulation (1993) – Janice H. Smith founded System Studies and Simulation, Inc. (S3) in 1993. With an extensive background in developing defense-related software, her main initial market was missile defense programs at Redstone Arsenal. In a short while, Srini R. Srinivas was brought in as a partner and CEO, allowing the firm to gain minority status (8-A). Facilities were at 4815 Bradford Drive in CRP. Over the first decade, S3 experienced growth averaging 20 percent per year. Activities were primarily in providing technical, programmatic, data management, and software development services.

In its second decade, System Studies and Simulation (S3) broadened its activities, establishing offices at several sites in the U.S. and overseas. Janice H. Smith, founder and CEO, was named Entrepreneur of the Year by the Huntsville/Madison Country Chamber of Commerce and UAH Business Fellow of the Year – both for 2008; the Chamber of Commerce also awarded S3 1st place as the 2008 Small Business of the Year in Technology. C. Thomas Houser joined S3 as President in 2010, leading further diversification into aviation services. As the firm continued to grow, headquarters and offices were relocated to 615 Discovery Drive in CRP West. Primary activities involved services in program acquisition, research and development, information technology, training, and logistics. As of 2013, S3 in Huntsville had about 150 employees but hundreds more in support of 40 Government locations worldwide.

Griffon Aerospace (1993) – Griffon Aerospace, based in nearby Madison, Alabama, is the only locally owned firm that produces aircraft and related products. Established by Larry A. French in 1993, the company was formed to design, develop, and manufacture manned aircraft kits using composite structures. With a facility at 901 Nick Fitchard Road, its initial kit was for the Lionheart, a six-seat aircraft powered by a Pratt & Whitney R-985 450-hp radial engine; this was first exhibited at the 1996 Oshkosh Experimental Aircraft Show and cover-featured in the October 1996 Kit Planes Magazine.

In 2002, Griffon Aerospace expanded its composite-forming capability into government contract work. Early efforts included a subcontract from Northrop Grumman for composite cryogenic tanks, and a contract from Lockheed to build a full-scale Crew Capsule. They designed a prototype 130-lb unmanned aircraft, which eventually became the U.S. Army's MQM-170A Outlaw. Other contracts included the MQM-171A Broadsword (a 600-lb UAV), and the MQM-170C (a much improved version of the Outlaw); their latest product is the Seahunter, a twin-engine, tactical-class UAV. As of 2013, Griffon Aerospace had produced over 3,500 unmanned aircraft being used by multiple nations. They also provide flight operation services, having been responsible for some 160,000 test and evaluation sorties at locations nationwide. Their main offices are at 101, 103, and 106 Commerce Circle; founder Larry A. French is the CEO / CTO.

STI Electronics (1993) – Located in Madison, Alabama, STI Electronics, Inc., designs, manufactures, and tests prototype circuit boards and assemblies for government agencies and the electronics manufacturing industry. The company also trains clients on designing and building these products, and is a distributor of fabrication supplies and equipment. It was founded as Soldering Technology International (STI), Inc., at San Dimas, California in 1982, by Jim D. Raby, a long-time ABMA and MSFC electronics engineer. The firm was relocated to 216 Palmer Road, Madison, in 1993, and several years later was renamed STI Electronics, Inc. It is owned by Mary Ellen Raby, with David E. Raby serving as President and CEO, and Jim Raby the Technical Director.

Internationally known for its training programs, STI Electronics incorporated state-of-the-art laboratories and manufacturing facilities for supporting this training; it then branched into using these facilities for engineering design and development, with a strong capability for producing prototype and small-to-medium-volume assemblies for classified hardware. The manufacturing laboratory encompasses 26,000 ft^2 of floor space, including a Class-1000 cleanroom. STI Electronics was selected by the Huntsville/Madison County Chamber of Commerce as the Small Business of the Year 2000 in the Business Services Category.

Schafer Corporation - Huntsville (1995) – Schafer Corporation is a highly respected scientific, engineering, and operational support company headquartered in Arlington, Virginia; it provides technical and management consulting and other services to the government and industry in the areas of defense, space, and energy. In 1995, Schafer Corporation opened a Huntsville Branch at 1500 Perimeter Parkway, then later relocated to 5030 Bradford Drive, CRP; Charles B. Chitwood, earlier the Director of MSFC, was the General Manager.

The operation in Huntsville became a part of Schafer's Aerospace Sector. Another Huntsville firm, 3D Research, was acquired by Schafer in 2010, and the operations combined at 360 Quality Circle. In 2013, Michael D. Griffin, previously Administrator of NASA and Eminent Professor at UAHuntsville, became Chairman / CEO of Schafer Corporation, which is now owned by

Citigroup Alternative Investments. William J. Marx was General Manager of the Huntsville operations with about 80 employees.

Aviagen (1995) – Aviagen is the world's leading poultry breeding company, developing pedigree lines for the production of broiler chickens under several brand names. The company is headquartered in Huntsville, with a number of wholly-owned operations and joint-ventures around the world, and a distribution network for customers in over 130 countries. Aviagen employs approximately 3,800 people globally, including 400 in North Alabama and 90 in Huntsville.

In 1995, Aviagen opened its North American office in Huntsville at 5015 Bradford Drive, CRP, and in 2000, this became the global headquarters for the company. In 2005, the parent Aviagen Group was purchased by the Erich Wesjohann (EW) Group, a family-run holding company based in Germany. Aviagen moved its global headquarters to a 27,000-ft^2 office building at 920 Explorer Boulevard in 2015. Jan K. Henriksen is CEO, Kevin McDaniel, President, and Jason M. Mack, VP of Operations.

Aviagen is involved in significant research activities. One of its many projects is in improving feed efficiency – the food conversion ratio – using continuous electronic measurements of feed and chick growth. A genomics project is focused on identifying naturally occurring markers within the genome of elite birds. In 2011, Aviagen opened a 15,000-ft^2, state-of-the-art veterinary diagnostic and research laboratory in nearby Elkmont, Alabama.

Gleason Research Associates (1995) – Thomas J. Gleason formed Gleason Research Associates (GRA) in Columbia, Maryland, in 1982. From the start, its mission was to provide specialized products and services to support the warfighter. In its first years, work centered on optical / electro-optical / IR / RF survivability enhancement, often in association with Army programs at Redstone Arsenal.

In 1995, Robert E. Yates, previously a leading engineer with MICOM, joined the Company and opened a GRA Huntsville Operations; the initial facility was at 2227 Drake Avenue. Yates became VP of GRA Huntsville Operations in 2003; two years later, Charles M. Vessels, an internationally known expert in explosive ordnance and also previously with MICOM, was hired. Over the next decade, GRA Huntsville Operations gave major assistance in the evolution of several missile systems, a major one being the AGM-114 HELLFIRE.

GRA was restructured in 2006: Huntsville was made GRA's headquarters, Bob Yates was appointed CEO, Chuck Vessels was named President, and founder Tom Gleason returned to being a researcher and serving as VP / Director of Maryland Operations. The Company's Huntsville facilities were relocated to 5030 Bradford Drive. As of 2013, GRA had about 130 employees in Huntsville; Yates died that year and was replaced by Vessels as the CEO.

Westar / QinetiQ - Huntsville (1996) – Westar Corporation, headquartred in Albuquerque, New Mexico, provided test and evaluation support to the Army Aviation Systems Command (AVSCOM) in Saint Louis. Missouri. In 1995, AVSCOM was merged with the Army Missile Command to form the Aviation and Missile Command (AMCOM). In 1996, Westar followed AVSCOM, and began an operation in Huntsville. After acquiring several other companies, Westar expanded to provide engineering and logistics services to a wide variety of DoD, NASA, and commercial customers. Their facilities were at 4950 Corporate Drive CRP; William B. Walheim was the VP/General Manager of the Huntsville operations.

In October 2004, Westar was acquired by the giant British defense firm QinetiQ, forming QinetiQ North America, Inc. (QNA). The former operation of Westar in Huntsville became the QNA Defense Solutions, headquartered at 890 Explorer Boulevard CRP West. As of 2013, QNA Defense had some 900 employees, with about 500 local; David A. Shrum was the EVP / General Manager. Their primary activities are in systems engineering, software integration, logistics management, training systems, information technology, and test and evaluation support.

SEI Group (1996) – Formed by Eloy J. Torrez in 1996, SEI Group, Inc. is an engineering, construction, and operation and maintenance (O & M) company. A privately held firm, SEI headquarters and primary offices were initially at 303 Williams Avenue in downtown Huntsville. From the start, the firm had a diversified range of technical and management expertise, including registered professional engineers, registered architects, security specialists, licensed contractors, computer-aided design technicians, and business and support staff. Torrez served as the President and CEO; a highly experienced engineer, he also led in many technical projects.

In 2002, *Hispanic Business Magazine* named SEI as the second fastest growing Hispanic business in the United States. Within a few years, SEI customers included the U.S. Army Corps of Engineers, the Naval Facilities Engineering Service Center, the Federal Emergency Management Agency, the U.S. Air Force, the Defense Threat Reduction Agency, and the Social Security Administration. In recent years, contracts have emphasized technical support and evaluation of energy efficiency designs, renewable power generation, and resource management technologies. The headquarters of SEI Group relocated to 689 Discovery Drive, CRP, in 2013.

Davidson Technologies, Inc. (1996) – Julian Davidson, a recognized early leader in ballistic missile defense, formed Davidson Technologies, Inc. (DTI) in 1996. DTI provides a broad range of capabilities in defense and space activities: systems engineering / analysis, modeling and simulation, software development, and program support / integration. Facilities are at 530 Discovery Drive, CRP. By the end of the century, DTI had about 180 employees.

VT Group - Huntsville (1997) – The British-based defense and service company, VT Group plc, had US operations, VT Aepco, that supported Army aviation. This was transferred to Huntsville in 1997; facilities for VT Group were at 9238 Madison Blvd, Madison, with Michael F. McClellan as Sr.VP Army Programs. Aepco also had a facility at 7500 Memorial Parkway NW. By 2010, there were about 215 employees, mainly working in Army facilities. The British defense company, Babcock International, then acquired VT Group plc, but the Huntsville operations were relatively unchanged. Then in 2012, Babcock sold the U.S. operations to the private investment fund, Resolute Fund II LP. As of 2013, the Huntsville operations continued as VT Group with about 135 employees.

CINRAM - Huntsville (1997) – CINRAM Group, Inc., a Canadian firm headquartered in Toronto, Ontario, was one of the world's largest producers of media-delivered services. Established in Montreal in 1969, the company started by mass-producing eight-track tapes and cassettes, and opened its first compact disc (CD) plant in 1987. The digital video disc (DVD, sometimes called digital versatile disc), came in 1995, and the next generation – the blu-ray disc (BD) – began in 2006. In 1987, CINRAM opened a Huntsville plant at 4905 Moores Mill Road, Chase / Lowe Industrial Park; Larry T. Tabb was VP of Huntsville Operations. The CINRAM facility was originally built for Disc Manufacturing, Inc. of Anaheim, California; this operated in Huntsville for a short time.

Within CINRAM's Huntsville 320,000-ft^2 facility, many highly automated injection molding machines use master mold-plates to stamp the digital audio or video tracks onto 120-mm (4.7-in) diameter plastic discs; other automated machines then inspect, label, and package the discs in cardboard sleeves ready for distribution. The original digital audio and video used in making the mold plates is highly competition sensitive; thus, the plant is operated under extremely tight security. Product demand is highly variable; thus, the number of employees varies considerably, sometimes up to 2,500 and more. In the past, increases in low-skill workers were often filled by persons from Jamaica and other countries with temporary U.S. work permits.

MEMS Optical / JENOPTIK Optical – Huntsville (1997) – An internally funded research activity in micro optics was started by Teledyne Brown Engineering (TBE) of Huntsville, Alabama, in the mid-1980s; this included a fabrication cleanroom foundry. In 1997, this was spun out as a commercial endeavor: MEMS (micro-electro-mechanical systems) Optical, LLC. In exchange for partial ownership, MEMS Optical obtained from TBE the intellectual property related to microoptics, optical foundry equipment, and existing microoptics customers. Rodney L. Clark led this new activity.

Over the next four years, MEMS Optical invested in building their own facility at 205 Import Circle in northwest Huntsville. Three fully equipped cleanrooms were opened; these were primarily designed for micro-machine fabrication, in particular photolithography, wafer bonding and glass and silicon etch technologies. In March 2001, the firm was incorporated as MEMS Optical, Inc., with Rodney Clark as the President and CTO. In 2005, ISO 9001:2000 registration for the design, development, and manufacture of optical and opto-mechanical components was obtained.

In 2006, MEMS Optical was acquired by a German firm, JENOPTIC AS of Jena, Germany, in 2006, but continued to operate as MEMS Optical in Huntsville. Then in early 2010, MEMS Optical and several other U.S.-based companies were merged to form JENOPTIK Optical Systems, Inc. Headquartered in Jupiter, Florida, but also with full facilities at 205 Import Circle in northwest Huntsville, this firm is a worldwide leader in the design, fabrication, and testing of optical and microoptical systems. Gregg T. Borek is the General Manager of Huntsville operations, and Tim R. Lindsey is the Microoptics Manager.

Avion Solutions (1997) – Headquartered in Huntsville and family-owned, Avion Solutions, Inc. provides a broad range of specialized engineering, software development, logistics, and technical services to the U.S. Army and U.S. Navy aviation communities. Its technical professionals include engineers, ex-military aviation experts, logisticians, analysts, and computer systems professionals, who provide expertise to support rotorcraft weapons systems, unmanned aerial systems (UAS), and associated subsystems. Formed at St. Louis, Missouri, in 1993, the company transferred to Huntsville during the Aviation Command Base Realignment and Closure (BRAC) initiative in 1997. First located at 7067 Old Madison Pike, Avion facilities are now at 4905 Research Drive, CRP. As of 2013, Avion had about 100 Huntsville employees; Chad B. Donald was the President, and Evan C. Wagner the VP and COO; Employees are also stationed at several Army and Navy aviation bases.

DRS Technologies - Huntsville (1997) – DRS Test & Energy Management (TEM), LLC, is the Huntsville operation of DRS (Diagnostic / Retrieval Systems) Technologies, Inc. DRS TEM is a leading designer, integrator, and producer of diagnostic, health management, rugged electronics, and energy management systems for combat vehicles and weapon systems of the U.S. ground forces. As of 2013, DRS TEM had about 220 employees at its main facility at 110 Wynn Drive, CRP; Timothy A. Smith was the VP / General Manager, Barton L. Rossnager the Engineering Manager, and William T. Sartor the Technology Manager.

Although officially formed in 1997, DRS TEM draws on experience as a direct descendant of Chrysler Corporation's Huntsville activities that opened in 1952, and, most recently, Chrysler Pentastar Electronics. The parent DRS Technologies, headquartered in Arlington, Virginia, is a leading supplier of integrated products, services, and support to military forces, intelligence agencies, and prime contractors worldwide.

General Dynamics C4 Systems - Huntsville (1997) – General Dynamics (GD) was formed in 1952, primarily through the combination of Electric Boat Company and Consolidated Vultee (Aircraft). GD grew organically and through acquisitions until the early 1990s, when they sold nearly all of their divisions except those in boat building and combat vehicles. Starting in the mid-1990s, they again began expanding by acquiring businesses in combat vehicles, information technology, and services. Headquartered in Falls Church, Virginia, they acquired many new businesses and are now the fifth largest (in sales) defense contractor in the world.

Through the years, General Dynamics has had sales and technical representatives in Huntsville, but not an operating unit. In 1997, General Dynamics C4 Systems, a division of their Information Systems and Technology segment, opened an engineering and production facility at 6000 Technology Drive in CRP. With Eric M. Nelson as the Site Manager and about 120 employees, it is described as GD's hub of designing, integrating, testing, fielding, training, and sustaining military command and control systems.

Miltec / Ducommun Miltec - Huntsville (1997) – Donald W. Miller, the former Chief Engineer at Lockheed Missiles and Space, founded Miltec Corporation in March 1997. Miltec's main facilities were at 678 Discovery Drive; they grew to some 120 employees by 2000. Activities included aerodynamics, propulsion, structures, guidance-navigation-and-control, seekers and sensors, lethality/warheads, and test and launch operations support. In addition to the Huntsville operations, Miltec opened facilities in nearby Guntersville, Alabama, and in Iuka, Mississippi.

In 2006, the firm was acquired by Ducommun Corporation, a California firm dating from 1849, and a leading provider in diverse electronic and aircraft structures markets. The Huntsville operations became Ducommun Miltec. As of 2013, Ducommun Miltec executives included J. Michael Stanfield, President; Scott K. McWhirter, General Manager; W. Bruce Beaufait, VP Advanced Development; William A. Koons, VP Programs; and Michael P. Dreessen, VP Engineering. There were about 130 employees. Much attention has recently been given to Miltec's development of nanosatellites for tactical applications by the U.S. Army.

Science and Engineering Services - Huntsville (1997) – Hyo Sang Lee formed Science and Engineering Services, Inc. (SES) at Columbia, Maryland, in 1990. SES Huntsville was started in a 63,200-ft^2 facility at 4015 Pulaski Pike NW in 1997 (this facility was later called SES-East). SES Huntsville activities include the design, development, integration, testing, and manufacture of various aviation sub-systems. Russell Chunn was the EVP, and, by the end of the century, there were near 400 employees, many working in Government facilities.

Garver Engineers - Huntsville (1998) – Dating from 1923, Garver Engineers, headquartered in Little Rock, Arkansas, is a national leader in civil engineering activities. In 1997, Garver opened a regional office at 218 Holmes Avenue in Huntsville, with Jerry D. McCarley as VP and District Manager. Garver provides engineering, land surveying, and planning for government, industry, and private clients. The company's design engineering includes water, sewer, and natural gas systems; road and highway systems; and water-front facilities. They are presently located in facilities at 5125 Research Drive, CRP.

Gray Research (1998-2008) – Gray Research, Inc. was a provider of engineering and business solutions to DoD and NASA customers. Founded by Ronald W. Gray in 1998, its initial facility was at 675 Discovery Drive CRP; Gray served as President and Michael D. Holmes was VP Operations. As it grew, its core capabilities included program management, information technology, space system analysis, test and evaluation, logistics, and facility engineering. In addition to Huntsville, major customers are in Colorado Springs, Colorado, and Cleveland, Ohio.

In 2008, the firm was acquired by MacAulay-Brown, Inc. (MacB), a large space and defense contractor with corporate headquarters in Dayton, Ohio, and

National Capital headquarters in Vienna, Virginia. For several years, Gray Research continued to operate as a subsidiary of MacB and had about 150 employees. Their main facility was at 655 Discovery Drive.

Digium (1999) – Digium, Inc. is a privately held communications technology company based in Huntsville. Founded as Linus Support Services by Mark A. Spenser in 1999, early experimentation with telephone call processing using a PC running Linux led to developing open-source software called Asterisk to act as a telephone system. As interest in Asterisk grew rapidly, the firm focused on supporting Asterisk and addressing digital hardware for the telecom market. In 2001, the company changed its name to Digium and began manufacturing its first telephony hardware interface card, the Wildcard. Most of the initial key personnel previously held positions with ADTRAN in Huntsville.

Over the next decade, Digium has become the dominant open-source, lower cost, alternative to proprietary communication providers. Switchvox is Digium's family of VoIP (Voice Over Internet Protocol) phone systems designed for small- and medium-sized businesses. In October 2007, Digium opened a new 60,000-ft^2 headquarters and engineering facility at 445 Jan Davis Drive in CRP West. As of 2013, key personnel included Danny J. Windham, CEO; Mark A. Spencer, CTO; and David R. Deaton, VP Engineering; there were about 120 employees.

Benchmark Electronics – Huntsville (1999) – Electronics, Inc., was formed in Clute, Texas in 1979, specializing in low-volume medical products. There was a management buyout in 1986, and the name changed to Benchmark Electronics, Inc. The firm had major growth, primarily through acquisitions. In 1994, the headquarters moved to Arlington, Texas. Benchmark purchased AVEX Electronics in 1999, including a large plant in Huntsville, Alabama. The parent Benchmark Electronics now has 19 plants in 9 different countries

Located at 4807 Bradford Drive in CRP, Benchmark's Huntsville Division has a facility of some 276,000 ft^2. Peter W. Cianfaglione was the Division General Manager. The operation has services to support the full life cycle for new and existing products: circuit design, board layout, prototype development, full production, electromechanical assembly, system integration, reliability testing, and end-of-life management

WestWind Technologies (1999) – Recognizing the business potential of U.S. Army and commercial aviation in Huntsville, local investors opened WestWind Technologies, Inc., in 1999. With headquarters at 2901 Wall Triana Highway adjacent to the Huntsville International Airport, WestWind has four facilities totaling 135,000 ft². This includes a 62,000-ft² hangar that can accommodate large fixed-wing aircraft (up to the C-130) and multiple rotary-wing aircraft simultaneously. The company's core competencies include avionics and mechanical systems engineering, electrical and sheet metal fabrication, electronics and mechanical assembly, and aircraft structural modification.

WestWind provides aircraft maintenance and upgrading in seamless operations from disassembly, integration, repair, reassembly, test flight, and return to service; helicopters have included the U-60 Blackhawk, the CH-47 Chinook, and the AH-64 Apache. Representative of WestWind's manufacturing is full-rate production of CH-47F Transportable Flight Proficiency Simulators. As of 2013, there were some 120 employees; William H. Jolly was the President / CEO, Roger T. Messick the COO, and Mark R. Cellarius the Engineering Manager.

Radiance Technologies (1999) – Founded by George M. Clark, Radiance Technologies, Inc. is employee owned by many of its engineers and scientists. Originally with facilities at 500 Wynn Drive, Radiance occupied a new headquarters building at 350 Wynn Drive in about 2007. Radiance has focused on government support, technology development and applications, and specialty manufacturing. Key products include electro-optical systems, radar power-distribution technologies, high-speed signal processing, and life-sensing systems.

As of 2013, Radiance had operations in nine states as well as South Korea. In Huntsville, a rapid prototyping facility opened at 7272 Governors Drive West. William C. Bailey, Jr., succeeded George Clark as President. Other leaders included John A. Dennis, EVP; Peter L. Weiland, Chief Engineer; and Lori A. Sisk, Manufacturing Operations Manager. Radiance had about 130 employees in Huntsville and some 500 employees worldwide.

Intuitive Research and Technology Corporation (1999) – Upon retiring from the U.S. Army, A.R. "Rey" Almodóvar and Harold R. Brewer formed Intuitive Research and Technology Corporation based on extensive experience and team collaboration. With Almodóvar as CEO and Brewer as COB / President, the firm provided analytical services, management, and system solutions and was an immediate success; facilities were at 5030 Bradford Drive. A native of Puerto Rico and a P.E., Almodóvar was soon a leader in the technical community and the Chamber of Commerce. By 2013, there were 250 employees in 22 locations locations.

Huntsville's Oldest Engineering Firm

As described in Chapter II, George Walter Jones formed the first firm in Huntsville offering surveying and engineering in 1885. Later named G.W. Jones & Sons, Inc., this was a family-owned and -operated enterprise; the firm had its 100th anniversary in 1985.

Although George Walter Jones did not hold a college degree, he had learned surveying and engineering from an uncle. When the Alabama Board of Registration of Professional Engineers and Professional Land Surveyors formed in 1935, founder Jones was "grandfathered" into official registration. From the start, the firm's offices were on Franklin Street in downtown Huntsville.

As the 1900s closed, a grandson, Raymond B. Jones was President, and there were about 50 employees. Recent work was mainly in the transportation, aviation, industrial, municipal, and utility sectors under Mark Yokley as President. In 2013, G.W. Jones and Sons was acquired by Nashville-based Littlejohn Engineering Associates.

Franklin Street Offices

HIGHER EDUCATION DIVERSITY

The influence of higher education on the technological evolution of certain areas in the United States is well recognized – for example the Boston / Cambridge concentration in Massachusetts and the San Francisco / Silicon Valley region of California. Although not as large or mature as these and several others, the Greater Huntsville community would rank well in comparison. Starting in the 1950s, the local government agencies and technical industries have evolved at almost an unbelievable rate, largely due to its high concentration of engineers, scientists, technologists, and other well-educated personnel in the area – a large portion of them receiving their undergraduate or graduate studies locally.

This section will provide brief descriptions of the higher-education institutions in the area, with an emphasis on their diversity.

Evolution in Science and Technology Education

During the last three decades of the 20th century, there was significant evolution in advanced science and technology education available in Greater Huntsville. This is primarily in the initiation of the University of Alabama in

Huntsville (1970), the Southeastern Institute of Technology (1976), and the improvements at Alabama A&M University.

University of Alabama in Huntsville – As described in the previous chapter, the University of Alabama, with its main campus in Tuscaloosa, had operated an Extension Center in Huntsville since 1950. By 1960, the operation had expanded to justify a campus, a resident faculty, and academic programs, but completion of degrees was under the control of the Tuscaloosa departments. Throughout the 1960s, undergraduate and graduate courses were offered in engineering, science, mathematics, business management, social sciences, and liberal arts. The first master's degrees based entirely on courses completed in Huntsville were offered in 1963, and, similarly, undergraduate degrees in a number of areas began in 1968.

In 1969, the Board of Trustees of the University of Alabama established the University of Alabama System, with three independent, autonomous operations at Tuscaloosa, Birmingham, and Huntsville. The University of Alabama in Huntsville (UAH, and later UAHuntsville) became fully operational with the appointment of Benjamin B. Graves as the first President in 1970. James L. Wilson was VP for Academic Affairs, and Joseph C. Dowdle was VP for Administration.

The initial academic organization included James M. Horner, Dean of the Faculty; the School of Humanities and Behavioral Sciences, headed by Jon G. Rogers; the School of Science and Engineering, under Nadeem F. Audeh; the School of Nursing and Allied Health Science, led by Kathryn M. Crossland; the School of Graduate Studies, under J. Edwin Rush, Jr.; and the Division of Continuous Education, directed by Raymond C. Watson, Jr. Jean M. Perreault headed the Library, and Nan. G. Hall was the Registrar. Somewhat later in the 1970s, the School of Primary Medical Care, under G. Gale Stephens, was added. In addition to the original Morton Hall, the 332-acre campus contained the Science and Engineering Building, the Humanities Building, the Graduate Studies Building, the Research Institute Building, the University Union Building, and the Library.

Southeastern Institute of Technology – As high-technology activities increased at the local governmental agencies and their supporting industries, they generated needs in advanced education that were not being met by the existing institutions. In the fall of 1975, John L. McDaniel, then Technical Director of the Army Missile Command, formed a group of representatives from Army organizations and defense firms to address the local educational situation and determine how needed changes might be implemented.

Their findings might be summarized as follows: (1) Although excellent for preparing entry-level personnel, the content of existing graduate programs was not suitable for a large segment of the professional community – there should be less fundamental theory and much more state-of-the-art, practical applications. (2) Courses should be available on both a credit and non-credit basis – attendees should be afforded the opportunity to use the courses in earning appropriate graduate-level professional degrees. (3) Strong use should be made of professional practitioners in developing and delivering the courses.

After several meetings and consideration of alternatives, it was eventually agreed that a completely new institute matched to these needs should be established. Southeastern Institute of Technology (SIT) was then formed in 1976.

For the next 28 years, this private, not-for-profit, non-traditional Institute served the Greater Huntsville technical community with professional-level courses and programs of the highest quality. Persons receiving this education filled senior positions in the local community and elsewhere, making significant contributions to Huntsville's technical evolution. Therefore, the section describing SIT has been given in detail.

Southeastern Institute of Technology (SIT), a private, not-for-profit, professional school in Huntsville, Alabama, was legally formed in early 1976. Its purpose was to provide professional-level continuing education and graduate degree programs in engineering, management, and applied science for career advancement, updating, and redirection.

The Institute was under the control of a Board of Directors, which, among other community leaders, included the Chairman of the Madison County Commission and the President of Athens College. This book's author, Raymond C. Watson, Jr. – with 30 years industrial and academic experience, and previously with UAH on the graduate engineering faculty and Director of their Continuous Education Division – was appointed to lead the development of SIT and then to serve as President.

During the first part of 1976, legal and financial arrangements were made, courses and programs were developed, a faculty of academically qualified professional practitioners was recruited, and initial approvals were obtained from the Alabama Department of Education (they would give final approval – equivalent to accreditation – after at least one year of operation).

On 31 August 1976, Southeastern Institute of Technology officially began its first classes. In due course, it was Approved and Licensed by the Alabama Department of Education under Title 16-46-1 through 10, Code of Alabama, authorizing the Institute to conduct educational programs and to award degrees.

SIT received 501(c)(3) status from the Internal Revenue Service, allowing contributions to the Institute to be tax deductible. It was also approved to receive tuition assistance from the Veterans Administration, U.S. Army and NASA Training Departments, and most local companies. Essentially from the start, the Institute was financially self-sufficient.

SIT facilities were in an office building at 200 Sparkman Drive in the Cummings Research Park. In addition to administrative offices, there were classrooms, a personal-computer laboratory, and a library with eventually about 10,000 volumes and over 100 journal and professional magazine series. SIT held a Secret-level facility clearance, allowing classified courses to be offered for cleared attendees.

SIT Facility

As a professional school, SIT differed from standard academic institutions in a number of important ways:

- SIT granted professional degrees, as contrasted with academic degrees. • Admission to degree programs required prior professional employment. • The SIT faculty was composed exclusively of professional practitioners – most holding doctorate degrees – teaching and advising on a part-time basis. • Application, rather than theory, was emphasized in all courses and programs.

While fully approved by the Alabama Department of Education, SIT's professional degree programs did not have nationally recognized accreditation. At that time, the Southern Association of Colleges and Schools (SACS) – the regional body for institutional accreditation – accredited only traditional academic schools. As an independent professional school with many non-traditional characteristics, SIT was not eligible for SACS accreditation. Program specializations included the following, listed by popularity order:

- Radar & Signal Processing • Optics & Electro-Optics • Advanced Computers & Software • Missile & Space Systems • Science & Engineering Management • Contracts & Procurement Management
- Cost Estimating & Analysis.

All of the specializations included courses in computer applications and higher mathematics / operations research. A series on radar systems, led by Stephen M. Gilbert, had the most-attended courses.

Typically, about half the students were pursuing degrees; requirements for degrees were comparable to those of the best of traditional graduate schools. Authorized degrees and their abbreviations were as follows:

- Master of Science (M.Sc.) • Master of Science in Engineering
(M.Sc.Engr.) • Master of Science in Management (M.Sc.Mgt.)
- Doctor of Science (D.Sc.) • Doctor of Engineering (D.Engr.)
- Doctor of Management (D.Mgt.).

In some periods, certain other degrees were available, including the Master of Business Administration (M.B.A.) and Doctor of Business Administration (D.B.A.). Although SIT did not offer undergraduate degree programs, select non-degreed persons with considerable prior education and senior-level experience might be admitted to a combined Bachelor's-Master's Program.

To provide for experimental research in certain specialized degree programs, arrangements were made with many, well-equipped laboratories of local government agencies and industries. Also, SIT had an affiliated Center for Applied Research (CAR) that conducted both unclassified and classified studies under non-profit contracts, subcontracts, and purchase orders; Charles R. Wyman was the Principal Researcher in many CAR activities.

Typical SIT Class Settings

As the operation matured, the technical programs were coordinated by William J. Barksdale, while Richard H. Shuford coordinated the management programs; Mary W. Downey was the Registrar / Office Administrator, and R. Coke Watson, III the Librarian.

During the 1980s, SIT had an instructional activity in South Africa, mainly giving graduate courses in radar, optical systems, and project management. Also, courses in operations research and engineering management were given in Egypt under a U.S. Government subcontract. In the 1990s, SIT was invited to affiliate with the New York Institute of Technology in forming a world-wide higher education institute, but the economics did not work out.

Over the years, SIT was well received, serving several thousand students and awarding hundreds of degrees – significantly contributing to the technological evolution of Greater Huntsville. In this same period, however, offerings of a similar content nature slowly became available from the local state schools.

In 2004, the SIT Board, acting on the recommendation of the President, decided that the original purpose of the Institute was no longer valid and duplication of effort was not in the community's best interest. SIT was placed in an inactive status, not accepting new students or offering classroom courses. The Institute, however, continued as an entity, allowing students who were near the end of their studies to complete requirements – mainly independent papers, theses, and dissertations.

Alabama A&M University – In the 1970s and 1980s, offerings in the physical sciences and mathematics were strengthened at Alabama A&M University (AAMU); degree programs in areas of engineering technology were developed, and its master's degree programs in computer science and business administration were the first in Huntsville. Offerings in physics were elevated to the master's degree level, and then to the doctorate level with specializations in optics and materials.

During the 1990s, Dean Arthur J. Bond – who held a doctorate in engineering from Purdue University and had many years of industrial experience – led the formation of the School of Engineering and Technology, with bachelor degree programs in Civil, Mechanical, and Electrical Engineering, as well as bachelor degree programs in Civil, Mechanical, and Electrical Engineering Technology. The engineering building at Alabama A&M University was named in honor of Dean Bond.

In the early 2000s, under the leadership of School Dean Vernell T. Montgomery, the B.S. degree programs in Civil Engineering, Electrical Engineering, and Mechanical Engineering became accredited by the Engineering Accreditation Commission of ABET; the B.S. degree programs in Electrical Engineering Technology and Mechanical Engineering Technology were accredited by the Engineering Technology Accreditation Commission of ABET; and the B.S. degree program in Computer Science was accredited by the Computing Accreditation Commission of ABET. The author of this book was privileged to have served for many years as Advisory Board Chairman for the School as well as for the EE Department.

Analytical and experimental research had been conducted for a number of years by faculty members and students in a Research Institute. Unfortunately, the administration allowed this Research Institute to be privatized and taken out of the AAMU control, consequently leaving the University without a facility security clearance – a vital status for performing most defense-sponsored research.

As of 2013, the College of Engineering, Technology, and Physical Sciences, with Chance M. Glenn, Sr., as the Dean, offered a variety of undergraduate programs. Graduate degrees are offered in computer science (M.S.), construction management (M.S.), materiel engineering (M.Eng.), industrial technology (M.S.), and physics (M.S. and Ph.D.). Vann R. Newkirk is Dean of the Graduate School, and Andrew Hugine, Jr. is the 11th President of Alabama A&M University,

Other Educational Institutes

As the technology community of Greater Huntsville diversified and enlarged, other educational institutes did also and will be noted here.

Calhoun Community College - Calhoun Community College is a two-year institution of higher learning; its main campus is located in Decatur, 20 miles west of Huntsville. Founded 1949, it is the largest of the 27 two-year schools comprising the Alabama Community College System. Calhoun is an open-admission, coeducational, comprehensive community college.

As of 2013, Calhoun serves just over 11,300 students and offers 49 associate degree programs and 52 career/certificate programs. Of Calhoun's 302 full-time employees, 133 serve on the college faculty. Over 80 percent of the faculty possess at least a master's degree, and approximately 15 percent hold a doctorate. The college president is Marilyn C. Beck.

The Associate of Science (A.S.) degree is the award conferred on students who wish to transfer to an Alabama senior institution and pursue a Bachelor's Degree. The Associate of Applied Science (A.A.S.) degree is the award conferred on students who wish to complete two years of education at the community college level and then enter the workforce.

Calhoun has a Huntsville campus located at the corner of Wynn Drive and Old Madison Road in CRP. A full curriculum of mathematics, science, and basic engineering is available in day and evening classes; credits on most of these courses may be transferred to bachelor's degree programs at UA Huntsville or Alabama A&M. Athens University functions as an upper-division institution, providing a direct continuation of Calhoun's academic programs.

Athens State College / University - What is now Athens State University is the oldest institution of higher education in Alabama, tracing its origins to Athens Female Academy that openen in 1822. Starting in 1931, Athens College operated under the control of the Methodist Church. In 1974 the State of Alabama accepted the control and the school became known as Athens State College, initially operating under the Alabama Department of Education.

Offering coursework at the junior and senior level, it became the only baccalaureate degree-granting institution of the Alabama College System.

In educational activities that might directly support the technology community, Athens State had long offered majors in mathematics, chemistry, biology, accounting, and general management. As the technology community enlarged, majors in computer science, management of technology, acquisition and contract management, logistics and supply chain management, and human resources management were developed. There were also minors in several other areas.

In 1998, the College became Athens State University. Recently, Athens State gained the status of an autonomous university status and has established its own Board of Trustees. As of 2013, there were about 3,500 students, pursuing 33 different majors, in three colleges: Arts & Science, Business, and Education. Robert K. Glenn was the President.

J. F. Drake State Technical College - The college was founded as the Huntsville State Vocational Technical College in 1961. From the start, the school was located at 3421 Meridian Street, near Alabama A&M. The school was renamed J. F. Drake State Technical Trade School in 1966, the name honoring Joseph Fanning Drake, a long-serving president of Alabama A&M. In August 1973, the school was given technical college status by the State Board of Education, training students for current and future employment in technical, industrial, and vocational careers. With the new status, the school was named J. F. Drake State Technical College and authorized by the Board to offer the Associate Degree in Applied Technology (A.A.T.) The institution earned membership in the Southern Association of Colleges and Schools in July 2012; the next year it was again renamed to J.F. Drake State Community and Technical College. As of 2013, there were some 1,500 students, about equally divided between full-time and part-time, and 16 certificate and two-year degree programs were offered. Helen T. McAlpine was the President.

Florida Institute of Technology - In 1958, Brevard Engineering College was founded by Jerome P. Keuper at Cape Canaveral, Florida, to provide continuing education opportunities for scientists, engineers, and technicians working for NASA at what is now Kennedy Space Center. The new school grew quickly, establishing a campus at Melbourne in 1961. The name was changed to Florida Institute of Technology (FIT) in 1966, and off-campus programs at locations of government agencies were started in 1972.

A program was initiated by FIT at Redstone Arsenal in 1976. Initially limited to government and contractor employees, the courses were all in the general area of business management and leading to the Master of Science (M.S.) degree. Classrooms were at Redstone Arsenal in Toftoy Hall on Patton Road. Later, admission was available to the public – but still on Redstone Arsenal – and courses in Systems Management and Information Systems were added. Recently, some classes have been given in Huntsville at 6767 Old Madison Pike.

As of 2013, available offerings from FIT in Huntsville / Redstone Arsenal include Master of Science degree programs in Computer Information Systems, Logistics, and a number of management disciplines. FIT also offers a Master of Business Administration degree program locally. Students may access other courses and programs via the Virtual Graduate Center through the Extended Campus distance learning program. Timothy J. White is the Director of the Huntsville / Redstone Arsenal Center.

Oakwood College / University - In 1896, the Seventh-day Adventist Church opened the Oakwood Industrial School on a 380-acre former slave plantation in northwest Huntsville; its mission was to provide faith-based manual training for the recently freed Black Americans of the South. By 1917, its curriculum had expanded to include academic courses, and it became Oakwood Junior College. Academic expansion continued, and it became Oakwood College in 1943. In 1958, it was accredited by the Southern Association of Colleges and Schools (SACS) to award both associate and baccalaureate degrees. From its beginnings, the Institute had used its land to cultivate food crops and raise cattle; in 1918, additional land was acquired, bringing the campus size to its current 1,186 acres.

Historically, Oakwood's degree programs that might support Greater Huntsville's technical community have included majors in mathematics, chemistry, biology, and accounting. In more recent years, these have expanded to include biochemistry, chemical engineering, computer science, computer networks, management information systems, organizational management, and communication media, as well as a full, nationally accredited Bachelor of Business Administration (B.B.A.) program.

To pursue engineering, there is a cooperative dual degree program in which the student spends approximately three years at Oakwood University and approximately two years at the University of Alabama in Huntsville (UAHuntsville). Following the successful completion of all requirements, the student will be awarded the Bachelor of Science degree in Applied Mathematics from Oakwood University as well as the Bachelor of Science degree in Engineering from UAHuntsville in one of their undergraduate areas.

In 2008, the Institute became Oakwood University, adding its first graduate degree: the Master of Arts in Pastoral Studies. As of 2013, Oakwood University had over 2,000 students; the President was Leslie N. Pollard, a 1978 graduate of Oakwood.

AGRICULTURAL DIVERSIFICATION

Industrialization in Greater Huntsville and Madison County – and the associated reduction in farming – started with the Cotton Mills of the 1930s and continued during the 1940s with the World War II chemical munitions arsenals. Going into the 1950s and the Rocket City Age, employment opportunities with the Government agencies and in the high-technology industries further reduced the role of agriculture in the Tennessee Valley.

While Alabama was known as *the* Cotton State in the century following the Civil War, the State's agriculture became more and more diversified during the 20th century. In the last decade of the century, poultry and eggs accounted for 65 percent of total agricultural sales, cattle and calves for 11 percent, nursery and greenhouse crops for 8 percent, and cotton for only 4 percent.

Although high in capital cost, mechanical cotton-harvesting equipment greatly decreased labor cost and increased land-use efficiency. Between 1948 and the late 1960s, mechanical harvesting of the cotton crop went from essentially zero to 96 percent of the crop. The machines reduced the man-hours required to produce a cotton crop from 125 hours per acre to 25.

As shown in the following table, Madison County was still a major producer of cotton, performing only second to neighboring Limestone County in acreage and bales produced. The statistics show the decline in Madison County farmland as well as other changes. The increase in average farm size results from the use of more efficient equipment.

	1929	1949	1979	1999
Number of Farms	7,178	5,004	1,213	973
Average Size, Acres	54	83	173	216
Total Farm Land	387,612	415,332	209,745	210,168
Percent of County	74	80	40	40
Cotton, Acres	129,780	109,400	45,200	44,400
Cotton, Bales	41,700	50,800	41,400	51,500
Cotton Yield (pounds/acre)	154	223	440	557
Corn, Acres	62,300	56,400	11,800	9,800
Soy Beans, Acres	-	5,400	88,100	27,500

Although the 1999 amount of land in cotton was only some 40 percent of the acreage in 1949, about the same number of bales were produced at the end of the century as were being produced at mid-century. A standard bale of cotton lint is considered to weigh 480 pounds; thus, with the acreage about halved, the yield – in pounds per acre – has increased by 150 percent. Therein is shown the influence of technology.

There are three non-environmental factors for increasing cotton yield: insect control, weed management, and seed genetics. The Tennessee Valley Research and Extension Center at Belle Mina in adjacent Limestone County and the Agriculture Department at Alabama A&M University have been involved in research in all of these factors.

In previous chapters, the detrimental effect of the boll weevil and the control by crop dusting using calcium arsenate was described. DDT, available after WWII, was quickly adopted, but this was forbidden in the mid-1970s and other techniques and insecticides were developed.

Weed control through the use of herbicides was described in the previous chapter. In 1970, the chemical glyphosate was discovered to be a highly effective, broad-spectrum systemic herbicide; sold my Monsanto under the trade name Roundup, it has been widely used to kill weeds, especially annual broadleaf weeds and grasses. Monsanto's last commercially relevant United States patent on glyphosate expired in 2000; consequently, more sources of glyphosate will be forthcoming, the cost lowered, and even more will be used.

Seed genetics has led to new plant varieties that overcome environmental and climate problems. One of these is cotton for fast emergence; it is marketed by Bayer CropScience under the trade name Stoneville. Another is seed for crops, such as soy, that are affected by glyphosate from nearby fields; these Roundup-resistant seed are marketed by Monsanto as Roundup Ready.

A very important development in seed genetics was insect-resistant cotton. Commonly called Bt cotton, it was introduced by Monsanto under the trade name Bollgard in 1996. This cotton produces an insecticidal protein from the naturally occurring soil bacterium, *Bacillus thuringiensis* – thus the name Bt. In Bt cotton, the insecticide is always present in the plant rather than applied in periodic spraying. Although the seed cost more, there are major savings through reduced spraying. Bt cotton was immediately adopted worldwide and is sometimes called the most important advancement in the history of agriculture.

Selected Bibliography / Additional Information for Chapter IV

"Alabama Supercomputer Authority," https://www.asc.edu/

"Apollo Applications Program," Wikipedia;
http://en.wikipedia.org/wiki/Apollo_Applications_Program

"Base Realignment and Closure," Wikipedia;
http://en.wikipedia.org/wiki/Base_Realignment_and_Closure

Caton, Jeffrey L., "Evolving Army Needs for Space Support," War College
Press; http://www.strategicstudiesinstitute.army.mil/pdffiles/PUB1262.pdf

"Chandra X-ray Observatory," Wikipedia;
http://en.wikipedia.org/wiki/Chandra_X-ray_Observatory

Cheek, F.A., "History of Ballistic Missile Defense," New Technology, Inc.,
report for BMDATC, 1983

"Compton Gamma Ray Observatory," Wikipedia;
http://en.wikipedia.org/wiki/Compton_Gamma_Ray_Observatory

"Gravity Probe B," Wikipedia; http://en.wikipedia.org/wiki/Gravity_Probe_B

Hardy, B. T. et al; "Bt Cotton"; U. S. Department of Agriculture, ARS-154, Jan.
2001

"HEAO Program," Wikipedia; http://en.wikipedia.org/wiki/HEAO_Program

Holbrook, Bill (Editor), "1999-2000 Industrial Directory, Nineteenth Edition"
Huntsville / Madison County Chamber of Commerce, 2000
"Hubble Space Telescope," Wikipedia;
http://en.wikipedia.org/wiki/Hubble_Space_Telescope

"Installation History," U.S. Army Aviation and Missile Life Cycle Management
Command; http://history.redstone.army.mil/ihist.html

Jones, Ray, "The Story of G. W. Jones and Sons: Our First century," Tennessee
Valley Press, 1986. Reprinted in *The Historic Huntsville Quarterly*, Vol. XVI,
Nos. 1 & 2 (Winter-Spring 1990);
http://historyconnections.info/hh/hhpics/pdf/hhq2/HHQ-Vol-XVI-12-
WinSpr90.pdf

"Karas, Thomas H. (Project Director), Office of Technology Assessment, U.S. Congress; *Ballistic Missile Defense Technologies*; U.S. Gov. Printing Office, 1985

"LAGEOS," Wikipedia; http://en.wikipedia.org/wiki/LAGEOS

"Origins of the International Space Station," Wikipedia; http://en.wikipedia.org/wiki/Origins_of_the_International_Space_Station

Reynolds, Gina, "A smashing success UAH has the biggest test gun in the country"; *Huntsville Times*, May 18, 2008 http://www.freerepublic.com/focus/news/2018260/posts

"Safeguard Program," Wikipedia; http://en.wikipedia.org/wiki/Safeguard_Program

"Southeastern Institute of Technology," Wikipedia; http://en.wikipedia.org/wiki/Southeastern_Institute_of_Technology

"U.S. Army Space and Missile Defense Command," Wikipedia; http://en.wikipedia.org/wiki/United_States_Army_Space_and_Missile_Defense_Command

U.S. Census of Agriculture, published every five years by the U.S. Department of Agriculture's National Agricultural Statistics Service; (1) http://agcensus.mannlib.cornell.edu/AgCensus/homepage.do;jsessionid=E763E75D11EAB5599E4159EF707035AC ; (2) http://www.agcensus.usda.gov/

Walker, James A., Frances Martin, and Sharon S. Watkins, *Strategic Defense: Four Decades of Progress*, Historical Office, U.S. Army Space and Strategic Defense Command, 1995

Walker, James A., Lewis Bernstein, and Sharon Lang; *Seize the High Ground: The Army in Space and Missile Defense*, U.S. Gov. Printing Office, 2003; ISBN 0-15-072308-6

Weitze, Karen J., et al, "Historical Assessment of Marshall Space Flight Center," EDAW, Inc., for NASA MSFC, Nov. 2003

Chapter V

PRESENT and BEYOND

During the first century (the 1800s) of Greater Huntsville, there had been only modest population and associated technology growth. Local technology evolution had accelerated at the start of World War II when government munition operations opened in the area; the city population was 13,050 and the county was 66,317 in 1940. A decade later – when the local rocket era began – the populations were 16,437 and 72,903, respectively. The roughly ten percent population increases in this decade came largely from the influx of engineers and related personnel who then instigated a revolution in technological development.

By 1970, the populations had increased to 139,282 in Huntsville and 186,540 in Madison County – a phenomenal growth rate averaging some 37 percent per year. It is estimated that at the start of the 21st century approximately half of Greater Huntsville's economy was derived from governmental and industrial high-technology activities. The official populations in 2000 were 158,216 persons in Huntsville and 276,700 in Madison County; by 2013, these populations had grown to 186,254 and 346,892 respectively. It is noted that the City of Madison – considered a part of Greater Huntsville – grew 56 percent in this 13-year period, reaching a population of 45,799 in 2013.

Chapter V of this book concerns the present era of Greater Huntsville – the 21st century. Previously covered were Chapter I, Origins and Maturing (the 19th century); Chapter II, the Manufacturing Era (first half of the 20th century); Chapter III, the Rocket City Era (the 1950s and 1960s); and Chapter IV, the Diversification Era (the 1970s 1980s, and 1990s). Chapter V will present an attempted summarization of the present technological status; inherent in this will be projections of the future technological evolution.

The present Greater Huntsville is a government-dominated metropolis – it has been such since the announcement in the 3 July 1941 *Huntsville Times* that the U.S. Army was taking over thousands of acres of farmland just southwest of the city. In short order, this land was used for constructing and operating chemical arsenals, and every available worker from miles around was hired to manufacture and store munitions for the ensuing World War II.

After the war, the Redstone Arsenal became the nation's center for Army rocket development, taking the largest share of technological employment – both government and private – in Greater Huntsville. With the opening of the Marshall Space Flight Center in 1960, a good portion of the local technical activities were space, rather than military, related, but it was still 'government work.' Only in the last three decades of the 20th century was there significant diversification of local technological activities, but government funding still dominated the local economy.

At the start of the 21st century, Redstone Arsenal – set on 37,910 acres in the heart of the Tennessee Valley in northern Alabama – contained 11.7 million ft^2 of building space; this includes state-of-the-art offices, laboratories, and classrooms, as well as outdoor test facilities and ranges. It remains the epicenter of some of the Nation's most advanced research and development in space and defense technologies. In addition to Redstone Arsenal, Greater Huntsville had over 10,000 acres in highly concentrated research and industrial parks containing high-technology firms.

NASA / SPACE ACTIVITIES

The National Aeronautics and Space Administration (NASA) is the United States government agency responsible for the civilian space program as well as for aeronautics and aerospace research. President Dwight D. Eisenhower established NASA in 1958, with a distinctly civilian (rather than military) orientation encouraging peaceful applications in space science.

In 2009, Charles F. Bolden, Jr., a retired U.S. Marine Corps Major General and former astronaut, became the 12th Administrator of NASA. A graduate of the U.S. Naval Academy and president of his class, Bolden initially trained as a test pilot, flew more than 100 sorties as a Naval Aviator during the Vietnam War, and has logged a total of more than 6,000 hours flying time. He became an astronaut in 1980, and logged over 680 hours in four Shuttle space flights. Appointed by President Barack Obama, Bolden was the first African American to head NASA on a permanent basis

MARSHALL SPACE FLIGHT CENTER

If the average person in the United States, or perhaps even the world, thinks of Huntsville, Alabama, it is usually as being the home of space systems – the Marshall Space Flight Center (MSFC), NASA's most technically diverse center. Although far more government employees and associated contractors are with local defense organizations, MSFC gets the media coverage and is generally regarded as being more "high technology" than the Army activities. MSFC Directors in the 2000s were as follows:

Arthur G. Stephenson	September 1998 - May 2003
David A. King	June 2003 - March 2009
Robert M. Lightfoot, Jr.	August 2009 - March 2012
Arthur E. Goldman (acting)	March 2012 - August 2012
Robin N. Henderson (acting)	August 2012 - September 2012
Patrick E. Scheuermann	September 2012 - present

The first decade of the 21st century was a tumultuous time for NASA and MSFC. On 1 February 2003, the Space Shuttle *Columbia* disaster occurred – the orbiter disintegrating during reentry, resulting in the death of its seven crew members. Flights of the other Shuttles were put on hold for 29 months. MSFC had responsibility for the Space Shuttle's propulsion systems, including the external tank. An extensive investigation indicated that the failure was caused by a piece of insulation that broke off the external tank during launch and damaged the thermal protection on the Orbiter's left wing. After detailed study at MSFC, no changes in the external tank and its insulation were made; reliance was placed on inspection of the orbiter's critical elements to be made prior to reentry on future flights. NASA retired the Space Shuttle in 2011, leaving America dependent on the Russian Soyuz spacecraft for manned missions.

In 2005, the Constellation Program became a NASA human spaceflight activity for returning to the Moon by 2020, with the planet Mars as the ultimate goal. Development of the program's booster rockets, Ares I and Ares V (roughly corresponding to the Saturn I and Saturn V of the 1960s), was a major responsibility of MSFC. In 2009, a review of NASA's human space flight plans by a prestigious committee led by Norman R. Augustine (retired Chairman and CEO of Lockheed Martin) judged the Constellation program to be so behind schedule, underfunded, and over budget that meeting any of its goals would be essentially impossible. Consequently, President Barack Obama removed the activity from the 2010 budget, therefore canceling the program.

Although the overall Constellation program was cancelled, work on the intended spacecraft continued under management by the Johnson Space Center (JSC). Called the *Orion* Multi-Purpose Crew Vehicle (MPCV), it is composed of two major elements: a service module and a crew module. The first unmanned, Earth-orbital test flight was successfully launched by a DELTA IV Heavy rocket (built by United Launch Alliance in nearby Decatur) on 5 December 2014. (MPCV missions to carry astronauts are not expected to take place until 2021 at the earliest.)

Government and contractor teams at MSFC provided critical support ahead of the MPCV's first flight. MSFC fabricated almost 1,000 pieces of *Orion* flight and ground-test hardware; conducted structural testing of the service and crew module elements; designed and built the stage adapter that connected the *Orion* to the DELTA IV rocket; and managed oversight of the launch abort system propulsion elements.

In early March 2011, NASA announced that MSFC would lead the efforts on the Space Launch System (SLS) – a new heavy-lift rocket that, like the SATURN V of the lunar exploration program in the late 1960s, could boost large, man-rated payloads beyond low-Earth orbit.

A Renewed MSFC

Because of the events during the prior decade and the new opportunity with the SLS Program, there was a significant reorganization within MSFC in the fall of 2011. The remainder of this section concerns capabilities, activities, and personnel of the renewed MSFC, although many of the activities had their start much earlier.

As of 2013, MSFC had about 6,000 people working in their facilities, divided approximately 2,400 civil service and 3,600 contractor employees, and an annual budget of around $2.5 billion. It occupies about 4.5 million ft^2 of space in Huntsville plus some 2.2 million ft^2 of manufacturing space at the Michoud Assembly Facility in New Orleans. MSFC presently has capabilities and projects supporting NASA's mission in three key areas: Space Vehicles (Lifting from Earth), International Space Station (Living and Working in Space), and Advanced Scientific Research (Understanding Our World and Beyond).

As previously noted, Patrick E. Scheuermann became MSFC's Director in 2012. Assisting Scheuermann in technical functions are L. Dale Thomas, Associate Director - Technical, and Andrew S. Keys, Center Chief Technologist. Other key executives in the Office of the Director include Terresa V. Vanhooser, Deputy Director; Robin N. Henderson, Associate Director; Michael L. Tinker, Deputy Chief Technologist; and James L. Reuter, Senior Executive for Technology & Integration.

Patrick Scheuermann

Since reorganizing in 2011, MSFC has operated with the following major functional organizations:

Engineering Directorate – Christopher E. Singer, Director
Science and Technology Office – Daniel M. Schumacher, Manager
Flight Programs & Partnerships Office – Joan A. Singer, Manager
Safety & Mission Assurance Directorate – Stephen F. Cash, Director
Space Launch Systems Program Office – Todd A. May, Manager
Michoud Assembly Facility – Roy W. Malone, Director

The existing technical activities of MSFC, as well as some of their future activities, are best presented by describing these functional organizations and their subunits. The names of organizational leaders are as given on the MSFC informal organizational chart as of 17 January 2014. Comprehensive descriptions may be found at the current Web Site: "Marshall Space Flight Center, Core Capabilities and Services".

Since this is a book on technological history, only the functional (technical) organizations are covered. It is well recognized, however, that the administrative and service organizations are vital to the overall operation. Within MSFC, these are Financial Office, Procurement Office, Diversity & Equal Opportunity Office, Information Office, Strategic Analysis & Communications Office, Chief Council Office, Human Capital Office, and Center Operations Office.

Engineering Directorate

Going into the 21st century, James W. Kennedy was the Director of MSFC's Engineering Directorate (he later became the eighth director of the Kennedy Space Center). Michael U. Rudolphi (2001-2007) followed Kennedy, and then Daniel L. Dumbacher (2007-2011) served as the Director.

Christopher E. Singer was named Director in May 2011; C. Preston Jones was the Deputy Director, and Paul K. McConnaughey and James E. Turner were Associate Directors. Chris Singer had joined NASA in 1983.

Lisa Watson-Morgan was Manager of the Office of the Chief Engineer, and Nelson C. Parker was the Deputy; this Office is responsible for assuring the excellence and technical success of activities in the Engineering Directorate as well as other operations, including the SLS.

Christopher Singer

As of 2013, the Engineering Directorate had about 1,400 civil service and 1,200 support contractor personnel. The primary functional units include the following:

Advanced Concepts Office

Headed by Mark N. Rogers (previously by Reginald A. Alexander), the Advanced Concepts Office supports the development and maturation of alternate and innovative concepts for spacecraft systems, subsystems, payloads, and transportation systems. The office performs complete and integrated systems analyses and independent assessments of potential concepts, as well as technology assessments and subsystem trades for current projects. C. Leslie Johnson, a popular science/science-fiction writer, is in this organization.

In April 2010, President Barack announced that the project of a human mission to an asteroid was a new goal for 2025. Obama added that such a deep space mission could serve as a springboard for a manned mission to Mars in the mid-2030s. The Advanced Concepts Office is examining the hardware and operational requirements for such missions, starting with the SLS booster and *Orion* spacecraft.

Materials and Processes Laboratory

The Materials and Processes Laboratory has projects ranging from basic materials research to fully integrated solutions for large, complex systems challenges. About 350 total civil service and contractor personnel are involved. Managed by Wendell R. Colberg with Surendra N. Singhal as Deputy, the laboratory has principal competencies in metals, composites, and ceramics, as well as additive manufacturing, materials and process modeling and simulation, space environmental effects, non-destructive evaluation, and fracture and failure analysis. The Materials and Processes Technical Information System (MAPTS), under Dennis E. Griffin, is a computer-based, Internet-available, information point for this and related data for MSFC applications.

The Laboratory has a major capability in the latest metal-joining technique: friction-stir welding. This process transforms metals from a solid state into a "plastic-like" state, and uses a rotating pin tool to soften, stir, and forge a bond between two metal plates to form a uniform welded joint free of defects. Jon Street is the resident specialist in welding technology.

Friction-Stir Welding Facility

In 2013, the Laboratory used additive manufacturing (also called 3-D printing) to produce an injector for a rocket engine. Designed by the Propulsion Systems Department, built by the Materials and Processes Laboratory, and tested in the Test Laboratory – all units of the Engineering Directorate – the total fabrication cost was about $5,000, while the same unit built by traditional processes cost over $10,000. A team led by Kenneth G. Cooper was responsible for the project. Chris Singer, director of the Engineering Directorate, noted, "This game-changing additive manufacturing process can eventually reduce the time and cost of producing complex parts by an order of magnitude."

Propulsion Systems Department

This department serves as the central activity for performing research, maturing technologies, and developing, testing, and sustaining propulsion systems for both Earth-to-orbit launch and in-space missions.

Led by Thomas J. Williams with MaryBeth B. Koelbl as Deputy, some 400 civil service and contractor personnel are engaged in the design, development, and integration of

Propulsion Research Laboratory

propulsion systems and components ranging from micro-thrusters to systems producing millions of pounds of force in the following categories:

- Liquid propulsion systems including integrated booster and upper-stage systems, spacecraft propulsion, reaction and attitude control systems, propulsion system health management, and cryofluid management for long-term propellant storage
- Advanced propulsion and power research and development including high-power electric propulsion, nuclear thermal propulsion, space nuclear power systems, nuclear surface power systems, and propellant-free systems such as solar sails
- Solid propulsion systems including solid boost propulsion, upper and in-space propulsion, separation and maneuvering systems, and launch abort motors
- Propulsion component design and development including advanced rotating machinery, injectors, nozzles, preburners and ignition systems, valves, actuators, and thrust vector control systems.

Much of the work is conducted in the 66,000-ft^2 Propulsion Research Laboratory, a cutting-edge facility divided into 11 key laboratories.

Test Laboratory

The MSFC Test Laboratory is spread over 400 acres, with capabilities for testing items from small components to full engine systems. With M. Ralph Carruth, Jr., as Supervisor and J. Matthew Hammond as Deputy, there are about 400 civil service and contractor personnel engaged in this Laboratory. They are involved in operating over 50 different facilities. Some of these facilities pre-date MSFC – carried over from ABMA – but others are state-of-the art for testing in support of the SLS development.

Test of Sub-Scale SLS Booster

The Propulsion Test Facilities cover solid, liquid, and hybrid tests for component, scale-model, subsystem, and system-level testing. Pressures and propellants include LOX, LH_2, CH_4, and RP-1. The Advanced Engine Test Facility has a two-position, tri-propellant test stand capable of evaluating full-up engine and vehicle stage systems, with a maximum thrust load of 7.5M lbf.

The Enviromental Test Facility has 27 high-vacuum chambers with the capability of simulating launch and high-altitude conditions. There are 14 thermal vacuum chambers and 8 thermal humidity chambers. The Structural Test Facility includes an 8-ft-diameter shell-buckling machine, a 3-Mlbf tensile test machine, and a 20-ft universal test fixture.

At the Dynamic Test Stand, vibration tests simulate dynamic loads to validate flight and ground hardware. Acoustic testing is conducted in a reverberation chamber, a progressive wave tube, and an anechoic chamber. The Aerodynamics Test Facility has a 14- by -14-in. tri-sonic wind tunnel with Mach number range of 0.2-5.0.

Mission Operations Laboratory

Led by Lewis N. Wooten with Angela L. Marsh assisting, the Mission Operations Laboratory is responsible for the Payload Operations Center (POC), also known as Huntsville Operations Support Center (HOSC). The POC functions 24/7 as the primary science command post for the International Space Station (ISS), and directs communications between researchers around the world and their onboard experiments. It also manages satellites and other scientific payloads. Ground systems work includes telemetry, voice, video, information management, and pay-

Huntsville Operations Support Center

load planning. There are about 500 civil service and contractor employees in the Laboratory.

Space Systems Department

This department is responsible for designing, developing, assembling, integrating, testing, and delivering flight, ground, prototype, and development products for human space flight programs. Led by J. Larry Leopard with Robert H. Champion as Deputy, the Space Systems Department is staffed with over 600 civil servants and contractor personnel. They provide expertise in systems engineering and integration; mechanical design, analysis, and fabrication; avionics and electrical design, analysis, and fabrication; software development and testing; and systems integration and testing for space systems used throughout NASA.

Spacecraft and Vehicle Systems Department

With Helen V. McConnaughey as the Supervisor and James E. Turner the Deputy, this department is responsible for the technical design, analysis, evaluation, verification, and integration in the development of spacecraft and supporting vehicles. The department has around 700 civil service and contractor employees who function in four activities: spacecraft and vehicle systems engineering and integration; stages systems engineering and integration; flight mechanics; and structural design and analysis.

Flight Programs & Partnerships Office

This office has a combined workforce of more than 500 civil servants and contractors; Judy (Jody) A. Singer is the Manager and Paul A. Gilbert the Deputy. The office is responsible for MSFC work in areas of human exploration projects and tasks; flight mission programs and projects; and International Space Station hardware integration and operations. The office also is tasked with creating and maintaining partnerships with other government agencies and international and commercial partners.

Jody Singer

The functional units and their supervisors are the International Space Station Office, Annette M. Sledd; Robotic Mission Programs Office, Dennon J. Clardy; and Human Exploration Development Office, J. Chris Cianciola.

Safety & Mission Assurance Directorate

The Safety & Mission Assurance Directorate, with Stephen F. Cash as Director and Steven D. Pearson as Associate Director, leads the entire MSFC workforce in its commitment to assuring mission success and making the workforce and the work of the center and the agency as safe as possible.

As of 2013, the functional units of the directorate and their supervisors were Safety and Quality Department, Peter W. Allen; Mission Systems Assurance & Technical Support Department, Harold W. Dean; and Vehicle Systems Department, David J. Spacek.

Stephen Cash

Space Launch System Program Office

Since March 2011, when NASA announced that MSFC would lead the efforts on a new heavy-lift rocket, the Space Launch System (SLS) has been at the center of local activities. As of 2013, the leaders of this organization included Todd A. May, Manager; John H. Honeycutt, Deputy Manager; W. Keith Hefner, Associate Program Manager; and Sharon Cobb, Assistant Program Manager.

Garry M. Lyles is the SLS Chief Engineer but, for checks and balances, he is actually on the staff of the MSFC Engineering Directorate.

Todd May

The SLS is to be upgraded over time with more powerful versions. The initial Block 1 version uses four RS-25 engines and an external tank for the LH_2/LOX fuel. Blocks 1 and 1B will also use two five-segment Solid Rocket Boosters (SRBs) based on the four-segment Space Shuttle SRBs. Block 1B will add an Exploration Upper Stage. Block 2 will replace the SRBs with new advanced boosters. The proposed Block 2 will have similar lift capacity and height to the original Saturn V.

With its heavy-lift capability, the SLS will allow astronauts aboard the *Orion* spacecraft to explore deep-space destinations, including near-Earth asteroids, the Moon, and ultimately Mars. The team of MSFC employees, support contractor engineers, and lead contractor Boeing are on schedule to develop the SLS by 2017 – the deadline that had been set by Congress.

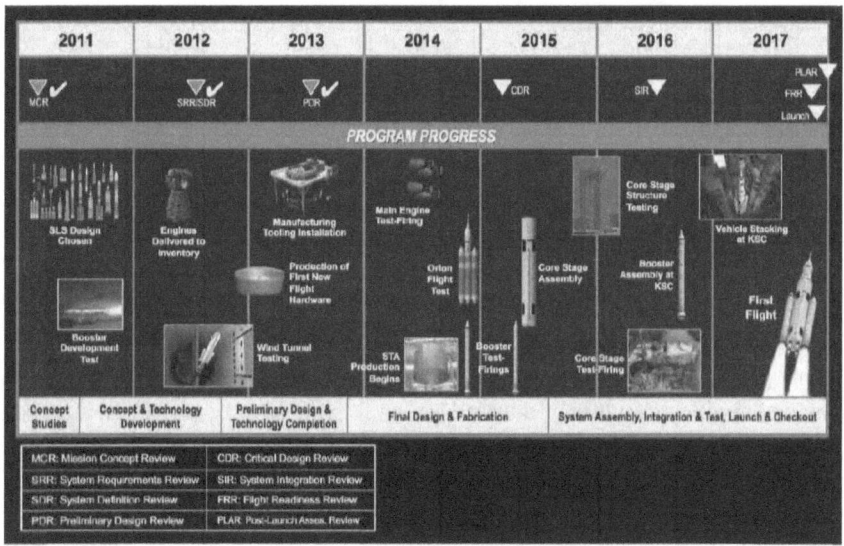

SLS Development Schedule

Science & Technology Office

From its beginning in 1960, MSFC has been involved in advanced space research; much of the early work can be directly attributed to Ernst Stuhlinger, then Director of the Space Sciences Laboratory and sometimes called the "brains" of the von Braun team. In recent years, the areas of research have expanded, ranging from near-Earth environments to deep-space astrophysics. Established in the 2011 reorganization, the Science & Technology Office was assembled from activities previously in many parts of MSFC.

Daniel Schumacher

With Daniel M. Schumacher as the Manager and Raymond G. Clinton, Jr., as Deputy Manager, the mission of this new office is to perform basic and applied research to achieve NASA's science and exploration objectives as well as giving oversight in the technology transfer among government and industry. Cynthia K. Ferguson is the Assistant Manager, and Melissa A. McGrath the Senior Science Advisor. Dan Schumacher was previously the director of MSFC's Science and Mission Systems Office, and "Corky" Clinton earlier served as deputy manager of this same organization.

The functional units of the Science & Technology Office are the Science Research Office, the Science & Space Technology Office, and the Technology Development and Transfer Office. More than 250 civil service and contractor personnel are involved.

Science Research Office

The Science Research Office (SRO) is the primary science organization at MSFC; James F. Spann is the SRO Manager and Michael R. Lapointe is Deputy Manager. (It is noted that Jim Spann has an audio presentation on Internet Archive entitled "Reconciling Faith and Science," relating his pilgrimage as a Christian believer in the scientific community.)

James Spann

There are over 100 NASA-employed scientists and technologists in the SRO with some 40 percent holding doctoral degrees; they are primarily located in the facility of the National Space Science and Technology Center (described elsewhere). At a given time, there are usually about 50 research projects underway, often pursued in cooperation with other organizations located in this facility.

The mission of the SRO is to provide continued support to the Nation and NASA through research and development efforts to expand scientific knowledge and in exploration of the Earth and its universe. The activities of the SRO are through three Research Offices: Earth Science, Astrophysics, and Heliophysics & Planetary Science.

Earth Science Office - Part of NASA's charter is to perform research to improve the understanding of the Earth – developing technologies to better observe the Earth system and transitioning them to operational agencies such as the National Oceanic and Atmospheric Administration (NOAA). Such activities are conducted by the Earth Science Office, managed by James L. Smoot. Specializations include observations of lightning and atmospheric convection, microwave remote sensing, studies of climate dynamics, and enabling use of NASA observations and modeling capabilities by the National Weather Service.

The Earth Science Office is a major participant in NASA's SERVIR (Spanish, "to serve") project. SERVIR brings satellite-driven technologies to address problems of disaster mitigation, ecosystems, biodiversity, weather, water resources, climate change, health, and agriculture in Central America, East Africa, and Hindu-Kush Himalaya. Daniel E. Irwin has been the Project Manager and Cynthia K. Ferguson the Project Director on the MSFC SERVIR Coordination Team.

The Earth Science Office's success in applications work is indicated by the diversity of the funding sources. Besides NASA, this includes NOAA, the DoD, the Centers for Disease Control and Prevention, National Institutes of Health, the World Bank, and the U.S. Agency for International Development. In these projects, the goal is to bring NASA capabilities, assets, and information to bear on significant problems that must be addressed by decision-makers at all levels of society.

Astrophysics Office - Managed by Nasser F. Barghounty, activities in the Astrophysics Office center on X-ray astronomy. With prior experience in all phases of the Chandra X-ray Observatory, the X-ray team members conduct technology development, build flight hardware, and participate in concept studies for numerous planned or potential high-energy astrophysics, heliophysics, and lunar or planetary missions – ranging from balloon and rocket experiments to probe- and facility-class X-ray observatories.

One very unusual project was the development of mirror modules for an X-ray instrument – the Astronomical Röntgen Telescope – that will be used in orbit by the Space Research Institute (IKI) of the Russian Academy of Sciences; Mikhail V. Gubarev and Brian D. Ransey, MSFC scientists in the Astrophysics Office were responsible for this project. The modules were built in the Precision Optical Fabrication Laboratory and tested in the Stray Light Facility, both at MSFC on Redstone Arsenal. (The work was funded under an agreement between NASA and the IKI; MSFC obtained the necessary export licenses.)

The Fermi Gamma-ray Space Telescope (FGST), initially called the Gamma-ray Large Area Space Telescope (GLAST), was launched in June 2008. The main telescope is complemented by the GLAST Burst Monitor (GBM); this can detect bursts of X-rays and gamma rays in the 8-keV to 3-MeV energy range. The GBM is managed by MSFC, and Charles A. Meegan is the Principal Investigator. Many new discoveries have been made in the initial period of operation; in May 2009, a burst was detected that is believed to negate some approaches to a new theory of gravity. The telescope and GBM have a life goal of 10 years.

As noted under the Compton Gamma Ray Observatory (CGRO) in the previous chapter, MSFC was responsible for the Burst and Transient Source Experiment (BATSE) using the CGRO. In nine years of operation, BATSE triggered about 8000 gamma-ray phenomena, of which some 2700 were strong bursts that were analyzed to have come from distant galaxies. Gerald J. Fishman, a senior scientist in the Astrophysics Office, was the Principal Investigator of a project to continue examination of data from BATSE and other gamma-ray projects. The 2011 Shaw Prize in Astronomy was shared by Fishman and Italian astrophysicist Enrico Costa. The Shaw Prize – widely regarded as the "Asia's Nobel Prize" – carries a monetary award of US$1million.

Heliophysics & Planetary Science Office – Under the leadership of Dennis L. Gallagher, researchers in the Heliophysics & Planetary Science Office (HPSO) are engaged in a number of interesting and beneficial analytical and experimental projects.

Heliophysics (from the French word "héliophysique") is all of the science common to the field of the Sun-Earth connections. Activities at the HPSO in this field are mainly in support of NASA Headquarter's research program to understand the origin and nature of solar activity and its effect on the space environment of the Earth. An objective is techniques to early predict the onset of solar activity and the extent of the disturbance to the Earth's ionosphere and subsequent electromagnetic communication. HPSO is a participant in the Solar Wind Electrons Alphas and Protons (SWEAP) investigation; in this, a probe will plunge directly into the atmosphere of the Sun and, for the first time, make measurements in this completely unexplored region of space. Jonathon W. Cirtain is the heliophysics team leader

Planetary science at the HPSO is at the leading edge of new knowledge of the Solar System's content, origin, evolution, and the potential for life elsewhere. Using flybys, orbiters, landers, and rovers, this involves one of the oldest of scientific pursuits: the observation and discovery of our solar system's planetary objects. Research areas of expertise include planetary interior modeling, and planetary atmosphere observations. A major objective is planetary sample analysis, returning samples from planetary bodies to NASA-supported laboratories. Barbara A. Cohen is the planetary team leader.

Specialists from HPSO serve as Project Scientists for NASA Headquarters' Lunar Mapping and Modeling Project (LMMP) and Robotic Lunar Lander Development Project (RLLD). LMMP is building a suite of tools and data products that support human lunar exploration activities including commercial missions. The RLLD designs small robotic landers for low-cost exploration and science missions to the Moon, Mercury, and near-Earth asteroids.

A prototype robotic lander – called Mighty Eagle – was developed at MSFC by a team from the HPSO, the Engineering Directorate, the Applied Physics Laboratory of Johns Hopkins University, and several local industries. D. Gregory Chavers was MSFC's Systems Engineer for the project. The prototype was completed in 2011, and, over the next two years, was tether-tested at TBE facilities then free-flight tested at MSFC. All testing was satisfactory, but funding has not been made available for building the full robotic lander.

Mighty Eagle Free Flight

Science & Space Technology Projects Office

Managed by Lawrence D. Hill and Randy M. Baggert as Assistant Manager, this Office is responsible for overseeing the planning, scheduling, resources, support requirements, management systems, and personnel for multiple programs and projects. Those include the Chandra X-ray Observatory, Hinode Project, James Webb Space Telescope, and the Astronomical Röntgen Telescope.

Technology Development & Transfer Office

The mission of this Office is to develop and mature a broad range of technologies that address the challenges of human and robotic space exploration. With Daniel J. Dorney as Manager, representative major activities include cryogenic propellant storage and transfer; an environmental control and life support system; next generation life support system; and a nuclear cryogenic propulsion stage. Terry L. Taylor is the Assistant Manager, responsible for MSFC's technology transfer program.

Michoud Assembly Facility

Although located in New Orleans, Louisiana, the Michoud Assembly Facility (MAF) is a functional unit of MSFC. With Roy W. Malone as Director and Michael H. Kynard as Deputy Director, MAF is operated by Jacobs Technology and houses a variety of occupants from both the commercial and government sectors. On an 832-acre site, the MAF is one of the world's largest indoor manufacturing facilities and has state-of-the-art equipment such as automated machining, automated fiber-placement machines, and friction-stir welding. It has direct access to a deep water port and rail, air, and an interstate highway transportation is nearby. The Space Launch System will be manufactured at Michoud.

Roy Malone

Other MSFC Activities

In addition to its official organizational structure, MSFC is involved in a number of important partnerships and collaborative activities. Several of these will be described.

International Space Station

The International Space Station (ISS), a partnership of the United States, Russian, European, Japanese, and Canadian Space Agencies, started being assembled in space with the delivery of Russian module *Zarya* in November 1998. This was followed the next month by the U.S. module *Unity* (also called *Node 1*) built by Boeing in MSFC facilities. As the 21st century started, Space Shuttle flights carried up supplies and additional items of equipment, including a portion of the solar power array.

A three-man crew arrived by a Soyuz spacecraft on 2 November 2000; since then, the ISS has been continuously occupied. In February 2001, *Destiny* arrived; this was the U.S. Laboratory also built by Boeing at MSFC. The *Quest* air lock was added to *Unity* in July 2001, allowing extra-vehicular activities. The 2005 NASA Authorization Act designated the U.S. segment of the space station as a national laboratory.

In October 2007, *Harmony* (or *Node 2*), also managed by MSFC, was added to *Destiny*; this provided connection hubs for European and Japanese modules and also gave additional living space, allowing an increase to six crew members. The final U.S. and Boeing-built element was delivered in February 2009; this was a truss that supported a full set of solar arrays and increased the available power to 30 kW. As of March 2010, Boeing had completed all of its contracted work on the U.S. segment of the ISS; NASA plans for this ISS configuration to be operated at least through 2020.

Completed ISS

As described earlier, the Payload Operations Center (POC), also known as Huntsville Operations Support Center (HOSC), of MSFC's Mission Operations Laboratory functions continuously as the primary command post for scientific work aboard the ISS.

With the retirement of the Space Shuttle in 2011, manned resupply flights to the ISS had to use the *Soyut* spacecraft. A number of different unmanned vehicles have also made resupply flights; these include the Russian *Progress* spacecraft, European Automated Transfer Vehicles, Japanese *Kounotori* vehicles, and the American *Dragon* and *Cygnus* spacecraft. MSFC has little or no involvement with these spacecraft. *Dragon* is a product of Space Exploration Technologies Corporation (SpaceX), and, under contract with NASA, is launched by a SpaceX *Falcon 9* booster to resupply the American portion of the

ISS; it is the only present unmanned resupply vehicle capable of returning a payload to Earth. *Cygnus*, an expendable spacecraft, is built by Orbital Sciences Corporation and launched by their *Antares* booster.

National Institute for Rocket Propulsion Systems

The National Institute for Rocket Propulsion Systems (NIRPS) is a collaboration between NASA, the FAA, the DoD, and industry. Founded in 2011, NIRPS serves as a steward of national propulsion capabilities, recognizing their vital role in national security, economic competitiveness, and the continued exploration of space. Operating from MSFC, L. Dale Thomas is the Director; other initial leaders included Rajiv Doreswamy (implementation manager) and Kendall K. Brown. There is a monthly meeting, with teleconference participants from across the U.S.

The NIRPS serves as an information source in deriving government and commercial spaceflight solutions, and assists with developmental and operational challenges as new propulsion technologies are released. The Institute's creation was in response to widely acknowledged concerns dating back more than a decade about the U.S. rocket propulsion base. U.S. leadership in rocket and missile propulsion is threatened by long-term industry downsizing, a shortage of new solid and liquid propulsion programs, limited ability to attract and retain fresh talent, and pressure on discretionary federal budgets.

Global Hydrology Resource Center

One of NASA's Earth Science Data Centers, the Global Hydrology Resource Center (GHRC) is a joint venture of MSFC and the Information Technology and Systems Center of the University of Alabama in Huntsville. Rahul Ramachandran is the Manager, and Helen T. Conover the Operations Manager. Established in 1991, the GHRC provides both historical and current Earth Science data, information, and products from satellite, airborne, and surface-based instruments. Some of these instruments were developed by MSFC and were described in the previous chapter. The GHRC acquires basic data streams and produces derived products from widespread instruments. The Center specializes in data involving the hydrological cycle, severe weather interactions, lightning, and convective processes.

National Space Science and Technology Center

In August 2000, the National Space Science and Technology Center (NSSTC) opened as a joint research venture between MSFC and the Space Science and Technology Alliance, a consortium of Alabama research universities. Located in a facility at 320 Sparkman Drive, the NSSTC was initially organized in seven research centers: Space Science, Global Hydrology & Climate, Information Technology, Advanced Optics, Biotechnology, Material Science, and Propulsion. The aim of the NSSTC was to foster collaboration in research between government, academia, and industry. In 2005, Martin P. Kress, former Deputy Director at NASA's Glenn Research Center in Cleveland, Ohio, was named NSSTC's Executive Director. It was planned that the 200,000-ft^2 facility would house approximately 450 people from MSFC, other government agencies, academia, and private industry.

NSSTC Building

Over the next decade, some excellent research work was done by personnel at the NSSTC; this was, however, essentially all through MSFC and UAH – the Alliance *per se* had little involvement. The extra cost was a burden to both MSFC and UAH; thus, the NSSTC as an organization was eventually abandoned. The building, owned by UAH and called Cramer Hall, still carries the name NSSTC and houses a number of research and related activities. These include the following:

- Science Research Office, MSFC - James F. Spann, Manager
- Global Hydrology Resource Center, MSFC - Rahul Ramachandran, Manager
- Center for Space Plasma & Aeronomic Research, UAH - Gary P. Zank, Director
- Earth System Science Center, UAH - John R. Christy, Director
- Alabama State Climatologist Office - John R. Christy, State Climatologist
- National Weather Service, Huntsville - Chris B. Darden, Meteorologist-in-Charge
- Von Braun Center for Science & Innovation, Inc. - Kimberly S. Webb, Acting Director

Although not directly in the NSSTC facility, the Severe Weather Institute and Research Laboratory (SWIRL) is in an adjacent building with a tornado/vortex-appearance entrance. The SWIRL is a part of the Earth System Science Center of UAHuntsville.

The Von Braun Center for Science & Innovation, Inc. (VCSI) was established in 2006 by business, government, and university leaders in the Huntsville community as a not-for-profit research for development organization. The plan for VCSI was, through collaborative activities, to blend local resources with national assets to provide science and technology innovation and solutions and to meet critical NASA and DOD mission requirements. After some initial success, the national economic downturn restricted sustainment. As of the end of 2014, VCSI no longer occupied offices in the NSSTC facility and apparently has ceased to exist.

Testing for the Next Generation Space Telescope

The next generation of large space telescopes is the James Webb Space Telescope (JWST), scheduled for launch in 2018. Named after James E. Webb, the second Administrator of NASA who played an integral role in the Apollo program, the JWST is considered a scientific successor to the Hubble; however, its primary emphasis is on infrared observation, rather than visible imaging. JWST is an international collaboration of 20 countries led by NASA, and managed by the Goddard Space Flight Center.

MSFC is responsible for JWST mirror technology development and environmental research. The Large Optics Test Facility – officially called the X-Ray Cryogenic Facility (XRCF) and described in the preceding chapter – is finding major use in the testing of mirrors for the JWST). The primary mirror is a 6.5-m (21.4-ft) diameter gold-coated beryllium reflector composed of 18 hexagonal segments,

Testing JWST Segments

which will unfold after the telescope is launched. MSFC engineers conduct round-the-clock cryogenic testing of groups of six segments, chilling them to -248°C (-414°F) to confirm that the mirrors will respond as expected to the extreme temperatures of space. Helen J. Cole is the Project Manager for JWST activities at MSFC; Jeffrey R. Kegley is the XRCF testing manager.

DEFENSE ACTIVITIES

Defense activities in Greater Huntsville are almost exclusively under the jurisdiction of the U.S. Department of Defense (DoD) and mainly in programs of the DoD's Department of the Army. After it was designated as the center for missile development in 1949, Redstone Arsenal's primary Army activity had been primarily in this field. The Army Missile Command (MICOM) was created in 1962, and was Greater Huntsville's largest defense organization for the next third of a century.

In 1995, as described in the preceding chapter, the DoD's Base Realignment and Closure (BRAC) brought major diversification in local defense activities with the merging of the Army Aviation Command (previously in Saint Louis) and the Army Missile Command. MICOM officially became the Aviation and Missile Command (AMCOM) in 1997; also formed at this time were the Program Executive Office - Aviation, and the Program Executive Office - Missiles and Space.

Going into the 21st century, the changes under BRAC 1995 were still being implemented. The former missile and aviation RD&E centers were merged in 2000, becoming the Aviation and Missile Research, Development, and Engineering Center (AMRDEC). Then in 2004, AMRDEC was assigned to the Research, Development, and Engineering Command (RDECOM) and became independent of AMCOM; AMRDEC, however, remained at Redstone Arsenal.

Other local military organizations carried over from the previous century included the Army Garrison - Redstone; the Missile and Space Intelligence Center; the Logistics Support Activity; the Redstone Test Center; the Army Engineering and Support Center (Corps of Engineers); the Fox Army Health Center (transformed from the Fox Army Hospital in 1996); and the Defense Acquisition University.

MAJOR CHANGES IN 2005

Although the BRAC 1995 had brought a major change in the largest Army organization in Greater Huntsville, the changes brought 10 years later by BRAC 2005 were much greater. The preliminary list of realignments and closings was released by the DoD on May 13, 2005. It recommended closing 22 major United States military bases and the "realignment" (either enlarging or shrinking) of 33 others.

On September 15, 2005, President George W. Bush approved the BRAC Commission's recommendations, leaving the fate of the bases to the United

States Congress. Congress had a maximum of 45 days to reject the proposal by passing a joint resolution of disapproval, or the recommendations would automatically enter into effect. Such a resolution failed to pass (95-324) in the House of Representatives (there was no vote in the Congress), thereby enacting the list of recommendations. The Secretary of Defense was required to begin implementing the recommendations by September 15, 2007 and to complete implementation no later than September 15, 2011.

Local Changes Under BRAC 2005

The local changes in Greater Huntsville's DoD organizations is estimated to have involved 4,700 direct military, civilian government, and embedded contractor personnel, as well as some 15,000 indirect jobs in the surrounding area. In addition, there was about $500 million of new military construction on Redstone Arsenal to support the BRAC relocations, and possibly even more to handle related industry enlargements.

The most significant of the BRAC 2005 relocations was that of the Headquarters, U.S. Army Materiel Command (AMC), bringing to Redstone Arsenal 1,350 positions with AMC plus 300 positions with the Army Auditing Command. With a four-star General as Commander, AMC has one of the largest budgets in the Army.

Two other Headquarters also relocated to Redstone Arsenal under BRAC 2005: the U.S. Army Space & Missile Defense Command (SMDC), with about 250 positions, and the U.S. Army Security Assistance Command (ASAC), with some 320 positions.

The largest number of incoming jobs was with the DoD's Missile Defense Agency (MDA). BRAC 2005 established the MDA headquarters at Fort Belvoir, Virginia, but realigned most of the MDA's mission and mission support activities to Redstone Arsenal, bringing more than 2,200 new military, government, and contractor positions.

Other incoming activities under BRAC 2005 were the Aviation Technical Test Center, with 323 positions; Rotary Wing Center, 50 positions; the 2nd Recruiting Brigade, 110 positions; and the 2nd Medical Recruiting Battalion, 14 positions.

Redstone Arsenal had only one significant loss under BRAC 2005. The Ordnance Munitions and Electronic Maintenance School (OMEMS) was relocated to Fort Lee, Virginia. This was a loss of 425 operating personnel and 1,100 temporary-duty student soldiers.

Von Braun Complex

Von Braun Complex

To accommodate the Army Materiel Command and the local activities of the Missile Defense Agency, the Corps of Engineers initiated a huge facility addition – the Von Braun Complex. Built at 5220 Martin Road in four phases, the last of these was completed in 2014.

Other Additions at Redstone Arsenal

In addition to the BRAC 2005 changes, there were several other significant additions to Redstone Arsenal in this period. The Headquarters, U.S. Army Contracting Command (ACC) was activated in March 2008. Coming with ACC to Redstone Arsenal was a subordinate, the Expeditionary Contracting Command (ECC). ACC and ECC combined have some 300 local personnel.

The Greater Huntsville area, now recognized as having one of the most significant concentrations of high-technology expertise in the nation, is attracting non-military governmental activities. The Bureau of Alcohol, Tobacco, Firearms and Explosives (ATF) moved its National Center for Explosives Training and Research into a new headquarters building on Redstone Arsenal that opened in October 2010.

The Federal Bureau of Investigation (FBI) is opening a new $132-million headquarters on Redstone Arsenal for its Terrorist Explosive Device Analytical Center (TEDAC) that is relocating from

FBI TEDAC Campus

Quantico, Virginia. The new TEDAC facility is a dedicated forensic and technical laboratory to support intelligence reporting, exploitation, and analysis capabilities critical to counter-IED operations.

MILITARY ORGANIZATION DESCRIPTIONS

All of the BRAC 2005 relocations were completed by September 2011. The following are descriptions of military organizations in the Greater Huntsville area (mainly on Redstone Arsenal) after this time. A group of five of these organizations is commonly called the U.S. Army Aviation and Missile Materiel Enterprise; these are AMCOM, AMRDEC, PEO Aviation, PEO Missiles & Space, and Army Contracting Command.

The order of the listing has no significance – it is mainly by age (when the organization or its direct predecessor first operated locally) and, generally, not by size or rank of the commanding officer..

Army Garrison - Redstone

Since it was first formed in 1941, Redstone Arsenal – as an organization, a U.S. Army post, and a census-designated place – has undergone many changes. On 1 October 2002, the Redstone Arsenal Support Activity's name was changed to the Army Garrison - Redstone, and it was placed under the U.S. Army Installation Management Command (IMCOM). Headquartered in San Antonio, Texas, IMCOM handles the day-to-day operations of U.S. Army installations around the globe.

Garrison-Redstone now provides the support services for all aspects of the following: 38,125 acres of land, 19 million ft^2 of building space, 72 tenant organizations, 350 private housing units, 200 miles of roads, 10 miles of riverfront with barge docks, a 7,200-ft airport, two railheads, 14 flag officers, over 100 Senior Executive Service appointees, and about 35,000 daily, on-base personnel (1,000 active military, 19,000 civil service, and 15,000 contractors). As of 2013, COL William L. Marks II was the Commander, and Kyle W. Crump the Command Sergeant Major.

Aviation and Missile Command

Under BRAC 1997, the Aviation and Missile Command (AMCOM) had been formally established at Redstone Arsenal with the merger of the U.S. Army Missile Command (MICOM) and the U.S. Army Aviation and Troop Command (ATCOM) on 1 October 1997. AMCOM was a major subordinate command of the Army Materiel Command (AMC).

As the 21st century started, AMCOM continued with the implementations initiated under BRAC 1995. The main organizational units included:

• Security Assistance Management Directorate (SAMD), • AMCOM Logistics Center (ALC), • U.S. Army Test, Measurement, and Diagnostics Equipment Activity (USADC), and • Missile Research, Development and Engineering Center (MRDEC), which included the • Redstone Technical Test Center. The • Redstone Arsenal Support Activity (Redstone Arsenal) also reported to AMCOM).

Over the next several years, there were a number of significant changes in AMCOM. In October 2000, MRDEC was merged with the RD&E aviation unit to form the Aviation and Missile Research, Development and Engineering Center (AMRDEC); then in June 2003, AMRDEC was organizationally assigned to the Research, Development, and Engineering Command (RDECOM), headquartered at Aberdeen Proving Ground, Maryland; RDECOM was also a major subordinate command of the Army Materiel Command.

In October 2002, the Redstone Arsenal Support Activity's name was changed to Garrison-Redstone; it was placed under the U.S. Army Installation Management Command (IMCOM) headquartered at Fort Sam Houston, Texas, and was no longer part of AMCOM.

In February 2003, the Aviation Center Logistics Command (ACLC), located at Fort Rucker, Alabama, was placed under the operational control of AMCOM; with this, AMCOM took over the maintenance and supply management of Fort Rucker's aviation fleet.

AMCOM Becomes AMLCMC

On 5 October 2004, the Army Materiel Command established the Aviation and Missile Life Cycle Management Command (LCMC), composed of all elements of the Aviation and Missile Command. This was formally activated 16 June 2005, with MG James H. Pillsbury, previously the AMCOM Commander, becoming the first LCMC Commander. The central part of AMLCMC's work involves acquisition and sustainment support of missile, helicopter, unmanned ground vehicle, and unmanned aerial vehicle weapon systems throughout their life cycle.

This change to an LCMC designation was a part of a major implementation across all of the DoD of Supply Chain Materiel Management Policy, dated 22 April 2004, reforming all of the supply chain under the concepts of Life Cycle Management (LCM). For simplification, however, AMLCMC is still more often referred to as AMCOM.

As part of the implementation process in AMLCMC, Program Executive Offices for Missiles and Space and for Aviation were activated in January 2005.

BG Samuel M. Cannon was the PEO Missiles and Space, and Paul Bogosian the PMO Aviation.

Major AMCOM Organizations

In addition to the ACLC operations at Fort Rucker, under BRAC 2005 AMCOM became responsible for two more maintenance and supply depots in other states: the Corpus Christi Army Depot in Texas (CCAD, a Center of Industrial and Technical Excellence for Rotary Wing Aircraft) in Texas, and the Letterkenny Army Depot in Pennsylvania (LEAD, a Center of Industrial and Technical Excellence for Air Defense and Tactical Missile Systems).

As previously noted, there was one significant loss to Huntsville / AMCOM due to BRAC 2005. The Ordnance Munitions and Electronic Maintenance School (OMEMS) – which had been located at Redstone Arsenal (under a variety of names) for more than 50 years – relocated to Fort Lee, Virginia in 2010.

As new commands came to Redstone Arsenal, AMCOM underwent other changes in dividing activities between organizations. The most recent major AMCOM units are as follows:

AMCOM Logistics Center (ALC) – The ALC develops, acquires, fields, and sustains logistics support for the Army aviation and missile systems and associated support equipment. As of 2013, B. Keith Roberson was the ALC Executive Director, and COL Christopher J. Brewer the Military Deputy Director.

Security Assistance Management Directorate (SAMD) – The SAMD primarily supports the U.S. Army Security Assistance Command (USASAC) in Foreign Military Sales (FMS). SAMD's major units are the Patriot Division, the Attack Systems Division, and the Non-Standard and Tactical Missile Division, and there are branches for a number of products and customers. As of 2013, Jeffrey K. Young was the Director.

Army Contracting Command Redstone (ACC-Redstone) – The ACC-Redstone provides support to AMCOM, the Missile Defense Agency, Redstone Arsenal-Garrison, the Space and Missile Defense Command, PEOs and Program Managers at Redstone Arsenal, PEOs and Program Managers at Redstone Arsenal, and industrial operations centers at the Letterkenny and Corpus Christi Army Depots. As of 2013, Rebecca E. Weirick was the Executive Director.

U.S. Army Test Measurement and Diagnostic Equipment Activity (USATA) – The USATA has the responsibility of performing the test, measurement, and diagnostic equipment (TMDE) calibration and repair mission for the Army, other DOD claimants, and reimbursable customers. Providing

measurement traceability through primary standards to the International System of Units, USATA oversees and supports over 700,000 items of TMDE worldwide. This mission is performed with a team of over 650 professionals stationed in 10 countries and 59 different support activities. The USATA facilities are in Building 5435 on Fowler Road. As of June 2015, Myra S. Gray is USATA's Executive Director.

Myra Gray

AMCOM Leadership

As of 2013, the leaders of AMCOM included MG Lynn A. Collyar, Commanding General; Mary C. Dickens, Deputy to the Commanding General; COL Donald R. Nitti, Chief of Staff; and Tod A. Glidewell, Command Sergeant Major. The following have been the Commanding Generals of AMCOM from its formation in July 1997 to the present:

MG Emmitt E. Gibson	July 1997 - July 1999
MG Julian A. Sullivan	July 1999 - Sept. 2001
MG Larry J. Dodgen	Sept. 2001 - Dec. 2003
MG James H. Pillsbury	Dec. 2003 - July 2007
MG James R. Myles	July 2007 - Sept. 2010
MG James E. Rogers	Sept. 2010 - June 2012
MG Lynn A. Collyar	June 2012 - June 2014
MG James M. Richardson	June 2014

It is noted that the Richardsons are a "two-star" family; Jim Richardson's wife, Laura, is also a Major General, serving as the Chief Legislative Liaison Officer in the Army Headquarters.

Program Executive Offices

In October 2006, 11 major Program Management Offices within the U.S. Army were designated as direct reporting units of the Assistant Secretary of the Army for Acquisition, Logistics, and Technology (ASA(ALT)). Two of these were the PEO Aviation and the PEO Missiles and Space. While reporting to ASA(ALT), they are operationally partnered with the Aviation and Missile Command (AMCOM). The programs assigned to the PEOs spans the full spectrum of the acquisition process from system development to acquisition, testing, production, product improvement, fielding, sustainment, and eventual retirement from the force.

Program Executive Office - Aviation

PEO Aviation (PEOA) is responsible for purchasing and managing the life cycle of all of the Army's aviation weapon systems and equipment. Dating from 1987, PMOA was transferred from St. Louis to Huntsville under the BRAC 1995 movements. It has an annual average budget of more than $7.5 billion, the largest procurement budget in the U.S. Army.

With about 1,500 military, civil service, and contractor personnel, PMO Aviation facilities are mainly in a campus around Building 5681 on Wood Road. BG Robert L. Marion is the Program Executive Officer. There are eight project offices:

- Apache Attack Helicopters
- Armed Scout Helicopters
- Cargo Helicopters
- Utility Helicopters
- Unmanned Aircraft Systems
- Aviation Systems
- Fixed Wing Aircraft
- Non-Standard Rotary Wing Aircraft.

Program Executive Office - Missiles and Space

The Program Executive Office, Missiles and Space (PEOMS) provides centralized management for all Army tactical and air defense missile programs and selected Army Space programs. The PEOMS was established in January 2005 with the merger of the PEO Air, Space and Missile Defense and the PEO Tactical Missiles.

PEOMS is located at 5250 Martin Road. As of 2014, BG L. Neil Thurgood is the Program Executive Officer Missiles & Space. There are about 1,800 military, civil service and contractor personnel. Functionally, there are eight Project Offices:

- Close Combat Weapon Systems
- Cruise Missile Defense Systems
- Counter Rocket, Artillery, & Mortar
- Integrated Air & Missile Defense
- Joint Attack Munition Systems
- Precision Fires Rocket & Missile
- Lower Tier Project Office
- Missile Defense & Space Systems

In addition, PEOMS manages two international cooperative programs: Medium Extended Air Defense System (MEADS) with Germany and Italy, and Multiple Launch Rocket System (MLRS) with Germany, Italy, UK, and France. MEADS, as well as the Patriot Missile System which is under the Lower Tier PO, are briefly discussed in a future section.

Aviation and Missiles Research, Development, and Engineering Center

In this era, the Aviation and Missiles Research, Development, and Engineering Center (AMRDEC) is the largest and most technically important activity of the U.S. Army on Redstone Arsenal. This had its start when the engineering functions of MICOM began being consolidated at the newly constructed McMorrow Laboratories in 1964. This became the Army Missile Laboratory in 1980, then the Missile Research, Development, and Engineering Center (MRDEC) in 1985.

When MICOM and the Army Aviation Troop Command were combined, the associated Aviation Research, Development, and Engineering Center (ARDEC)

was officially combined with MRDEC to form the AMRDEC in 1999. In 2004, AMRDEC was placed under the new U.S. Army Research, Development and Engineering Command (RDECOM), a

AMRDEC Main Complex

major subordinate command of the Army Materiel Command.

With William C. McCorkle, Jr. continuing to serve as the Director, AMRDEC organized and expanded over the following years. In 2009, COL John T. Wright was the Military Deputy Director, Phillip W. Hodges the Associate Director for Aviation and Missile Systems, MG (U.S, Army, Retired) James R. Snider the Associate Director for Aviation Technology, and Robin B. Buckelew the Associate Director for Missile Technology. The complex had expanded to have some 1.5 million ft^2 of space.

McCorkle retired at the end of 2009, and in 2014, James B. Lackey became the Executive Director of AMRDEC. At that time, there were about 3,000 civil service and some 5,000 contractor employees; the annual operating budget was $2.4 billion. AMRDEC has 11 Directorates: • Aviation Engineering, • Aviation Applied Technology, • Aeroflightdynamics, • Advanced Science & Technology, • Engineering, • Software Engineering, • System Simulation & Development, • Applied Sensors, Guidance & Electronics, *James Lackey* • Propulsion & Structures, • Weapon Sciences, and • Technical Management.

AMRDEC has some of the most outstanding facilities for research, development, and testing; the major ones include the following: • Advanced Prototyping Laboratory, • Advanced Simulation Center, • Aero-Optic

Evaluation Center, • Air Defense Radar Operations Facility, • Radar Frequency Technology Facility, • Antenna Range, • Army Missile Optical Range, • Battlefield Automation Laboratory, • Composite Structures Manufacturing Facility, • Ducted Rocket Test Facility, • GPS Satellite Simulation Facility, • Gel Propellant Rheology Facility, • Liquid/Gel Rocket Test Facility, • Microfabrication Laboratory, • Laser Countermeasures Laboratory • Physical Sciences Complex, • Redstone Aviation Propulsion Test and Research Facility, and • Warfighter Protection Laboratory. Wind Tunnels at NASA Langley and Moffett Federal Airfield, and special facilities at Fort Eustis are also used and have AMRDEC personnel.

There is a particular capability in tactical missile system simulation and development. Specific activities are shown in the following illustration:

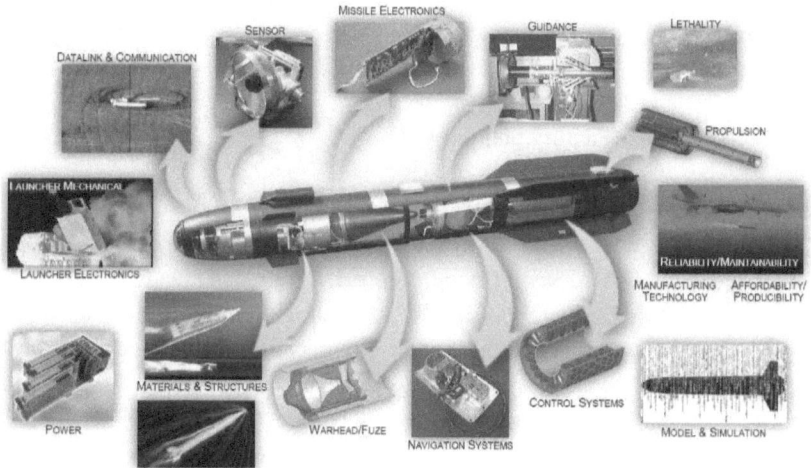

Specializations in Tactical Missiles

AMRDEC serves as the DoD lead for Rotorcraft Science and Technology and the DoD lead for Gel Propellants. In addition, AMRDEC has one of the few Capability Maturity Model (CMM) Level 4 software engineering facilities in the Army; major credit for this achievement is given to William A. Craig, the Director of Software Engineering.

William Craig

It is important to note that the RDECOM, of which AMRDC is a part, is the largest command under AMC. RDECOM has about 15,000 scientists, engineers, and other professionals developing technology solutions to existing and future needs of the U.S. Army.

Redstone Test Center

The last major change in the BRAC 2005 was the merger of the Redstone Technical Test Center (RTTC) and the Aviation Test Center (ATC) previously at Fort Rucker. In August 2009, the merger of these two operations formed the Redstone Test Center (RTC). Commander of the RTC was COL Steven W. Kihara, previously at ATC and an Army Test Pilot; the Technical Director was T. David Byrd, previously the Director of RTTC.

David Byrd

Aviation testing was not entirely new to the Center; testing of aviation hardware had been done since the Army Aviation Command merged with MICOM to form AMCOM in October 1997. However, for the broadened activities, new facilities costing over $46 million were constructed; included was a 128,278-ft^2 Rotary Wing Center Hangar Facility and the RTC's new headquarters buildings on Martin Road.

Rotary Wing Center Hangar Facility

The RTC is a unit of the Army Test and Evaluation Command (ATEC), headquartered at Aberdeen Proving Ground, Maryland, and is a major subcommand of AMC. In July 2013, COL Patrick H. Mason took command of the RTC; David Byrd remained the Technical Director. Major operating units of RTC are the System Engineering Directorate, Aviation Flight Test Directorate, Missiles and Sensors Test Directorate, and Environmental and Component Test Directorate.

Primary testing capabilities include the following: • Aviation Flight, • Climate, • Component & Surveillance, • Dynamic, • Electromagnetic Environmental Effects, • Field Sensors, • RF Controlled Environment, • Laboratory Sensors, • Missile Flight, • Propulsion, • Subsystems, and • Telemetry. Most of these capabilities are located on Redstone Arsenal; for certain testing, RTC personnel are stationed at four other government testing facilities.

Missile & Space Intelligence Center

The Missile and Space Intelligence Center (MSIC) is a major unit of the Defense Intelligence Agency (DIA). Operating under the jurisdiction of Department of Defense, the DIA is the main foreign military espionage organization of the United States. As one of the principal members of the U.S.

Intelligence Community, DIA informs national civilian and defense policymakers about the military intentions and capabilities of foreign governments and non-state elements, while also providing department-level intelligence assistance and coordination to individual military service intelligence components and the warfighter.

From a modest start with only six employees in 1956, the missile intelligence activities on Redstone Arsenal have continuously grown in employees, facilities, capabilities, and responsibilities. Reformed as the Missile and Space Intelligence Center (MSIC) in 1985, it was an original unit of the DIA.

With main facilities in the 200,000-ft^2 Richard C. Shelby Center for Missile Intelligence on Fowler Road, MSIC now has some 650 civil service and military personnel; several hundred contractor personnel also work in MSIC facilities. The Center includes high- and medium-bay exploitation work areas, and specialized areas such as the Radio Frequency, Electro-Optics/Infrared, and Signal Analysis Laboratories.

MSIC efforts involve applying complex scientific and technical methods to determine the characteristics, performance, operations, and vulnerabilities of threat weapons. Analysis areas include electronics systems, missile systems, directed energy weapons, and communications. Analysis methods include signals analysis, aerodynamic assessments, integrated modeling and simulation, computational fluid dynamics, radio frequency and infrared signature assessments, and high-performance computing.

Clyde E. Walker became the MSIC Director in 2000, and served until 2007. Other leaders in this period were Thomas E. Richardson, Technical Director, and R. Leslie Stamps, Staff Director and Chief, Office for Program

Management. In 2007, Pamela J. McCue was appointed as Director; she served until 2013, then moved up to the DIA as a Deputy Director. Tom Richardson filled the position of Director, but retired in 2014; Mark A. Clark was then made Director. In this period, Randall P. Jones was the Chief Scientist.

MSIC's overall mission is to supply scientific and technical all-source intelligence on surface-to-air missiles

Mark Clark   (SAMs), short-range ballistic missiles (SRBMs), anti-tank guided missiles (ATGMs), missile defense systems, directed-energy weapons (DEWs), selected space programs and systems, and relevant command, control, communications, computers, intelligence, surveillance, and reconnaissance (C4ISR).

There had long been a conflict between MSIC and the National Air and Space Intelligence Center at Wright-Patterson Air Force Base in Dayton, Ohio, as to who had responsibility for ballistic missile intelligence analysis. In 2010, the DIA created a Defense Intelligence Ballistic Missile Analysis Committee and MSIC was made the lead integrator. This was designed to ensure the incorporation of disparate analysis efforts from across the defense community.

The latest facility of MSIC is the Explosive Ordnance Exploitation Complex. Hidden away in the far reaches of the Arsenal's test area, this 25,000-ft^2 facility will eventually house about 250 engineers, analysts, and test specialists on the Foreign Military Exploitation team. Work there includes the disassembly, analysis, and testing of foreign missile systems and related hardware – activities that were previously conducted in a number of older buildings in various areas.

The new complex includes seven stand-alone facilities and two open storage buildings housing hazardous and non-hazardous materials. Hundreds of nearby acres are used for open-air testing. MSIC estimates that it performs more than 900 test operations per year.

U.S. Army Space and Missile Defense Command

The U.S. Department of Defense has nine Unified Combat Commands – six geographically organized and three organized on a functional basis. The U.S. Strategic Command (USSTRATCOM), organized in 2002, is one of the functional-based commands. It is charged with missile defense; space operations (such as military satellites); information operations (computer network attack and defense); global C4ISR; global strike and strategic deterrence (the U.S. nuclear arsenal); and combatting weapons of mass destruction.

The Army Space and Missile Defense Command (SMDC) was formed under an earlier Unified Combat Command in October 1997, then was placed into the Strategic Command in 2002. At that time, SMDC had the following major components: the SMDC Headquarters and the Force Development Integration Center in Arlington, Virginia; the U.S. Army Space Command in Colorado Springs, Colorado; and the Space and Missile Defense Technical Center (SMDTC), the Space and Missile Defense Battle Lab (SMDBL), and the Space and Missile Defense Acquisition Center (SMDAC), all three based in Huntsville.

The U.S. Army Space Command (ARSPACE), earlier organized in 1988, was a subordinate command. The commander was a dual-hatted leader; in addition to the duties of SMDC commander, he also served as the commander of the ARSPACE; consequently, the combination was officially called SMDC / ARSPACE, but is commonly referred to as simply SMDC.

Under BRAC 2005, the Headquarters of the SMDC / ARSTRAT was relocated to Huntsville; facilities are at the Von Braun Complex on Redstone Arsenal. As of 2013, LTG David L. Mann was the Commanding General. Others in the Command Group were Ronald E. Chronister, Deputy to the Commander; COL Dewey A. Granger, Chief of Staff; BG Gregory S. Bowen, Deputy Commander for Operations; and James N. Ross, Command Sergeant Major. The Headquarters has about 250 personnel.

David Mann

SMDC Technical Center

An important SMDC operation in Huntsville is the Technical Center (SMDTC); this activity researches, develops, tests, and integrates capabilities for material solutions in tactical space technologies, high-altitude systems, missile defense technology, cyberspace, directed energy, counter improvised explosive devices, and related areas. As of 2014, the Director of the Technical Center was Jewell (Julie) W. Schumacher, and the Deputy Director was COL Alan T. Statham.

Primary operational units of the SMDTC are • Space and Cyberspace Directorate, • Emerging Technology Directorate, • Rapid Transition Directorate, and • Reagan Test Site (in the Kwajalein Atoll). Major TC laboratories include the Advanced Concepts Lab, Army High Energy Laser Lab, and Advanced Measurements Optical Range.

Julie Schumacher

One of the TC developments was a system of nano-satellites allowing unprecedented battlefield communications.

SMDC Future Warfare Center

Also in Huntsville is the SMDC Future Warfare Center (FWC). With offices in Huntsville; Colorado Springs, Colorado; and Fort Eustis, Virginia; the Center is responsible for building future space and missile defense forces. FWC operational units include the • Space and Missile Defense Battle Lab, • Directorate of Combat Development, • Directorate of Training and Doctrine, and • Decision Support Directorate. As of 2014, Laurence (Larry) H. Burger was the FWC Director.

Larry Burger

Within the FWC Decision Support Directorate is the Information and Computational Engineering Division, which includes ● Cyberspace Initiatives, ● Cyber Technology Laboratory, ● Advanced Research Center (ARC – discussed elsewhere), ● Simulation Center, ● High Performance Computing Distribution Center, and ● Extended Air Defense Simulation (EADSIM).

More than 300 customers in the DoD and 17 friendly nations use EADSIM for experiments, exercises, test, analysis, and operational planning; this highly successful simulation was initially developed for the U.S. Government by Teledyne Brown Engineering (TBE) in the late 1980s, and has been continuously upgraded by TBE and others since then.

U.S. Army Materiel Command

The U.S. Army Materiel Command (AMC) is the Army's primary provider of materiel readiness – technology, acquisition support, materiel development, logistics power projection, and sustainment – to the total Army force. AMC often described itself as follows: If a Soldier shoots it, drives it, flies it, wears it, eats it, or communicates with it, AMC provides it. Under BRAC 2005, AMC Headquarters was relocated from Fort Belvoir, Virginia, to Huntsville; their facility is at the Von Braun Complex, Redstone Arsenal.

When first relocated to Redstone Arsenal in 2008, AMC was commanded by GEN Ann E. Dunwoody, the first woman in U.S. military history to achieve the rank of four-star General. She retired in August 2012, and was succeeded by GEN Dennis L. Via.

GEN Ann Dunwoody

As of the beginning of 2014, other key leaders were LTG Patricia E. McQuistion, AMC Deputy Commanding General and Redstone Arsenal Senior Commander; John B.

GEN Dennis Via

Nerger, AMC Executive Deputy to the Commanding General; BG Edward M. Daly, Deputy Chief of Staff; and James K. Sims, AMC Command Sergeant Major. At the Headquarters, there are about 1,350 civilian and military AMC personnel, plus some 300 personnel of the Army Auditing Command.

There are 10 special staff sections, ranging from Chaplain to Public & Congressional Affairs. Only one of these has a highly technological involvement, the Chief Technology Officer (CTO). The CTO serves as AMC's technical authority and has management oversight for research, development, and integration of science, technology, and engineering products and services.

Patrick O'Neill

This position was formed in 2010, and filled at that time by Grace M. Bochenek. As of 2015, Patrick O'Neill was appointed to this position.

AMC has ten major subordinate commands, three of which – the Army Aviation and Missile Command, the Army Security Assistance Command, and the Army Contracting Command – are also headquartered on Redstone Arsenal. A fourth major subordinate command of AMC – the Army Research, Development, and Engineering Command – is headquartered elsewhere but its largest operating unit – the Aviation and Missile Research, Development, and Engineering Center – is on Redstone Arsenal.

AMC operates the research, development and engineering centers; the Army Research Laboratory; depots, arsenals, ammunition plants, and other facilities; and maintains the Army's Prepositioned Stocks, both on land and afloat. The AMC is the DoD's Executive Agent for the chemical weapons stockpile and for conventional ammunition. AMC's maintenance depots and arsenals overhaul, modernize, and upgrade major weapons systems.

U.S. Army Security Assistance Command

A major subordinate command of the Army Materiel Command, the U.S. Army Security Assistance Command (USASAC) is often referred to as "The Army's Face to the World." It develops and manages security assistance programs and Foreign Military Sales (FMS) cases to build partner capacity, support Coordinating Committee for Multilateral Export Controls (COCOM) engagement strategies, and strengthen U.S. Global partnerships.

As of 2014, USASAC was Commanded by MG J. Mark McDonald, with Robert L. Moore, Deputy; COL Lawrence W. Fuller, Chief of Staff; and SM Rodger W. Mansker, Command Sergeant Major. With headquarters at 4402 Martin Road in the Von Braun Complex, the 350 personnel of USASAC annually manage about 4,600 FMS cases valued at more than $134 billion. Sales to overseas customers includes the same "total package" of quality material, spare parts, training, publications, technical documentation, maintenance support, and other services that AMC provides to U.S. Army units. Originally formed in 1975, USASAC co-leads the Army Security Assistance Enterprise, which is composed of more than 40 organizations.

U.S. Army Contracting Command

In October 2008, the U.S. Army Contracting Command (ACC) was formed as a major subordinate command under the Army Materiel Command. (This was a new action, not a part of BRAC 2005.) The ACC performs the majority of contracting work for the U.S. Army; in FY-2013, it awarded more than 22,000 contract items valued at about $20.1 billion.

Headquartered at 106 Wynn Drive in the Cummings Research Park, MG Theodore C. Harrison III is ACC's Commanding General; Michael Hutchison, Deputy; COL Martin Zybura, Chief of Staff; and David M. Puig, Command Sergeant Major. Overall, ACC has more than 5,500 military and civilian personnel at six major contracting centers and a total of 117 locations worldwide.

The ACC has two subordinate one-star commands – the Expeditionary Contracting Command (ECC) serving locations outside the continental United States, and the Mission and Installation Contracting Command (MICC) serving continental locations. The MICC is headquartered at Joint Base San Antonio-Fort Sam Houston, Texas, and the ECC is headquartered in Huntsville.

The ECC is co-located with the ACC in Huntsville at 106 Wynn Drive, CRP. ACC and ECC combined have some 300 local personnel. As of 2013, the ECC is commanded by BG Michael D. Hoskin with Bryan R. Samson, Deputy and Director of Contracting Operations.

Logistics Support Activity

A direct report to the AMC, the Logistics Support Activity (LOGSA) was transferred to Huntsville in BRAC 2005. LOGSA was initially housed in a number of spread facilities on Redstone Arsenal. It is now consolidated in Buildings 3303, 3305, and 3307, near Redstone Gate 10.

Logistics Information Warehouse

LOGSA provides integrated logistics data, information and expert analysis in support of Army operations worldwide. It is the home for the Logistics Information Warehouse (LIW), the Army's official storehouse for collecting, storing, organizing and delivering logistics data. LIW processes a tremendous amount of information daily, providing services to over 65,000 individual users and 150 direct trading partners around the world.

As of 2013, LOGSA was commanded by COL Charles B. Salvo. It had some 800 civil service, military, and contractor employees engaged in four major functional areas:

- Logistics Engineering Center
- Global Support Center
- Soldier Support Center
- Enterprise Integration Center.

Activities include providing logistics courses, preparing logistics documents and publications, and the centralize collection of equipment manuals.

Missile Defense Agency

As described earlier, the Department of Defense has eight Unified Commands. One of these is the U.S. Strategic Command (USSTRATCOM). Established in 1992, one of the responsibilities of this Command is early warning of, and defense against, missile attack, as well as long-range strategic attacks.

On 13 December 2001, President George W. Bush announced that the U.S. was withdrawing from the ABM Treaty; the cornerstone of Cold War arms control efforts, this treaty had been in effect for 30 years and was actually no longer applicable. Restrictions on developing and testing BMD technologies were reduced, and the BMDO – a unit of the Strategic Command – was renamed the Missile Defense Agency (MDA). This is a research, development, and acquisition agency with a workforce including government civilians, military service members, and contractor personnel in multiple locations across the United States.

Under BRAC 2005, MDA headquarters was established at Fort Belvoir, Virginia, but most of the MDA's mission and mission support activities were assigned to Redstone Arsenal, bringing to Greater Huntsville more than 2,200 additional military, government, and contractor personnel joining about 1,000 who were already here. The initial central facility – six-story, 840,000-ft^2 building in Von Braun Center Phase 3 – was occupied in October 2011. Another building for the MDA was incorporated as Phase 4 in the planning; was completed in 2014, this includes a five-story, 225,000-ft^2 administrative facility.

MDA Central Facility in Von Braun Complex

The Huntsville operation is the new center of gravity for the test, integration, and fielding activity. It is located in the cradle of American rocket technology and is responsible for guiding the programs essential to the overall success of the MDA mission. Expected to reach about 5,000 personnel, it will be the largest concentration of missile defense specialists in the world. The National Capital Region office maintains a headquarters focused on global strategy and overall program development. It is a liaison between the Department of Defense, the White House and Executive Branch, and Congress.

Missile defense technology being developed, tested, and deployed by the United States is designed to counter ballistic missiles of all ranges – short, medium, intermediate and long. Ballistic missiles have different ranges, speeds, size, and performance characteristics; therefore, the BMD System is an integrated, "layered" architecture that provides multiple opportunities to detect and destroy enemy missiles and their warheads before they can reach their targets.

James Syring

In November 2012, Vice Admiral James D. Syring was appointed as Director of the MDA, becoming the 9th Director of the MDA and its predecessor BMDO. In this capacity, he oversees the MDA's worldwide mission to develop a capability to defend deployed forces, the United States, allies, and friends against ballistic missile attacks. BG Kenneth E. Todorov serves as the MDA Deputy Director. There are major staff offices for Operations, Acquisition, Advanced Technology, Engineering, and C4ISR. Keith Englander is the MDA Chief Engineer.

Some 30 MDA program offices are in Huntsville. Major programs include Ground-Based Midcourse Defense (GMD), Terminal High Altitude Area Defense (THAAD), Global Deployment, US-Israeli Cooperative Development, Targets and Countermeasures, Integration Synchronization and Change, and Facilities, Military Construction, and Environmental Management. As of 2014, the following firms are listed by MDA as prime contractors: BAE Systems, Boeing, Booz Allen, CSC, General Dynamics IT, L-3 Communications, Lockheed Martin, Northrop Grumman, Orbital Sciences, Parsons (Sparta), Raytheon, and Teledyne Brown Engineering.

U.S. Army Engineering and Support Center

The Engineering and Support Center of the U.S. Army Corps of Engineers (USACE) is located at 4820 University Square, adjacent to Oakwood University in northwest Huntsville. As of 2013, the Center was commanded by COL Robert J. Ruch, and there are about 840 military, civil service, and contractor personnel. The Center typically executes more than 6,000 contracts valued at $1.6 billion annually in engineering, construction, and technical services to support a very diverse global customer base that includes USACE divisions and districts, federal agencies, and military installations worldwide.

The Center is responsible for the Army range and training land program; electronic security systems; medical facilities; environmental and munitions; and utility monitoring and control systems. The Center is also home to many technical centers of expertise: energy savings and performance contracting; facility systems safety; heating, ventilation, and air conditioning; installation support; facilities reduction; facilities repair and renewal; centralized furnishings; and operation and maintenance engineering enhancement. It is also assigned standardization in medical facilities; child and family services; sports and fitness facilities; fire and emergency facilities; and training ranges.

Other Local Defense Organizations

Several other local U.S. Army activities that are contributing to Greater Huntsville and the national defense should be noted. These include the Defense Acquisition University, the Fox Army Health Center, the Network Enterprise Center - Redstone, the U.S. Army 2nd Recruiting Brigade, and the U.S. Army 2nd Medical Recruiting Battalion.

Defense Acquisition University South

The Defense Acquisition University (DAU) is a Department of Defense establishment responsible for the training and career development of the more than 150,000 military and civilian members of the Defense Acquisition Workforce.

In the fields of acquisition, technology, and logistics, DAU provides the following services: • Acquisition certification and leadership training; • Mission assistance to acquisition organizations and teams; • Online knowledge-sharing resources; and • Continuous learning.

With its Headquarters at Ft. Belvoir, Virginia, DAU operates through five regional campuses. DAU South is the regional campus serving the southern half of the nation. As of 2013, Mark D. Lumb is the Dean of DAU South. With offices and classrooms located in Huntsville at 7115 Old Madison Pike (near Gate 9 of Redstone Arsenal), DAU South also has satellite campuses at Eglin AFB, Florida, and Warner Robins, Georgia.

Fox Army Health Center

The first government hospital in the area was built in 1942, one of the first brick-and-mortar facilities for use by both Redstone and Huntsville Arsenals; this was on what was commonly called Squirrel Hill. Through the years, this gradually grew into the Fox Army Hospital, a 127,000-ft^2 facility located at 4100 Goss Road opened in 1977. It transformed from a hospital to a primary health care center in January 1996.

The Fox Army Health Center (FAHC) is under the Southeastern Regional Medical Command headquartered in San Antonio Command, Texas. As of 2013, COL David J. Carpenter Jr, MD, is the FAHC Commander. With about 230 military and civilian employees, the average daily workload includes 210 patients and 1,230 prescriptions filled.

The FAHC provides health care in the following: Active Duty Physicals; Behavioral Health; Dental Services; Immunizations; Neuropsychology; Occupational Medicine; Optometry; Pathology; Pharmacy; Physical Therapy; Preventive Medicine; Primary Care; and Radiology.

FAHC services are available for active duty personnel and their families, civil service personnel, and contractor personnel working on Redstone Arsenal. Services are by appointment only. The Center is closed on Saturday, Sunday, and all federal holidays. Fox is an ambulatory care clinic and does not have an emergency room. Telephone: 256-955-8888 or 800-223-9531. Advice Nurses are available during non-duty hours.

U.S. Army Network Enterprise Center - Redstone

The Network Enterprise Center (NEC) - Redstone manages and maintains command, control, communications, computer information management (C4IM) services for Redstone Arsenal. NRC-Redstone, located at 3400 Ajax Road, is under the 93rd Signal Brigade, which commands all NECs in the eastern United States. Altogether, there are 39 NECs throughout the CONUS. This Brigade is a part of the Network Enterprise Technology Command/9th Signal Command, a subordinate of the U.S. Army Cyber Command headquartered at Fort Belvoir, Virginia. As of 2013, Ricky A. Prince is the Director of NEC-Redstone.

U.S. Army 2nd Recruiting Brigade

The U.S. Army 2nd Recruiting Brigade was relocated to 3435 Snooper Road Redstone Arsenal under BRAC 2005. The Brigade, which is a part of the U.S.

Army Recruiting Command, is responsible for recruiting operations for the active Army and Army Reserves in the eight Southeastern states, Puerto Rico, and the U.S. Virgin Islands. More than 2,000 soldiers and civilians are assigned to the Brigade. In recent years, an average of some 17,000 new soldiers per year were recruited by the Brigade for active duty forces, and more than 2,800 new soldiers were added to the Army Reserve. As of 2013, COL David Stewart was the Brigade Commander.

U.S. Army 2nd Medical Recruiting Battalion

The 2nd Medical Recruiting Battalion, relocated to Redstone Arsenal under BRAC 2005 and located at 3435 Snooper Road, has the mission of recruiting quality healthcare professionals to preserve the Army's fighting strength. With a staff of about 100 soldiers, the Battalion currently manages the medical recruiting operations of three recruiting Army Companies (in Nashville, Tennessee; Orlando, Florida; and Atlanta, Georgia), they have 18 recruiting stations in 10 states and two U.S. territories. As of 2013, the Battalion Commander was LTC Nancy Parson, RN, and Derise C. Miller was the Command Sergeant Major. The Battalion is a unit of the United States Army Medical Recruiting Brigade, headquartered at Fort Knox, Kentucky.

Progress in Laser Weapon Systems

Before leaving the descriptions of various defense organizations on Redstone Arsenal, a mention should be given to an activity that is receiving considerable attention in the media: a battlefield laser-weapon system.

The 11 July 1934 issue of the *New York Times* quoted Nikola Tesla as saying that he had invented a "death ray" that could "drop an army in its tracks and bring down squadrons of airplanes 250 miles away." Although no technical disclosure or demonstration of Tesla's device ever came about, it generated the interest of military researchers in a number of countries in electromagnetic radiation, and thus contributed to the technology that eventually became radar.

In 1960, the invention of the laser (an acronym for 'Light Amplification by Stimulated Emission of Radiation') rekindled thoughts of a Tesla-type weapon. The beam-forming capability (beam quality) was largely achieved through basic characteristics of the laser, but the power of the beam was miniscule – initially a few watts or perhaps a kilowatt at best for initial laser-based systems. Early research in lasers and their applications under agencies at Redstone Arsenal has been described in preceding chapters.

Along came the Strategic Defense Initiative (SDI or 'Star Wars') in 1984 – especially with Edward Teller proposing new ideas for megawatt-power beams and directed-energy weapons. Such weapons, however, were not forthcoming. One report indicates that since 1960, the Government has spent over $6 billion on directed energy, but until very recently there has been little to show for it.

Potential advantages of laser-based high-energy weapons are significant. While large guns and missiles have about reached their limit in capability, laser weapons have potential for increases in range, accuracy, lethality, reliability, and cost-effectiveness. Drawbacks include the limitations in beam generators, electrical power sources, and atmospherics.

In the past several years the U.S. has made significant advances in high-energy lasers. In 1996, the U.S. Space & Missile Defense Command and the

HEL Beam Director

Israeli Ministry of Defense entered into an agreement to produce a cooperative Tactical High-Energy Laser (THEL). For testing, a beam director system was built at the HEL test facility at White Sands Missile Range. Within five years, a THEL testbed in Israel shot down 28 artillery rockets and five artillery projectiles. More recently, THEL programs were pursued separately in the U.S. and Israel.

In the U.S., a central program is DARPA's High Energy Liquid Laser Area Defense System (HELLADS). The goal is to develop a 150-kW laser weapon system that is ten times smaller and lighter than current lasers of similar power under research. In May 2015, HELLADS demonstrated sufficient laser power and beam quality to advance to a series of field tests.

The U.S. Navy recently made public the successful demonstration of a 30-kW Laser Weapon System (LaWS) mounted atop the bridge of the USS *Ponce*.

In Huntsville, a high-energy laser for battlefield defense against incoming mortars and tactical missiles is being developed by Boeing under the management of the SMDC Technical Center. Presently involving a 10-kW solid-state laser mounted on a mobile unit, the HEL Mobile Demonstrator is being tested at White Sands Missile

HEL Mobile Demonstrator

Range. The power will soon be increased to 50 kW. For a number of years, Gerald T. Wilson has served as the government technical lead in most of the U.S. Army's high-energy laser activities.

Public announcements of laser-weapon demonstrations have also been made by several other countries. In 2012, *Izvestia Daily* reported that the Russian Defense Ministry was testing an aircraft-based laser weapon, and in early 2015, the China Academy of Engineering Physics reported the success of a laser-based defense system against low-flying drones. The Iron Beam air-defense system developed by Israeli contractor Rafael Advanced Defense Systems was announced at the Singapore Air Show in August 2015; this is said to be capable of destroying short-range rockets, artillery, and mortars at ranges up to 7 km. Thus, it appears that practical laser weapons are finally on the move.

RELATED NON-MILITARY ORGANIZATIONS

As previously noted, non-military government agencies are now relocating to the Greater Huntsville area, finding highly qualified personnel and gaining better access to related military organizations. Three of these – the Hazardous Devices School, the National Center for Explosives Training and Research, and the Terrorist Explosive Device Analytical Center – are operated by the U.S. Department of Justice.

Hazardous Devices School

Opened on Redstone Arsenal by the Federal Bureau of Investigation (FBI) in 1971, the Hazardous Devices School (HDS) is where the nation's 2,600 bomb technicians are trained and certified – the only world-class institution of its kind in the U.S. In addition to bomb technicians, more than 7,500 first responders

FBI Hazardous Devices School

have completed the basic course alone.

Following a $25 million improvement completed in 2004, the nearly 300-acre HDS campus at 7010 Redstone Road includes 14 mock training areas that mirror real-life environments: bus and airline terminals, homes, apartments, a church, a warehouse, a bank, a strip mall, and a gas pipeline. An "Anarchist Bookstore" is next door to the U.S. Army Recruiting Office, and the movie theater marquee features "The Hurt Locker."

Training activities at the HDS include the following topics: • Fundamentals of explosives; • How to recognize, assess, and render safe hazardous devices; • Post-blast investigations (what exploded and why); • Decontamination and disposal procedures; • Basic electronics; • Fragment analysis; • Protective clothing; and • Robotic machines. Recent world-wide events have led to cutting-edge counterterrorism bomb training on how to respond to suicide attacks, large vehicle bombs, weapons of mass destruction, and mortar attacks. As of 2013, Jeffrey C. Warren was the HDS Director.

National Center for Explosives Training and Research

The Bureau of Alcohol, Tobacco, Firearms and Explosives (ATF) moved its National Center for Explosives Training and Research (NCETR) to Redstone Arsenal in 2010. The new headquarters and operations building at 3750 Corporal Road has 83,500 ft^2 of classrooms, laboratories, offices, a mock courtroom, and a full suite of audio-video facilities.

Explosives Training

NCETR also has than 1,000 acres of explosive ranges, including a small town where would-be terrorists regularly plant explosives in the homes, post office, church, and airline terminal. Classroom lessons about bomb suits, robots, and methods of dealing with an explosive device are applied to real-life scenarios in the buildings of this town.

With Carl J. Vasilko as Director, there is a full-time staff of ATF agents and instructors, as well as a number of part-time personnel. NCETR consists of the Explosives Enforcement and Training Division, the Explosives Research and Development Division, and the Fire Investigation and Arson Enforcement Division. ATF is the only U.S. Government agency with fire and arson investigation as part of its core mission, and the programs that support that mission are located at NCETR.

Terrorist Explosive Device Analytical Center

The latest high-technology activity in Huntsville is the Terrorist Explosive Device Analytical Center (TEDAC) of the FBI. Located in a new $125-million facility at 4940 Fowler Road on Redstone Arsenal, the TEDAC is a dedicated

FBI TEDAC Laboratory

forensic and technical laboratory for intelligence reporting, exploitation, and analysis capabilities related to improvised explosive devices (IEDs).

TEDAC was established in 2003 to support military, homeland security, international partners, intelligence, and law enforcement communities in analyzing and exploiting intelligence related to terrorist improvised explosive devices. Previously operating at Quantico, Virginia, the relocation was to update

TEDAC's laboratories and place the personnel adjacent to the outstanding science and technology capabilities in Greater Huntsville.

When fully operational in 2015, the center will be staffed by electronics experts, engineers, explosives specialists, and intelligence analysts. The FBI and its partners at Redstone Arsenal – alongside the FBI's Hazardous Devices School and the ATF's National Center for Explosives Training and Research – join to advance TEDAC's mission by enhancing collaboration and further strengthening a government-wide approach to countering IEDs and other explosives threats.

TECHNICAL INDUSTRIES

With the significant addition of aviation programs under BRAC 1995, and the explosive increase in local defense programs from BRAC 2005, there was major growth in numbers of defense-related industries in Greater Huntsville during the first period of the 21st century. Adding to BRAC-related growth, several other government activities were begun here; also, MSFC gained new missions in high-thrust propulsion and science research activities, further building the industrial community.

By the beginning of the new century, Greater Huntsville had more than 10,000 acres in 11 specific research parks. These are shown in the following figure; city limits are heavily shaded:

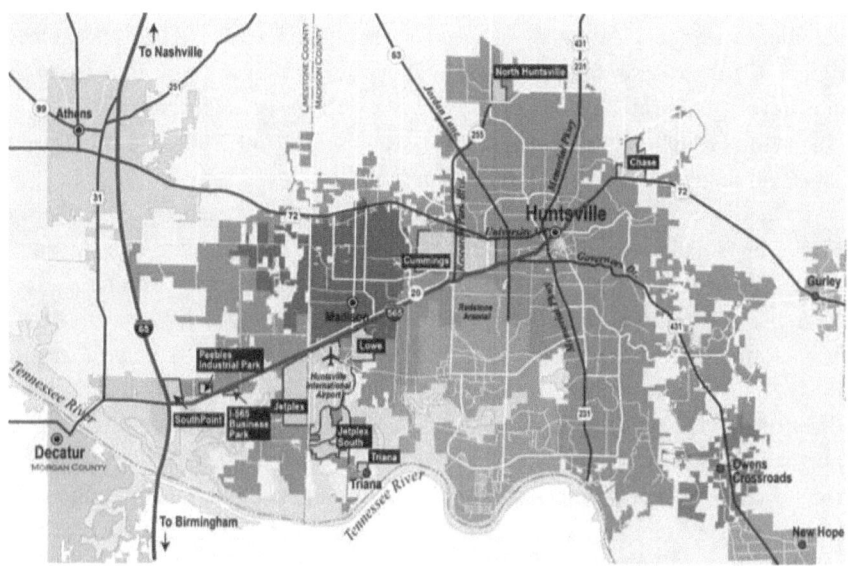

Greater Huntsville Research and Industrial Parks

INDUSTRY CLASSIFICATIONS

With the multitude of changes, the general presentation format used in the earlier chapters became too voluminous for this chapter on the present status. Thus, the format will change and with this certain classifications will be used.

Size Classification

Here a brief discussion of size in local industries might be beneficial. Since many firms in Greater Huntsville have sales to government agencies, they are often subject to governmental regulations.

In the Small Business Act of July 30, 1953, Congress created the Small Business Administration (SBA) under the Department of Commerce. The SBA's function was to aid, counsel, assist, and protect, insofar as possible, the interests of small business concerns. Among other functions, the charter stipulated that the SBA would ensure small businesses a "fair proportion" (23 percent) of the total value of all the government's prime contracts – this does not guarantee any portion of each such contract. Since its founding, the SBA has delivered millions of loans, loan guarantees, contracts, counseling session, and other forms of assistance to small businesses.

Two special types of contracts available from many federal agencies to small businesses should be noted. The Small Business Innovation Research (SBIR) is a highly competitive program that encourages domestic small businesses to engage in Federal Research/Research and Development that has the potential for commercialization. The Small Business Technology Transfer (STTR) is another program that expands funding opportunities in the federal innovation research and development arena. Under STTR, the proposing industry must have a nonprofit research institute as a partner. Both SBIR and STTR are often used for initial funding by start-up small businesses.

In addition to number of employees, other methods used by the SBA in classifying small companies include annual sales (turnover), value of assets, and net profit (balance sheet); these are used alone or in a mixed definition. The SBA establishes small-business size standards on an industry-by-industry basis. The North American Industry Classification System (NAICS) plays a major role. (Small Business Size Standards by NAICS Industry can be found at the Electronic Code of Federal Regulations, Title 13, Chapter I, Part 121.) As a rough guide, firms engaging in research and development in the physical, engineering, and life sciences – NAICS 541712 and used by a large number of local industries – might be classified as a small business if they have 500 or less employees.

The total number of employees of a firm – regardless of their location – is used in determining the size classification. In 2006, however, SBA records show that there were about 18,000 "small businesses" in the U.S. with over 500 employees; this accounted for about half of all the personnel employed at all small business.

NAICS Sectors

The North American Industry Classification System (NAICS) is used by business and government to classify business establishments according to type of economic process activity in Canada, Mexico, and the United States of America. This is totally a self-classification process; each establishment classifies itself to an industry according to its primary business activity. The first two digits designate the business sector, the third digit designates the subsector, the fourth digit designates the industry group, the fifth digit designates the NAICS industries, and the sixth digit designates the national industries. NAICS versions are released every five years.

The following is a 2012 NAICS listing designated by most industrial firms in Greater Huntsville. This may be used to determine the activity or activities claimed by local organizations. It is noted that some sectors are not represented (these include Agriculture - 11, Retail Trade - 42, and Arts and Entertainment - 71); also, all local industries have not claimed a NAICS listing. To simplify the list, some of the portions include only the first five digits. A full NAICS list may be found at the following URL: [http://www.census.gov/cgi-bin/sssd/naics/naicsrch?chart=]

The continued diversity of manufacturing activities in Greater Huntsville is clearly shown in this listing. A corresponding chart for the 1980s-1990s is not available, but it is estimated that such a chart would not include more than half the activities presently existing. While the technology firms of Greater Huntsville are commonly thought of as emphasizing analytical activities, there is certainly a further increasing of manufacturing activities.

32541 Pharmaceutical Manufacturing
32619 Plastics Product Injective Molding Manufacturing
32721 Glass Manufacturing
32791 Abrasive Products Manufacturing
33131 Primary Metal Manufacturing
33141 Nonferrous Metal (except Aluminum) Smelting and Refining
33142 Copper Rolling, Drawing, and Extruding
33231 Plate Work and Structural Product Manufacturing

33232 Sheet Metal Work Manufacturing

33251 Hardware Manufacturing

33271 Machine Shop and Precision Manufacturing

33281 Electroplating, Plating, Polishing, and Anodizing

33291 Metal Valve Manufacturing

33411 Computer and Electronic Product Manufacturing

33421 Telephone and Communication Apparatus Manufacturing

33422 Radio/Television Communication Equipment Manufacturing

33429 Communications Equipment Manufacturing

33431 Audio and Video Equipment Manufacturing

33441 Circuit Board and Electronic Component Manufacturing

33451 Search, Navigation, Guidance System Manufacturing

33531 Electrical Equipment and Component Manufacturing

33611 Motor Vehicle and Component Manufacturing

33621 Motor Vehicle Body and Trailer Manufacturing

33631 Motor Vehicle Engine and Engine Parts Manufacturing

33637 Motor Vehicle Metal Stamping

33641 Aerospace Systems and Component Manufacturing

33911 Medical and Dental Equipment Manufacturing

42444 Poultry and Poultry Product Wholesalers

42399 Miscellaneous Durable Goods Merchant Wholesalers

42499 Miscellaneous Nondurable Goods Merchant Wholesalers

51120 Software Publishers

51711 Wired Telecommunications Carriers

51821 Data Processing, Hosting, and Related Services

541330 Engineering Services

541370 Land Surveying

541380 Testing Laboratories

541511 Custom Computer Programming Services

541512 Computer Systems Design Services

541513 Computer Facilities Management Services

541519 Other Computer Related Services

541711 Research and Development in Biotechnology

541712 Research and Development in the Physical, Engineering, and
        Life Sciences (except Biotechnology)

541720 Research and Development in the Social Sciences

56142 Telemarketing and Contact Centers

92711 Space Research and Technology

Firm Size and Numbers

Information on industries in Greater Huntsville – actually all of Madison County – that design or manufacture products, or provide research and development services, is collected by the Chamber of Commerce of Huntsville / Madison County (CCHMC) and periodically published in their *Industrial Directory*. Hiroko Sedensky compiles and edits this document; the Twenty-fifth Edition is dated 2014. Using this latest edition – primarily based on information gathered in 2013 – there is the following distribution of 644 technical firms in Greater Huntsville based on number of local employees:

Number of Employees	Number of Firms
1000 or More	15
Between 200 and 999	37
Between 75 and 199	71
Between 25 and 74	172
Between 5 and 24	231
Between 1 and 4	<u>118</u>
	644

GREATER HUNTSVILLE TECHNICAL FIRMS

In view of the variables in broad size classifications, it was decided to use in this document the following for industries in Greater Huntsville: Large, 1,000 or more local employees; Medium, between 200 and 999; and Small, 199 or less. In every case, it is the number of local employees (those in Greater Huntsville) that determined the listed size classification. It is the obligation of every firm to annually report this number to the CCHMC.

As in similar lists in previous chapters, the name Huntsville (or other local town) in the firm's title indicates that this is a local operation of an outside owner; in most such cases, the location of the headquarters is given.

Large-Sized Technical Operations

As of 2013, there were, or had recently been, 15 technical firms in Greater Huntsville falling in the large-size category. Because of space limitations, these are the only firms given even brief descriptions. Thirteen of these were described in earlier chapters; thus, they are mainly listed here as Updated; two were opened in the 21st century. Another (Remington) has not yet opened, and another (Polaris) plans to open in 2016; both of these are listed as Expected.

Brief Descriptions of Large-Size Firms

The following brief descriptions of these firms are given in the order the year of initial opening in Huntsville. As previously noted, Huntsville in the name indicates a local operation of a firm with headquarters elsewhere.

Chrysler / Siemens / Continental - Huntsville (1952-2003, Updated) – In 2000, the Chrysler Huntsville complex had about 700,000 ft^2 of space and almost 3,000 employees in their facilities at 100 Electronics Boulevard, adjacent to the Huntsville International Airport and the Intermodal Transportation Center. This highly automated electronics plant was sold to Siemens VDO in 2003, ending Chrysler's 51 years of operation in Huntsville. Siemens operated the facility for three years and then sold it to Continental AG. The plant, with about 1,300 hourly and salaried employees (who were the highest paid among Continental's U.S. factories), operated for three more years; then, following a dispute with the UAW, the plant closed in November 2009. As of 2013, the facility remained vacant. (See Remington later in this section.)

Teledyne Brown Engineering - Huntsville (1953, Updated) – In 1999, Allegheny Teledyne, then owner of subsidiary Teledyne Brown Engineering, split into three independent companies; Teledyne Technologies (more often simply called Teledyne) was one of these. The new Teledyne, with Robert Mehrabian as Chairman, CEO, and President, was initially composed of 19 companies; Teledyne Brown Engineering (TBE), with about 1,300 employees, was one of these. Teledyne headquarters was soon established at Thousand Oaks, California. Like its predecessor, the new Teledyne also very aggressively acquired new companies; by 2011, there were over 100 subsidiaries and other affiliated firms.

With the start of the 21st century, TBE had about 1,300 employees. Raymond C. Watson, Jr. (author of this book) was the initial Acting President. In July 2001, James M. Link

Janice Hess

(LTG USA, Retired) was named President and served until 2008. Rex D. Geveden, previously an MSFC official, served as President until 2013, then Janice L. (Jan) Hess, earlier TBE's EVP and CFO, became President. Stephen J. Kuffner was the EVP and Roy E. Rice was the Chief Engineer. In 2004, TBE started recognizing highly accomplished

Stephen Kuffner

Roy Rice

(mainly prior) employees as Engineering Fellows; the author of this book is proud to be among those selected.

In diversification endeavors, a number of very different activities were started by TBE. Included were a number of environmental programs led by John R. Yanosky; one activity involved developing mobile systems for destroying U.S. Army chemical weapons and another was monitoring the dismantling of former USSR nuclear-powered submarines in Severodvinsk, Russia. In July 2011, TBE was awarded a contract from the United States Special Operations Command to design, develop, test, manufacture, and sustain the Shallow Water Combat Submersible (SWCS), a manned combat submersible vehicle specifically designed to insert and extract Special Operation Forces in high-threat areas. Earle W. Presson was the SWCS project's Lead Engineer.

By 2013, TBE employment had decreased to about 900, but there were good prospects to again bring this to above 1,000. NAICS: many, including 33251, 33411, 33429, 33451, 33641, 54138, 541511, 541512, 541513, 541712, 92711

The Boeing Company - Huntsville (1962, Updated) – The Army's Space and Missile Defense Command was created in 1997; its initial major program was National Missile Defense (NMD) managed out of Huntsville. In April 1998, Boeing won the contract for serving as the NMD Lead Systems Integrator, tasked to oversee and integrate systems from other major defense subcontractors. NMD was renamed Ground-based Midcourse Defense (GMD) in 2002. The system consists of radars and ground-based interceptor missiles that would collide with and destroy incoming warheads in space. Norman E. Tew was Boeing's GMD Program Manager. NAICS: 33641

As of 2013, there were about 2,600 Boeing employees in Huntsville; Anthony R. (Tony) Jones (LTG U.S. Army, Ret.) was the VP and Local Site Executive. Boeing is the largest aerospace company in Alabama and one of the state's largest employers. Current company operations in Huntsville include the Ground-based Midcourse Defense (GMD) program and other missile defense work: the Standard Missile-3 Block IIB program, the Arrow Weapon System, the Avenger, and the Patriot Advanced

Tony Jones

Capability-3 seeker. Huntsville-based programs that support NASA include Exploration Launch Systems, which oversees design and development of the new Space Launch System, and the International Space Station. The site also hosts Boeing's Rotorcraft Support business, and provides engineering support for commercial and commercial-derivative airplanes.

Boeing Complex

In April 2013, Boeing Huntsville headquarters began its relocation to a leased complex in the Redstone Gateway, a new center just outside Gate 9 of Redstone Arsenal. This complex will eventually consist of three buildings with a total of 363,000 ft^2 in office and engineering space. Boeing will retain its company-owned facilities at the Jetplex location.

In December 2013, Boeing announced that, as part of restructuring its Research and Technology organization, it would be opening five research centers, including one in Huntsville. The Huntsville center is planned to specialize in simulation and decision analytics, and metals and chemical technology; it will employ between 300 and 400 highly qualified personnel.

Northrop Grumman Corporation - Huntsville (1963, Updated) – In 1994, Northrop Corporation acquired Grumman Aerospace, forming Northrop Grumman Corporation. At that time, the combined companies had about 100 local employees. By 2000, the Huntsville Operations of Northrop Grumman had over 500 employees. Daniel Montgomery was the Director of Huntsville Operations at 620 Discovery Drive; here the activities were mainly in systems engineering, software development, and information technology services. Harry Quandt was the Director of the Engineering Center at 935 Explorer Boulevard, producing electronics for precision guided munitions.

Kevin Campbell

As of 2013, Kevin T. Campbell (LTG, U.S. Army, retired) was the VP / Corporate Lead Executive for Huntsville, and Kelley G. Zelickson the VP for Air and Missile Defense. Their Corporate Campus was at 301 Voyager Way; other facilities were at 915, 917, and 935 Explorer Boulevard, 330 Wynn Drive, and on Bat Road in Redstone Arsenal. There were over 1,200 Northrop Grumman employees in Greater Huntsville. NAICS: 33641

Lockheed Martin Corporation - Huntsville (1963, Updated) – By the start of the 21st century, Lockheed Martin had considerably expanded in Huntsville. The Space and Strategic Missiles unit led by Robert A. Drolet at 4800 Bradford Drive had about 450 employees. Harold Browning was the Director of the

Lockheed Martin Information Systems unit at 4000 South Memorial Parkway; this had some 150 employees. Celia A. Lang was the Program Director of the Space Operations Integrated Technology Solution at 4800 Bradford Drive with about 100 employees. In addition, Lockheed Martin Technology Services built test equipment at 4912 Research Drive, Charles D. Garner was the Huntsville Technical Director of the Michoud Operations, and Lockheed Martin Astronautics had a facility at 620 Discovery Drive.

John Holly

In 2009, Lockheed Martin Space Systems Company established a new Missile Defense Systems operating unit based in Huntsville. John W. Holly (MG U.S. Army, Retired) was named to lead the new operating unit as the VP for Missile Defense Systems. Activities under Holly included support to the Missile Defense Agency's Targets and Countermeasures Program, the tri-national Medium Extended Air Defense System (MEADS), and the production facility in Courtland, Alabama.

As of 2013, Lockheed Martin had over 1,100 employees in Greater Huntsville. This included approximately 800 employees at Lockheed Martin Space System Company's Missile Defense Systems campus on Bradford Drive, which now contains nine buildings. NAICS: 33641, 541712

Intergraph Corporation - Huntsville (1969, Updated) – In 2000, following burdensome lawsuits with Intel, Intergraph discontinued its hardware business, selling the Intense3D graphics accelerator division to 3Dlabs,Inc, and the workstation and server division to Silicon Graphics. James Taylor was named CEO and led Intergraph into providing enterprise engineering software and geospatially powered software to businesses, governments, and organizations around the world. Intergraph then operated through two divisions: Process, Power & Marine (PP&M) and Security, Government & Infrastructure (SG&I).

R. Halsey Wise became president and CEO of Intergraph in 2003, and in 2006, the company became a subsidiary of Intergraph Holding Company, owned by Hellman & Friedman and Texas Pacific Group. In 2008, Intergraph was listed as one of the hundred largest software companies in the world.

Intergraph Headquarters

In July 2010, Intergraph was acquired by the Swedish firm Hexagon AB; founded in 1992, and headquartered in Stockholm. Hexagon is a leading global provider of integrated design, measurement, and visualization technologies. With the acquisition, Ola Rollén became CEO of Intergraph. Intergraph's two divisions, PP&M and SG&I, continued to operate under the leadership of Gerhard Sallinger and John K. Graham, respectively. As of 2013, Intergraph remains in Huntsville; with over 1,300 employees, they announced a significant enlargement of facilities with a 250,000-ft^2 headquarters.

Intergraph Process, Power & Marine (PP&M) is the leading global provider of engineering software for the design, construction, and operation of plants, ships, and offshore facilities. Intergraph Security, Government & Infrastructure (SG&I) is the leader in smart solutions for emergency response, utilities, transportation, and other global challenges. Intergraph Government Solutions (IGS) is an independent subsidiary for SG&I's U.S. federal business. NAICS: 541511, 541512

SAIC (Science Applications International Corporation) - Huntsville (1973, Updated) – As the 21st century started, SAIC Huntsville had some 600 employees; about 70 percent in work for the DoD, mainly on missile and aviation programs, and the remaining 30 percent were on NASA contracts, mainly information technology programs of MSFC. Most SAIC employees on these contracts worked in the government facilities. A new headquarters facility for Huntsville SAIC was opened at 6725 Odyssey Drive. There was also an SAIC facility at 300 Voyager Way. John H. Gully became the General Manager in early 2009.

In the early 2000s, SAIC had about 39,000 employees nationwide and was the largest employee-owned research and engineering organization in the United States. The firm went public in 2006, and moved its headquarters from San Diego to McLean, Virginia in 2009. In May 2013, SAIC announced that it was splitting into two separate, publically traded corporations: a new SAIC – a smaller company retaining the original name and focused on federal government contracting, and Leidos – a new, larger company focused on energy, security, and health care markets.

Before the split, Huntsville operations had about 2,300 employees. Essentially all remained with the reformed SAIC; around 600 work on its two

John Gully

major NASA programs, and the other 1,600 support a variety of Army programs on Redstone Arsenal. One NASA program is running the worldwide network linking all NASA centers and affiliate sites, and the other program runs all of NASA's organizational networks managing administrative activities (finance, personnel, property, etc.); both programs are based in Huntsville. Essentially all of the previous Huntsville activities remained in the new

SAIC. Senior Vice President John Gully, based in Huntsville, heads the U.S. Army and U.S. Air Force customer group in the new SAIC. NAICS: 541712

Dynetics, Inc. (1974, Updated) – Dynetics is an employee-owned engineering, applied science, and information technology company headquartered in Huntsville. Its primary customers are the United States Department of Defense, the United States Intelligence Community, and National Aeronautics and Space Administration (NASA). Dynetics opened a 160.000 ft² corporate headquarters facility at 1002 Explorer Boulevard in 2006. Marcus J. Bendickson led the company as the CEO.

Marc Bendickson

Dynetics entered the space propulsion business by purchasing Orion Propulsion in 2009. Dynetics also entered the space hardware business by building, in cooperation with MSFC, the FASTSAT (Fast Affordable Science and Technology Satellite). Its space activities further expanded with a NASA SLS contract for upgrades to the F-1 engine.

Dynetics Campus

Dynetics opened its newest facility, the Solutions Complex at 1004 Explorer Boulevard, in 2012. This 226,500-ft² building expanded their capabilities in research, development, and production for aerospace, cyber, and defense products in Huntsville. David A. King, earlier the Director of NASA MSFC, was named the Dynetics President.

As of 2015, Dynetics had about 1,400 employees with about 1,000 in Huntsville and the remainder in seven other locations. Bendickson, who had been with the Company for 40 years, stepped aside, becoming the Chairman. The technical leadership was then as follows:

From left to right: Steve Cook, VP Corporate Development; Jonathan Whitcomb, VP Enterprise Solutions; Greg Lester, President; David King, CEO; Tom Baumbach, CTO; and Mike DeMaioribus, EVP Business Operations. Other officers were Sallie Patterson, VP Contracts, and Randy Reynolds, VP Finance. NAICS: 33451, 33641, 541511, 54171

CAS / Wyle CAS Group - Huntsville (1979, Updated) – By the turn of the century, CAS had grown to become the largest programmatic contractor and the second-largest defense consulting and services contractor in Greater Huntsville. With over 500 employees, its headquarters were at 5021 Technology Drive in CRP with operations at several other cities.

Wyle CAS Group Facility

In September 2006, CAS was acquired by EDO Corporation, a large defense firm involved in hardware and services. CAS, Inc. continued as an EDO subsidiary, and moved to 100 Quality Circle, Thornton Research Park. Then in December 2007, EDO was bought by ITT Corporation, an international, multi-discipline firm; CAS, Inc. continued as an IIT subsidiary.

The ownership changes continued; in September 2010, Wyle Laboratories, a privately held company of El Segundo, California, purchased CAS. It was named the Wyle CAS Group, joining with Wyle Huntsville, a division having long-standing operations in this city. As of 2013, the Group had about 1,200 employees; John R. Jordan was the President. In March 2014, the testing business in Huntsville was sold, leaving Wyle CAS with near 1,100 employees. NAICS: 541712

SCI / Sanmina-SCI Corporation - Huntsville (1980, Updated) – In December 2001, Sanmina, a world-wide electronics manufacturing company of San Jose, California, bought SCI Systems, a leading Huntsville industry since 1961; the combined firm became Sanmina-SCI Corporation. Although Sanmina was only about half the size of SCI at the time, it was in a better cash position. In the next several years, the firm consolidated many of former SCI national and worldwide operations. In November 2012, the company changed its name to Sanmina Corporation. At the same time a division, SCI Technology, Inc., was formed; headquarters and primary facilities are in Huntsville. The long-standing facility at 8600 South Memorial Parkway was eventually closed.

As of 2013, there were two operating companies in the facilities at 13000 South Memorial Parkway: Sanmina Huntsville and SCI Technology (SCI). The total employment for the two operations was about 1,350. Work at the Huntsville facilities includes consulting, design,

Sanmina-SCI Factory

engineering, logistics, new product introduction, assembly, machining, and fabrication to produce printed circuit boards, backplanes, cables, injection-molded plastics, enclosures and frames, and optics. Sanmina provides manufacturing to support to OEMs primarily in the communications, industrial and medical instrumentation, multimedia, enterprise computing and storage, and automotive technology sectors. SCI's work is primarily in the defense and aerospace sector sectors. NAICS: 33411, 33422, 33411, 33451

Qualitest Pharmaceuticals / Vintage Pharmaceuticals - Huntsville (1983, Updated) – Vintage Pharmaceuticals had been formed at Huntsville by William S. Propst in 1983; the firm had a 780,000-ft^2 facility in Chase Industrial Park, and an auxiliary facility in Charlotte, North Carolina, called Qualite Pharmaceuticals. In 2000, the U.S. Food and Drug Administration (FDA) charged that Vintage Pharmaceuticals had, on one drug, assigned expiration dates that did not conform to FDA requirements. In 2006, after extended proceedings, the charge was settled with fines to Vintage as well as to both William S. Propst, the CEO, and William S. Propst, Jr., the President; also, the Propsts were restricted from holding Vintage management positions for several years. This led to Vintage being merged into Qualitest and this group being put up for sale. In September 2007, funds advised by Apax Partners, an independent investment firm in Great Britain, acquired Qualitest Pharmaceuticals; the transaction was reported to be worth close $1 billion. The headquarters remained in Huntsville, but a Britisher, F.M.J. Urwin, was assigned as President and CEO; there were about 750 Huntsville employees.

In 2010, Qualitest was acquired for about $1.2 billion by Endo International, plc, a global healthcare company headquartered in Dublin, Ireland,

Qualitest Pharmaceuticals Complex

and U.S. headquarters in Chadds Ford, Pennsylvania. The firm continues in Huntsville and Charlotte as Qualitest Pharmaceuticals. In late 2011, Qualitest moved its commercial functions to a new facility at 4940 Research Drive, CRP. As of 2013, Daniel Carbery is the Vice President and General Manager; there were about 1,350 employees in Huntsville and 200 in Charlotte. The firm had a portfolio of more than 600 products and was the No. 6 ranking supplier of generics in the Nation. NAICS: 32541

ADTRAN, Inc. (1986, Updated) – ADTRAN has two operating divisions – Carrier Networks and Enterprise Networks – and their three major product categories remain Carrier Systems, Business Networking, and Loop Access. The company is a supplier of local-loop access and deployment products for fiber, DS3, T1/E1, wireless T1/E1, Digital Subscriber Line (HDSL, HDSL2, HDSL4,

SDSL, SHDSL), Frame Relay, Asynchronous Transfer Mode (ATM), Integrated Services Digital Network (ISDN), and Digital Data Service (DDS) digital services. ADTRAN supplies multi-service access platforms and Integrated Access Devices (IADs) for converged voice and data networks. Their typical new-product life cycle is 18 months.

During the major growth years, Mark C. Smith served as Chairman and CEO, and Howard A. Thraikill was President and COO. In 2005, both Smith and Thraikill retired, and were succeeded, respectively, by Thomas R. Stanton and Danny J. Windham. James E. Matthews was

ADTRAN Corporate Campus

CFO and Kevin W. Schneider was the CTO. Windham resigned in 2007, and the positions of COO and President were vacated. As of 2013, the top executives at ADTRAN were Stanton, Matthews, and Schneider, while on the operations side were James D. Wilson, Raymond R. Schansman, and Michael K. Foliano, leaders respectively of the Carrier Networks Division, Enterprise Networks Division, and the Global Operations. R&D was under Eduard Scheiterer. There were about 1,550 employees. NAICS: 33421

CINRAM Group, Inc. - Huntsville (1987, Updated) – A Canadian firm, CINRAM Group, opened a Huntsville plant at 4905 Moores Mill Road, Chase / Lowe Industrial Park, in 1987. Highly automated machines were used to mass produce, package, and distribute CVDs and related home entertainment products. With product demand highly variable, the number of employees – particularly those with low skills – varied widely, sometimes up to 2,500 and more.

In June 2012, CINRAM filed for Chapter 15 bankruptcy in the United States, and the Huntsville operation was purchased by Najafi Companies, a private investment firm based in Phoenix, Arizona. Except for ownership – now by an American

CINRAM Facility

organization – the Huntsville operation continues with the CINRAM name. As of 2013, the plant was managed by Fredrick M. Rudolph; there was an average of about 1,450 employees. NAICS: 42399

Toyota Motor Manufacturing, Alabama, Inc. - Huntsville (2001) – As Toyota Motor Corporation of Japan – founded by Kiichiro Toyota in 1937 – made plans for expanding automobile and truck production in the United States, representatives of the City, County, and State successfully promoted the advantages of placing an engine manufacturing plant in this area. Toyota Motor Manufacturing, Alabama, Inc. (commonly called TMMAL), was formed in June 2001. A car-and-truck engine factory was started on a 334-acre site at 1 Cotton Valley Drive, off Pulaski Pike, in north Huntsville. The plant opened in April 2003, with 350 employees.

Haruaki Hoshino was the initial President of TMMAL; James T. Bolte has served as President since April 2009. W. Wesley Woods set up the initial manufacturing, and has been the General Manager of Quality and Engineering from the start.

TMMAL Complex

The initial operations produced the 5VZ-FE V6 engines and 2UZ-FE V8 engines. Production of the 1GR-FE V6 engines began in 2005, the 3UR-FE V6 engines in 2007, and the 1UR-FE V8 engines in 2008. In 2011, lines for 4-cylinder AR-2.5 and AR-2.7 engines were opened. It is the only Toyota plant globally to produce all of these types of engines under one roof. As the engines lines were added, facilities have undergone major expansions, reaching 1.13 million ft^2 and employing about 1,200 people by the end of 2013. The Huntsville plant has an annual production capacity of more than 500,000 engines. NAICS: 33631

Verizon Wireless - Huntsville (2007) – Verizon Wireless (branded as Verizon) is a wholly owned subsidiary of Verizon Communications Inc. Verizon Communications began operations in April 2000, resulting from the merger between Bell Atlantic and GTE. As of June 2000, Verizon Wireless was the largest mobile network operator in the United States.

Headquartered in Baskin Ridge, New Jersey, Verizon opened a Call Center operation at Huntsville in 2007. Located at 475 Quality Circle in Thornton Research Park, the facility handles all aspects of customer service including technical support, customer retention, and quality service delivery. It also serves as Verizon's Alabama headquarters. Verizon began local operations with about 400 employees, and reached over 1,200 in 2013. Jeremiah Knight was the initial operations manager. In 2014, Huntsville Verizon was upgraded, doubling the network capacity. NAICS: 56142

Remington Arms Company, LLC - Huntsville (Expected) – Founded in 1816 and headquartered in Madison, North Carolina, Remington Arms Company, LLC (also called Remington Outdoor Company) is America's oldest continuously operating gun manufacturer. In the last several years, Remington has made significant strides in its commercial products for shooting sports, and in winning highly competitive military and law enforcement contracts. Remington is the only American manufacturer of both guns and ammunition.

In early 2014, Remington announced that they had selected Huntsville as the site of a new plant to develop and manufacture certain hand-held gun products. The plant – expected to have about 2,000 employees – will be located in a former Chrysler electronics facility on Wall-Triana Highway and Electronics Boulevard, near Huntsville International Airport; that plant was most recently occupied by the German firm, Continental AG.

Polaris Industries, Inc. – Huntsville (Expected) – In January 2015, Polaris Industries, headquartered in Minneapolis, Minnesota, announced plans to build a new production facility in Huntsville. Located in a 600,000-ft^2 facility to be built on a 453-acre site west of the city in the Greenbrier vicinity, the plant will focus on the development and production of off-road vehicles.

The facility will have multiple assembly lines equipped with state-of-the-art technologies. This will support several core processes, including vehicle assembly, chassis and body painting, welding, fabrication, and injection molding. At full capacity, the site is expected to employ at least 1,700 people.

Mid-Sized Technical Operations

In this section, high-technology companies in Greater Huntsville with local operations employing between 200 and 999 persons are listed alphabetically. These are given in two parts: those existing prior to the 21st century, and those with operations starting in Huntsville during the 21st century. Many of these firms have web pages giving additional information.

Mid-Sized Operations Dating from 20th Century

Most of those in this group are updated from previous chapters; thus, only basic information is given here. Readers may check prior chapters for details.

AI Signal Research, Inc. (1990, Updated) – 2001 Nichols Drive. ASRI received a NASA Exceptional Space Act Award and the Marshall Space Flight Center Software of the Year Award in 2009. As of 2013, Anthony J. Jones was President/CEO, and Jen-Yi Jong was Chairman and R&D Director. 315 local employees. NAICS: 541712

BASF Corporation - Huntsville (1974, Updated) – 9800 Kellner Road in Lowe Industrial Park; foreign owned, Ludwigshafen, Germany; U.S. Headquarters, Florham, New Jersey. Mark A. Todd, Site Manager. 430 employees. NAICS: 336312

Benchmark Electronics Inc. (1999, Updated) – 4807 Bradford Drive. As of 2013, Peter W. Cianfaglione was the Division General Manager, and there were about 320 employees. NAICS: 33441

Camber Corporation (1990, Updated) – In the early 2000s, Camber greatly expanded. In December 2008, New Mountain Capital – a firm dealing in private equity and growth capital investments – acquired ownership of Camber. By 2013, there were about 360 employees in Huntsville, and some 1,800 more in 100 other locations. Recent officials included John G. Lord, President, and Gene N.. Kakalec, EVP. NAICS: 541712

COLSA Corporation (1980, Updated) – COLSA continues with major operations in Huntsville – including operating the Army's Advanced Research Center – and six other facilities nationwide. As of 2013, executives include Francisco J. Collazo, Chairman and CEO; Richard Amos, President; Barry Gosnell, VP / COO; Ivan Garcia, VP / CTO; and John Welt, VP Advanced Research Center. There were 521 local employees. NAICS: 541712

Computer Sciences Corporation - Huntsville (1966, Updated) – CSC acquired Nichols Research Corporation in 2000; among the newly acquired activities was the services support to the Alabama Supercomputer Center in Huntsville. In 2010, CSC consolidated many of its Huntsville employees into its new 140,000-ft2 headquarters in Bridge Street Town Centre, at Old Madison Pike and Research Park Boulevard. As of 2013, Douglas Allen was the General Manager and there were about 500 local employees. NAICS: 54151

Davidson Technologies Inc. (1996, Updated) – Founder Julian Davidson died in 2013, and his wife, Dorothy Davidson, took over as Chairman / CEO. The Davidson Center for Space Exploration at the U.S. Space & Rocket Center is named for Julian and Dorothy Davidson. 230 employees. NAICS: 541712

DRS Test and Energy Management - Huntsville (1997, Updated) – In 2008, DRS Technologies was acquired by Finmeccanica S.P.A., the leading technology company in Italy, but continues under an American board. Timothy A. Smith was the VP / Site Manager 214 employees. NAICS: 33441

ERC, Inc. (1988, Updated) – See previous chapter. 600 employees nationally; local number not given. See previous chapter. NAICS: 541712

Jacobs Technology - Huntsville (1989, Updated) – 1500 Perimeter Parkway; Corp. Headquarters: Tullahoma, Tennessee. The Jacobs Engineering & Science Services and Skills Augmentation (ESSSA) group is the largest provider of support to NASA MSFC. Randy Lycans, General Manager. 643 employees. NAICS: 541712. Jacobs Information Technology Support Services (ITSS) is a separate operation; Gerald Burnett, General Manager, 100 employees. NAICS: 541541

LG Electronics Alabama, Inc. - Huntsville (1981, Updated) – Korean owned; technical and customer service and repair parts distribution for LG's North American market. KyuMoon Yu, President. 360 local employees. NAICS: 33431

Navistar Diesel of Alabama, Inc. - Huntsville (1999, Updated) – 646 James Record Road; Corp. Headquarters Lisle, Illinois. Lance Fulks, Plant Mgr.; Charles Sibley, Plant Mgr. 200 employees. NAICS: 33631

Parsons / SPARTA - Huntsville (1979, Updated) – In 2004, SPARTA established a company-owned 97,000-ft^2 facility at 401 Diamond Drive in Thornton Research Park; William D. Goodner was the Huntsville Operation Manager. Cobham, plc, a British defense company, acquired SPARTA in January 2008, changing the firm's name to Cobham Analytic Solutions. This operation was then sold to Parsons Corporation in October 2011. SPARTA, headquarteed in Lake Forest, California, now operates as a wholly-owned subsidiary of Parsons. Michael Byers and Philip Bolger, Huntsville Opns. Mgrs. 400 employees. NACS: 541712

Phoenix Industries (1973, Updated) – A non-profit organization. See previous chapter and a future section on non-profit organizations. 670 employees. NAICS: 42499

PPG Aerospace - Huntsville (1969, Updated) – 1719 Highway 72, Corp. Headquarters: Pittsburgh, Pennsylvania. Plant devoted to aircraft transparencies. Brian K. Pollock, Plant Manager. 750 employees. NAICS: 336413

QinetiQ North America, Inc. - Huntsville (1986, Updated) – 890 Explorer Boulevard. In October 2004, Westar was acquired by the giant British defense firm QinetiQ, forming the Defense Solution of QNA; Corp. Headquarters McLean, Virginia. As of 2013, QNA Defense Solutions had some 900 employees, with about 500 local; David A. Shrum was EVP / General Manager. NAICS: 541712

Factory-of-the-Future

Raytheon - Huntsville (1989, Updated) – In 2009, Raytheon selected the previous site of Thiokol on Redstone Arsenal for developing a "missile factory of the future" to produce Standard Missile 3 (SM-3) and its ship-defense SM-6. Using the latest robotics and computer-controlled tools, the first SM-6 left the plant in February 2013, and the first SM-3 came out the following June. Angel Crespo is Plant Manager and Randy Stevenson is Director of the Weapon Integration Center. Raytheon's Huntsville administration is at 401 Jan Davis Drive, where Kevin Byrnes is VP and Michael Ring is COO. There were 560 local employees. NAICS: 33641

Science and Engineering Services - Huntsville (1997, Updated) – In 2008, SES added Manufacturing West in the former facilities of Dunlop Tire Plant at 248 Dunlop Boulevard, occupying about 740,000 ft^2, of which 29,000 ft^2 is in hangar space. Edward. J. Sinclair, CEO; Russell C. Chunn Vice Chairman; and Paul Bogosian EVP. 760 employees. NAISC: Many, main ones 33251, 33271, 33451, 33531. 33641. 541511, 541519

Tec-Masters, Inc. (1988) – See previous chapter. 205 employees. NAICS: 541511, 541712

Mid-Sized Technical Operations Opened in 21st Century

Listed alphabetically in this section are the mid-sized technical operations that opened in Greater Huntsville after the start of the century. Some are locally owned new starts, and others are local operations of outside firms.

AAR Aerostructures & Interiors / AAR Integrated Technologies - Huntsville (2007) – Earlier Summa Technologies; see previous chapter. 140 Sparkman Drive; Corp. Headquarters Wood Dale, Illinois. Kevin Lawler, VP/GM A&I. Melissa Bosley, VP/GM IT. 235 employees. NAICS: 33641

AECOM Technology Corporation - Huntsville (2014) – 675 Discovery Drive; Corp. Headquarters Los Angeles, California. Previously URS Corporation, dating in Huntsville from early 2000s, was acquired by AECOM. Cynthia Crutchfield, VP; Mary Bissell, Director. 268 local employees. NAICS: 541712

ASRC Federal - Huntsville (2007) – 350 Voyager Way. In May 2007, Analytical Services was sold to ASRC Federal Holding Company of Beltville, Mayland. ASRC is itself a subsidiary of Arctic Slope Regional Corporation – an Alaska Native Corporation owned by approximately 11,000 Iñupiat shareholders. Mark M. Vaughn, Site Executive; Daniel Lambert, VP Operations; about 225 local employees. NAICS: 54171

Avocent / Emerson - Huntsville (2000) – 4991 Corporate Drive. Avocent Huntsville Corporation was formed in 2000 by the merger of Cybex Computer Products Corporation of Huntsville and Apex, Inc., of Redmond, Washington. Avocent's headquarters and major facilities were in Huntsville; John R. Cooper was President, and Doyle C. Weeks was EVP. In October 2009, Emerson Electric, a Fortune 500 corporation based in St. Louis, Missouri, acquired Avocent and made it a wholly-owned subsidiary of their Emerson Network Power division. As of 2013, officials were Stephen C. Hassell, President; Steven T. Geffin, VP Research; about 350 local employees. NAICS 334119

CGI - Huntsville (~2010) – 7067 Old Madison Pike; Canadian owned, U.S. Headquarters in Fairfax, Virginia. CGI has 68,000 professionals in 40 countries providing end-to-end IT and business process services. The Huntsville operation, under Scott Hill VP/GM, has about 275 employees. NAICS: 541712

Chucagh Government Solutions, LLC - Huntsville (~2009) – 4955 Corporate Drive; Corp. Headquarters Anchorage, Alaska. Mel Lynch, President. Representing more than 2,500 Aleut, Eskimo and Indian shareholders, Chugach Alaska Corporation was established in 1972 to serve the interests of the Alaska Native people of the Chugach region. About 400 employees. NAISC: 541712

Digital Fusion, Inc. - Huntsville (2002) – 4904 Research Drive; Corp. Headquarters Tampa Florida. 230 employees. NAICS: 541712

Hart & Cooley, Inc. - Huntsville (2000) – 4910 Moores Mill Road, Chase Industrial Park; Corp. Headquarters Holland, Michigan. Larry McAfee, Plant Manager. 250 Employees. NAICS: 33232

HudsonAlpha Institute for Biotechnology (2008) – See future section on non-profit organizations. 160 employees. NAICS: 541711, 541712

IBM Corporation - Huntsville (~2000*) – 6767 Old Madison Pike. *IBM has had various operations in Huntsville since the early 1960s. Essentially no information on the present operation is available. Michael Hawthorne, Client Executive; 200 employees. NAICS: 541511

Kennametal Firth Sterling - Huntsville (2000) – 7300 Madison Boulevard; Corp. Headquarters Latrobe, Pennsylvania. After Allegheny Technologies Incorporated (ATI) was reformed in 1999, the Huntsville portion became Firth Sterling – a name dating from the early 1900s. In 2012, this was sold to Kennametal, a large supplier of industrial materials. Huntsville products are primarily high-quality tungsten and tungsten-carbide powders. Stephen Meuler, Opns. Mgr. 245 employees. NAICS: 33141

MATSU Alabama, Inc. - Huntsville (2003) – 9650 Kellner Road, Lowe Industrial Park. Canadian owned. Motor vehicle metal stamping. Robert Todd, General Manager. 263 employees. NAICS: 33611

Torch Technologies, Inc. (2002) – 4035 Chris Drive; employee owned; founded by William Roark and Daniel Holder. Provides system engineering, applied science, modeling and simulation, and information technology. Scott Parker, COO; John Watson, President. 275 employees. NAICS: 541712

United Launch Alliance - Decatur (2006) – Although located in Decatur, Alabama (20 miles west of Huntsville), this firm is included because of its importance to the local technical community.

United Launch Alliance (ULA), LLC, is a joint venture of Lockheed Martin and Boeing, providing spacecraft launch services to the U.S. government. ULA headquarters and engineering are in Centennial, Colorado, and large manufacturing facilities are in Decatur and Harlingen, Texas. As of 2014, booster rockets built at Decatur were Atlas V, Delta II. Delta IV, and Delta IV Heavy; Cindy Nafus was the ULA Site Lead. NAISC: 33641

EFT-1 Boosters for Delta IV Heavy in Decatur

West Corporation - Huntsville (2000) – 5000 Bradford Drive; Corp. Headquarters Omaha, Nebraska. Teresa Stewart, Operations Director; 500 local employees. An in-bound center, taking technical, service, and sales calls from client's customers. NAISC: 561422

Wyle Laboratories / NTS - Huntsville (1949) – In 2010, Wyle purchased CAS, Inc., a Huntsville-based defense firm, and established the Wyle CAS Group as a part of the Huntsville operations. (See CAS / Wiley CAS earlier in this chapter.) In March 2014, California-based National Technical Systems, Inc. (NTS) acquired the testing business of Wyle in Huntsville. This was renamed NTS with Keith B. Wilson as VP and General Manager. NAICS: 541380

Yulista Aviation, Inc. - Huntsville (2007) – 631 Discovery Drive; Corp. Headquarters Anchorage, Alaska. An Alaska Native Corporation operating in Huntsville, providing aviation maintenance and modification work. Darrell Harrison, President; 360 local employees. NAICS: 33441, 33451, 33641, 541380, 54171

Smaller Technical Operations

For purposes of this book, technical firms with less than 200 local employees are classified as smaller operations. The previously cited publication, the CCHMC's 2014 *Industrial Directory*, lists 539 firms having between 1 and 199 local employees. While it is recognized that many of these firms are making important technical contributions, it is impractical to even present a full list.

Smaller Technical Firms Listing

The following is an alphabetical list of technical operations presently having between 75 and 199 local employees, or being in operation for many years, or having received some special recognition. Most of those that have not been included in previous chapters are given some minimal description. Many of these firms have web pages and might be visited for more complete information.

Advanced Systems Development, Inc. (2001) – 103 Quality Circle. Minority owned; IT solutions. Richard Bennett, President. NAICS: 54 1511

AEgis Technologies Group (1988) – See previous chapter. NAICS: 541511, 541512, 541712

Amtec Corporation (1988) – See previous chapter. NAICS: 33641, 541712

A-P-T Research, Inc. (1993) – 4950 Research Drive. Employee owned; system analysis and software. J. Thomas Pfitzer, President / CEO. NAICS: 541712

Aranea Solutions, Inc. (2002) – 5030 Bradford. Drive. Woman owned; data analysis. Dwayne Bradford, President; Kimberly Colette, CEO. NAICS: 541512

ATI Specialty Alloys and Compounds / ATI Wah Chang - Huntsville (1960) – 7600 Madison Blvd.; Corp. Headquarters Pittsburgh, Pennsylvania. Jarod Golden, Opns. Mgr. See Wah Chang in Chapter III; renamed in 2014. NAICS: 33141

ATK - Huntsville (2001) – 620 Discovery Drive; Corp. Headquarters Minneapolis, Minnesota. Earlier ATK Thiokol, then ATK Propulsion, then ATK Aerospace. In early 2015, Orbital Sciences and ATK Aerospace merged forming Orbital ATK, headquartered in Dulles, Virginia. NAICS: 33641

Aviagen, Inc. (1995) – See previous chapter. NAICS 42444

Avion Solutions, Inc. (1992) – See previous chapter. NAICS: 541511, 541512, 541712

BAE Systems- Huntsville (2003) – 308 Voyager Way; Corp. Headquarters London; air/missile defense. Randall Bounds, Dir. Bus. Dev. NAICS: 541712

Baron Services, Inc. (1989) – See previous chapter. NAICS: 33451

BASF Corporation (1974) – See previous chapter. NAICS: 33631
Bastion Technologies, Inc. (1982) – See previous chapter. NAICS: 741712

BCF Solutions - Huntsville (2003) – 2404 Commerce Court; Corp. Headquarters Arlington, Virginia. Brady Foster, CEO; Chand Gupta, Presidebt/COO; Adam Mungo, Director Def, & Space Div. NAICS: 541712

Booz, Allen & Hamilton - Huntsville (1986) – See previous chapter. NAICS: 541712

Brown Precision, Inc. (1986) – See previous chapter. NAICS: 33271

CFD Research Corp. (1987) – See previous chapter. NAICS: 541511, 541711

Chandler/May, Inc. (1990) – See previous chapter. NAICS: 33541

Computing Technologies, Inc. - Huntsville (1992) – See previous chapter. NAICS: 51711, 541512

deciBel Research, Inc. (2002) – 325 Bob Heath Drive. Jeff Gronberg, President; Enrico Poggio, Chief Scientist. Specializing in high-fidelity modeling of advanced radar systems. NAICS: 541511, 541519, 541712

Delta Research, Inc. (1978) – See previous chapter. NAICS: 541712

DESE Research, Inc. (1982) – See previous chapter. NAICS: 541512, 541712

DHS Systems LLC - Huntsville (2002) – 5855 Endeavor Way; Corp. Headquarters Orangeburg, New York. Fabricating rapid-assembly shelters. Jerald McAbee, GM/VP Opns. NAICS: 33621

Digium, Inc, (1999) – See previous chapter. NAICS: 33421, 33429

Ducommun Miltec (1997) – See previous chapter. NAICS: 541712

Excellance, Inc. (1975) – See previous chapter. NAICS: 33611, 33621

FPMI Solutions, Inc. (2004) – 245 Business Park Blvd., Madison. Human capital management services. Stephen Moffett, President; Clyde Blandford, EVP. NAICS: 541511, 541512, 56142

Future Research Corporation (1995) – 675 Discovery Drive. Minority owned; software and computer graphics. Jesse Nunn, CEO; James Williams, VP Opns. NAICS: 51711, 541511, 541712

Garver Engineers – Huntsville (1998) – See previous chapter. NAICS: 541330, 541370

GE Intelligent Platforms - Huntsville (2000) – 12090 Memorial Parkway; Corp. Headquarters Charlottesville, Virginia. Expertise in automated and embedded computing. NAICS: 33531

General Dynamics C4 Systems - Huntsville (1997) – 6000 Technology Drive; Corp. Headquarters Falls Church, Virginia. Specializing in command and control systems. Eric Nelson, Site Mgr. NAICS: 33422

General Products Partners, Inc. (1970) – See previous chapter. NAICS: 33232, 33271

Gleason Research Associates (1982) –See previous chapter. NAICS: 541712

G.W. Jones & Sons Engineers, Inc. (1895) – See previous chapter. NAIS: 54133, 54137

HudsonAlpha Institute for Biotechnology (2005) – 601 Genome Way; a not-for-profit organization. James R. Hudson, Jr., Co-Chairman. See description in later section. NAICS: 541711, 541712

InfoPro Corporation (1985) – See previous chapter. NAICS: 541511

Integration Innovation, Inc. (2007) – 689 Discovery Drive. Michael Wicks, President/CEO; Kenneth Shannon, COO; Walter Strankman, VP Operations. Training and technology development services. NAICS: 541511

KAYA Associates, Inc. (2004) – 101 Quality Circle. Minority owned; facility planning, installation management, and IT services. John Prince, CEO; Scott Kowerduck, President. NAICS: 541512

KBM, Inc. (1982) – 700 Boulevard South. Woman owned; support to R&D, IT, and electronics manufacturing. Daniel Knight, VP Pgm. Dev. NAICS: 33411, 33422, 541380, 541511, 541512, 541513, 541712

Kratos Defense and Security - Huntsville (2007) – 4904 Research Drive; Corp. Heasdquarters, San Diego, California. Opened with the acquisition of Madison Research (see precious chapter). Richard Selvaggio, President, Weapons Systems Solutions; Jennifer Ricker, Sr.VP. NAICS: 541712

LogiCore Corporation (2002) – 360 Quality Circle; Miranda Bouldin, President/CEO. I ntegration and life-cycle support. NAICS: 541519, 541712

MacAulay-Brown, Inc. - Huntsville (2008) – Formerly Gray Research. See previous chapter. NAICS: 541712

ManTech Nexolve Corpration - Huntsville (2007) – 655 Discovery Drive; Corp. Headquarters Fairfax, Virginia. James Moore, Div.VP. High-technology engineering and scientific services. NAICS: 541712

MCR, LLC – Huntsville (1998) – 700 Boulevard South; Corp. Headquarters McLean, Virginia. Integrated program management services. Ronald Duke, COO. NAICS: 541712

Mentor Graphics - Huntsville (2000) – 5030 Bradford Drive; Corp. Headquarters Wilsonville, Oregon. Henry Potts, VP. Graphics software for electroncs products. NAICS: 541511

Millennium System Services, Inc. (2000) – 555 Sparkman Drive; Amy Cassibry, President/COO. Logistics, management, and programmatic services. NAICS: 33641

MJLM Engineering & Technical Services - Huntsville (2003) – 4825 University Square; Corp. Headquarters Houston, Texas. Minority owned. Russell King, Site Leader. NAICS: 541380, 541511, 541512, 541513, 541712

Mitchell Plastics - Huntsville (2005) – 1519 Highway 72 East; Corp. Headquarters Kitchener, Onterio, Canada. Rebecca Branca, Gen. Mgr. Injection molding for automotive interiors. NAICS: 32619

National Copper & Smelting (1973) – 3333 Stanwood Boulevard; John Guthrie, President. Precision manufacturing of copper tubing. NAICS: 33142

Nektar Therapeutics - Huntsville (1992) – 1112 Church Street; Corp. Headquarters San Carlos, California; Anthony Sanders, VP Mfg./ Site Mgr. Industry leading pulmonary and PEGylation platforms. NAICS: 32541

NEXTEX, Inc. (1995) – 201 Next Tech Drive. James Treummer, President/CEO; John Roberts, CEO. Manufacturing services, from design through mid-volume production. NAICS: 33411

Norris Cylinder Company - Huntsville (1948) – See previous chapter. NAICS: 33231

PALCO Telcom Services (1986) – 2914 Green Cove Road; Janice Migliore, CEO/Chairperson; Randolph Piechocki, President. Repair / refurbishing of telecommunication network components. NAICS: 33421

Parker Hannifin Corporation - Huntsville (1961) – See Chapter III. NAICS 33291

PeopleTec, Inc. (2005) – 4901 Corporate Drive; woman owned. Terry Jennings, CEO; Doug Scalf, President. Engineering, modeling, and cyber security. NAICS: 51821, 541511, 541512, 531513, 54151, 541711, 541712

Plastic Fusing Fabricators, Inc. (1981) – 3455 Stanwood Blvd; Terry Ragsdale, President. Custom containment solutions using polyethylene pipe. NAICS: 32619

Pratt and Whitney / Aerojet Rocetdyne - Huntsville ('1983) – 555 Discovery Drive. Frederick D. Bachtel, Gen. Mgr. Alabama Operations. The Pratt & Whitney Rocketdyne (PWR) formed in 2005. June 2013, acquired by GenCorp, forming Aerojet Rocketdyne. Corp. Headquarters Sacramento, California. NAICS: 336414

Qualis Corporation (1993) – 689 Discovery Drive; Elizabeth Morard, President /CEO; Thad Mauldin, VP/COO. Engineering, test, and program support services. NAICS: 541712

Qualtech NP - Huntsville (1983) – 125 West Park Loop; Corp. Headquarters Idaho Falls, Idaho. A business unit of Curtis Wright Flow Control. C. Wiley Finley, Sr. Dir.; Ronald Golub, Engr. Mgr. NAICS: 541712

Quantum Research International, Inc. (1987) – See previous chapter. NAICS: 541712

Radiance Technologies (1999) – See previous chapter. NAICS: 541712

Ridgeview Industries - Huntsville (~1990) – 15990 Chaney Thompson Road; Corp. Headquarters Grand Rapids, Michigan. Steven Coste, Plant Mgr. Steel stamping, welding, and assembly for automotive industry NAICS: 33637

S3 (System Studies and Simulation) (1993) – See previous chapter. NAICS: 33641

Saint Gobain Grains & Powders - Huntsville (1955) – See previous chapter. NAICS: 32721, 32791

Schafer Corporation - Huntsville (1995) – See previous chapter. NAICS: 741912

Schwarze Industries - Huntsville (1974) – See previous chapter. NAICS: 3361126

Siemens PLM Software, Inc. - Huntsville (1998) – 675 Discovery Drive; Corp. Headquarters Plano, Texas. William McClure, VP; Daniel Staples, Director, Solid Edge Software Development. NAICS: 451712

Sigmatech, Inc. (1986) – See previous chapter. NAICS: 541712

Sikorski Aircraft Corporation - Huntsville (2000) – 5025 Bradford Drive; Corp. Headquarters Stamford, Connecticut; UTC Subsidiary. Theresa Barton, Opns. Mgr.; Rick Isaacs, Programs Integrator. NAICS: 33641

Strata-G Solutions, Inc. (2004) – 4906 Research Drive; Devin Whitaker, Founder / CEO. Integration and innovation expertise for warfighter needs. NAICS: 541712

Tabor Machine Company, Inc. (1976) – 12529 Memorial Parkway S. Carol Tabor, President. Precision maching. NAICS: 332322, 332710, 332999

Technology Service Corp Phase IV - Huntsville (1985) – See previous chapter. NAICS: 33441, 33451, 33641

Telcordia Technologies (1996) – 7800 Madison Bvld; Lori Opp, Exec. Dir. Telecommunications software and services. NAICS: 511210

Tetra Tech EMC, Inc. - Huntsville (1989) – 101 Quality Circle; Corp. Headquarters Camarillo, California. Civil engineering services. NAICS: 541712

T-H Marine Supply, Inc. (1975) – See previous chapter. NAICS: 32619, 33992

Triedum Corporation (2005) – 655 Discovefy Drive. Minority owned. Van Sullivan, President; Kurt Lessmann, VP; Lewis Hundley, VP. Technology R&D to analysis and design. NAICS: 51821, 541511, 541512, 541712

Tyonek Manufacturing Group - Madison (2004) – 229 Palmer Road, Madison; Headquarters, Anchorage, Alaska. Newman Sufflebarger, President. Prototyping and low- to high-rate production. NAICS: 33641

United Plating, Inc. (1973) – 3400 Stanwood Blvd., James Caudle, CEO; Michael Fann, President. Electroplating, painting, and anodizing. NAICS: 33281

VT Group (1997) – See previous chapter. NAICS: 541712

Watring Technologies (1997) – 2120 Meridian Street. Minority owned; Dale Watring, President/CEO. System integration and process automation hardware. NAICS: 541712

WestWind Group, Inc. (1999) – See previous chapter. NAICS: many, including 33251, 33271, 33411, 33441, 33451, 33531, 33641

OTHER TECHNICAL OPERATIONS

Three other general categories of organizations are important in the technical evolution of Greater Huntsville. These are non-profit firms, public service operations, and university research institutes.

Technical Non-Profit Operations

There are many organizations in Greater Huntsville that are certified as non-profit or not-for-profit, but only a few of these are technical organizations. Two of the best known are the HudsonAlpha Institute for Biotechnology and the Phoenix Industries operated by the Huntsville Rehabilitation Foundation. While these are extremely different organizations, both have a significant role in Greater Huntsville's technological evolution.

HudsonAlpha Institute for Biotechnology (2008) – James R. Hudson, Jr. and Lonnie S. McMillian., recognizing the potential of biotechnology for Greater Huntsville, led in initiating the not-for-profit HudsonAlpha Institute for Biotechnology in 2008. Richard M. Meyers was hired as President and

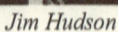

Jim Hudson   *Lonnie McMillian*

Research Director; Meyers was previously the Director of the Department of Genetics at the Stanford School of Medicine.

In short order, this Institute became the cornerstone of a 152-acre Biotech Campus at 601 Genome Way in West Cummings Research Park. Although Milton Cummings, the forward-thinking founder of this research park, has long-since been deceased, he would have been greatly pleased with this new developmental thrust.

Richard Meyers

The Institute laid out the new campus in a pattern symbolizing the DNA's double helix structure and built a 270,000 ft^2 central facility. This houses both a Genome Sequencing Center and Genomic Services Laboratory. A large portion of the area is in efficient and flexibly reconfigurable research space for wet chemistry, freezer farms, vivarium, biology stability testing, tissue culture, cell culture, microbiology, analytical chemistry, and bioscience clean laboratories. Its directive combines academic research, education, and incubation for applied research companies.

As of 2014, James (Jim) Hudson was Chairman; Richard Myers, President/Director; and O'Neal Smitherman, EVP; and there were 160 employees   Under the leadership of Jane Grimwood and Jeremy Schmutz, the Genome Sequencing Center is one of the few centers in the world performing *de novo* sequencing of plants and specializing in applying genomic

HudsonAlpha Institute

techniques to understand how plants function in response to environmental stimuli – highly important for research in food production and renewable biofuels.

Phoenix Industries (1969) – Since 1969, the Huntsville Rehabilitation Foundation – a 501(c)(3) organization – has done business as Phoenix Industries of Huntsville; this is the totally self-supporting industrial side of the Foundation's comprehensive community-based vocational rehabilitation programs. Through contracts with commercial firms and government procuring units, Phoenix provides work opportunities to individuals who need rehabilitation.

Bryan Dodson

As of 2013, Phoenix Industries had about 570 employees, at least 75 percent of whom had a vocational disability. H. Bryan Dodson has served as President since 1981; David D. Perez is Senior VP for Vocational Services, Wesley C. Tyler is VP of the Manufacturing Division, and Tim M. Stickley is VP for the Services Division. The main facility is at 2939 Johnson Road, SW.

Manufacturing and service work includes heavy-duty sewing for military application, apparel sewing for government procurement units, other light sewing work, electronic assembly / disassembly, parts sorting and inspection, mail room operations, and custodial and grounds maintenance services.

Sewing Flags at Phoenix Industries

Public Service Operations

The Greater Huntsville area has several public service activities that are certainly contributors to the area's technical evolution. Described here are the Huntsville Utilities, the Huntsville Hospital System, the U.S. Space & Rocket Center, and the Alabama Supercomputer Center. Most of these have been described earlier, and thus an updating is included.

Huntsville Utilities – Owned by the City of Huntsville, Huntsville Utilities (HU) is composed of three separate operations: HU Water System, HU Gas System, and HU Electric System, each with a Board appointed by the City Council. The earlier chapters give the history of these operations – the water system being the first in Alabama. The three systems each have technical operations but share top management, customer services, meter reading, and billing. Officially organized in 1940, as of 2013, Huntsville Utilities had some 625 employees, divided about equally between administration and technical; Stacy S. Cantrell was the overall Engineering Superintendent. The main offices were located at 112 Spragins Avenue in downtown Huntsville.

Huntsville's first elementary water works were built in 1823; the initial supply source was the Big Spring. The City of Huntsville acquired the privately owned system in 1858. Today, the HU Water System is supplied from several wells and the Tennessee River and uses two treatment plants, each with a capacity of 48 million gallons per day (MGD). The average combined production is about 35 MGD, supplying some 91 thousand local customers including Redstone Arsenal. There are about 1,000 miles of water mains and 52 million gallons in storage tanks.

A plant for producing manufactured gas opened in Huntsville by a private investor in 1856; the operation remained relatively unchanged until acquired by Alabama Gas Corporation in 1946, and converted to shipped-in liquefied propane. The City of Huntsville bought the gasworks system in 1950. Today, the HU Gas System distributes natural gas purchased from a number of suppliers. Two major pipelines bring the gas to Huntsville from both onshore and offshore sources. The system serves close to 50 thousand local customers.

Electrical power was introduced into the area by a private firm, the Huntsville Electric Company, in 1887; the electrical energy came from a local steam-powered generating plant. After several ownership changes, Alabama Power Company purchased the Huntsville operations in 1915. Shortly thereafter, Huntsville was connected into an Alabama Power network that delivered power generated at hydroelectric plants on several rivers. As part of the emerging Huntsville Utilities, the City of Huntsville acquired the electrical system in Madison County from the Alabama Power Company in 1940, and still provides electrical power for all of Madison County. HU also began the still-existing practice of buying power from the Tennessee Valley Authority (TVA).

As of 2014, the HU Electric System had 97 substations driven through 4 independent feeder routes from TVA, and 4,572 miles of distribution lines. There were close to 177 thousand customers, with an average monthly peak-

power demand of 1,016 megawatts. A Dispatch Center, staffed 24/7, monitors all of HU's electric system, as well as gas and water distribution, and supports crews in the field.

In efforts toward increasing renewable energy development, 30 HU customers have solar panels connected through TVA's Green Power Providers Program.

HU Dispatch Center

Huntsville Hospital System – The second largest hospital in Alabama, Huntsville Hospital System has a central campus at 101 Sivley Road. It is a 941-bed publicly owned organization that serves as the regional referral center for north Alabama and southern Tennessee.

Huntsville Hospital Central Campus

Dating from 1895, Huntsville Hospital was deeded over to the City of Huntsville in 1961. In recent years the hospital has expanded its service throughout the region with the development of Huntsville Hospital Health System. As of 2014, this was the fifth largest publicly owned hospital system in the nation with nine hospitals, more than 1,800 beds, and 12,000 employees. David S. Spillers is the CEO and Philip W. Bentley, Jr., is the Board Chairman.

While there are hundreds of nurses and technicians using the latest equipment throughout the facilities, Huntsville Hospital more than qualifies as a high-

technology operation through its Clinical Laboratory. In early 2015, the Clinical Laboratory – the eighth largest hospital-based laboratory facility in the U.S. – began functioning in its new facility in Twickenham Square, connected to the hospital via a walking

Clinical Laboratory

bridge and a high-speed specimen transfer network. With a staff of over 200 technicians and supporting personnel working 24/7, the expanded facility is capable of handling close to 8,000 specimens daily. More than 750 physicians rely on the laboratory for assistance in diagnosing patient conditions. Vicky McClain is the Laboratory Services Director.

The U.S. Space & Rocket Center – While serving as the greatest tourist attraction in Alabama, the U.S. Space & Rocket Center is also an important technical operation. Commonly called the Space Center, it is owned by the State of Alabama and operated by the Alabama Space Science Exhibit Commission. Opened in 1970, Edward O. Buckbee was the founding Director and served until 199. The Space Center is located at Exit 15 on Interstate 565, adjacent to Redstone Arsenal.

The newest addition to the Space Center is the Davidson Center for Space Exploration, named after Julian Davidson, founder of Davidson Technologies. Opened in January 2008, the 68,000-ft^2 (6,300-m^2) building was designed to house the SATURN V and other space exploration exhibits. The vehicle is elevated above the floor surface with separated stages and engines exposed, forming the ceiling of a huge meeting room.

Central Hall of Davidson Center

Following Buckbee, Michael Wing served as the Director for two years, and almost financially destroyed the operations. Larry R. Capps was the CEO

Deborah Barnhart

during a recovery and growth period between 1999 and 2011; a NASA Educator Resource Center was opened during Capp's tenure. In December 2012, former Navy Captain and Vanderbilt Doctorate Deborah E. Barnhart was appointed the CEO; since then, she has implemented many additions and improvements.

As of 2013, with 584,000 visitors, the Space Center was recognized as the top paid tourist attraction in Alabama. It serves as NASA-MSFC's Visitor Center and is a Smithsonian Affiliate. The Space Center has been the setting for a number of feature movies. (See Chapter III for background information.)

Supercomputer Center and Network –The Alabama Supercomputer Center (ASC) in Huntsville and the associated statewide Alabama Supercomputer Research and Education Network are both under the Alabama Supercomputer Authority (ASA), a public-service corporation headquartered in Montgomery. (See the previous chapter for background information.) After first becoming operational in 1988, the Center and Network continued without significant problems through the 1990s. Ralph D. Fulmer, II became the CEO of the ASA in 2002. From the beginning, a professional services contractor operated the ACS; this effort was rebid in 2004, and the contract awarded to the existing contractor, Computer Sciences Corporation (CSC).

During the 2000s, the supercomputer clusters were continuously changed to improve the performance. A new system, a Cray XD1, was installed in 2004; this had 144 processors running at 2.2 GHz, 240 gigabytes of memory, and 7 terabytes of shared disk. This system was decommissioned in early 2009.

Central Control

Further upgrades at the ASC facility involved SGI Altix 350, SGI UV, and DMC supercomputers. First installed in 2004, the SGI Altix 350 now has1800 CPU cores, 10.1 terabytes of memory, and 225 terabytes of disk space. The SGI Ultraviolet 2000 (named UV), was installed at the end of 2012; it has twelve processor cores, a single large compute node with 256 processor cores, and 4 terabytes of memory. The Dense Memory Cluster (DMC), installed in 2008, was built at ASC using component units from many vendors; in its present configuration it has 1800 CPU cores, 10.1 terabytes of memory, and 225 terabytes of disk space.

From the Cray X-MP up to the present configuration, the CPU processing power of ASC computers has grown over 180,000 fold, the memory capacity by a factor of 447,000, and the disk capacity over 20,000 times.

During the 2000s, the network constantly expanded in both the number of organizations connected as well as the bandwidth – from T1 links to DS3 connections, to OC3 connections, to 10-gigabit fiber-optic connections. Atlanta and Dallas are the Internet egress points for commodity internet as well as the research and education-specific internet, National Lambda Rail (NLR ceased operations in March 2014). Now called the Alabama Research and Education Network (AREN) with about 225 client organizations, there are over 1000 items of network equipment, and it extends into every county in the State.

As of 2013, John M. Wade was the CEO of the ASA. In Huntsville, K. Wayne McVay was the Facility Manager, and David R. Ivey the CSC Program Manager.

University Organized Research

The University of Alabama in Huntsville – As described in Chapter III, the initial activities in Huntsville of the University of Alabama included contracted research work. A Research Institute was funded by the State Legislature in 1961, and formal research activities have grown continuously since then. UAH faculty, staff, and students presently conduct research in association with world-class experts from NASA's Marshall Space Flight Center and important Department of Defense partners. It also performs cutting-edge research for the National Science Foundation, the National Oceanic and Atmospheric Administration, the Department of Energy, and the National Institutes of Health.

UAH ranks 16th nationally in universities with DoD funded research, and 15th in NASA research. Their research capabilities in Astrophysics,

Cybersecurity, Data Analytics, Logistics and Supply Chain Management, Optical Systems and Engineering, Reliability and Failure Analysis, Rotorcraft and Unmanned Systems, Severe Weather, Space Propulsion, and many other areas are nationally and internationally known. Recently, UAH researchers have averaged about $80 million per year in contracts and grants. Rayford (Ray) B. Vaughn, Jr., is the UAH Vice President for Research and Economic Development.

Ray Vaughn

Alabama A&M University – For many years, the faculty and departments of Alabama A&M University have received grants for research in agriculture and many areas of science, technology, engineering, and management (STEM). A formal Research Institute operated for several years, but was then taken private. Founded in 2014, the Research, Innovation, Science, and Engineering (RISE) Foundation at AAMU was established to enhance contractual opportunities, technology transfers, and faculty and student research in the STEM areas. A 501(c)3 subsidiary holding a facility security clearance, RISE is an extension of the AAMU's scientific impact on the surrounding community and an agile organization to swiftly pursue opportunities with government, industrial, and academic organizations.

Chance Glenn

Chance M. Glenn, Sr., is the President/Executive Director of RISE. Paul B. Ruffin and R. Barry Johnson, both with highly recognized R&D experience, serve as Senior Advisors.

ASSOCIATIONS OF PROFESSIONAL SOCIETIES

"Birds of a feather flock together." This saying is certainly true of the technical professionals in Greater Huntsville, and the names of local technical societies give a good representation of the multitude of disciplines functioning in the area. The first such local society has not been identified, but it is likely that the Textile Engineering Society had a chapter here in the 1930s, then the American Chemical Society possibly came in the early 1940s. With the advent of rocket development in the late 1940s, chapters of many technical societies were started here.

In 1969, the Huntsville Association of Technical Societies (HATS, new renamed The Association) was established for informing, coordinating, and networking on matters of concern to technical professionals and Huntsville area nonprofit organizations. HATS started with chapters of eight national and local societies and has grown to 36 member societies.

The largest local chapters, sections, and societies (including a number that are not in HATS) and representing an estimated 20,000 individuals, include the following:

ACTE - Association for Career and Technical Education, Alabama Chapter

AFA - Air Force Association, Tennessee Valley Chapter

AFCEA - Armed Forces Communications and Electronics Association, Huntsville Chapter

AHS - American Helicopter Society, Redstone Chapter

AIAA - American Institute of Aeronautics and Astronautics, Greater Huntsville Section

AIChE - American Institute of Chemical Engineers, North Alabama Section

APICS - Association for Operations Management, Tennessee Valley Chapter

ASA - Alabama Solar Association

ASCE - American Society of Civil Engineers, Alabama Section, Huntsville Branch

ASME - American Society of Mechanical Engineers, North Alabama Section

ASPA – Army Space Professionals Association, Rocket City Chapter

ASPE - Alabama Society of Professional Engineers, (NSPE)

ASQ - American Society for Quality, Huntsville Section

AUSA - Association of the United States Army, Redstone-Huntsville Chapter

HAL5 - Huntsville Chapter of the National Space Society

HASBAT - Huntsville Association of Small Business in Advanced Technology

HEOS - Huntsville Electro Optical Society

IEEE - Institute of Electrical and Electronics Engineers, Huntsville Section

ISA - International Society of Automation, North Alabama Section

INCOSE - International Council On Systems Engineering, Huntsville Chapter

ITEA - International Test & Evaluation Association, Rocket City Chapter

NCMA - National Contract Management Association, Huntsville Chapter

NDIA - National Defense Industrial Association, Tennessee Valley Chapter

NEACA - Northeast Alabama Craftsman's Association

NSBE - National Society of Black Engineers, (NSBE-NAAE)

NSCH - National Space Club of Huntsville, Huntsville Chapter

PMI - Project Management Institute, North Alabama Chapter

SAME - Society of American Military Engineers, Huntsville Post

SCEA - Society of Cost Estimating and Analysis, Greater Alabama Chapter

SOLE - Society of Logistics Engineers, Tennessee Valley Chapter

SRE - Society of Reliability Engineers, Huntsville Chapter

SSS - System Safety Society, Tennessee Valley Chapter

STC- Society for Technical Communication, North Alabama Chapter

VBAS - Von Braun Astronomical Society

WID – Women in Defense, Tennessee Valley Chapter

It is noted that many large national societies do not have local representation except as associated with higher education institutions; thus, it is likely that there are a number of other professional associations that are not on this list.

In addition to the listed associations, there are a number of affiliated working groups and other activities that regularly meet in Greater Huntsville. One of the largest and most active is the Space and Missile Defense Working Group (SMDWG), an affiliate of the NDIA.

AGRICULTURE – STILL IMPORTANT

As it has for two centuries, agriculture still has a major role in Alabama. According to statistics from the Alabama Farmer's Federation for 2010-2011 (the latest available), total farm receipts were over $5 billion for the year. The top farm commodities for cash receipts were (1) poultry, (2) cattle and calves, (3) greenhouse and nursery, (4) cotton, and (5) soybeans. At the start of the 21st century, Alabama had 46.0 thousand farms with a total of 8.9 million acres (average 193 acres per farm); in 2010, the farm number had increased to 48.5 thousand and 9.0 million acres (186 acres per farm). Alabama also had some 22 million acres in vital forest land.

The Tennessee Valley Research and Extension Center (TVREC) is a 755-acre experimental farm located at Belle Mina in adjacent Limestone County; it is Jointly operated by Auburn and Alabama A&M Universities. Although TVREC research projects include corn, soybeans, and wheat, it is for its cotton research that the Center is nationally recognized.

TVREC at Belle Mina

The TVREC was an early convert to precision agriculture technologies and has a major effort in the evaluation of geospatial navigation tools for efficient application of plant nutrients; they have a full inventory of auto-steer tractors, yield monitors, and light-bar guidance systems. As of 2010, GPS-based technology was being used by 34 percent of Alabama farmers in precision agriculture, reducing chemical costs while increasing yields. A new administrative building at the TVREC was opened in 2012, made possible by a $1 million donation from the Alabama Farmers Federation

Although Greater Huntsville has become one of the principal technology development centers of America, Madison County remains one of the State's leading agriculture areas. As previously noted, Madison County had a total population of near 347,000 in 2013; of these, some 113,000 (about 33 percent)

lived in rural areas, many of them engaged in full- or part-time farming. The Madison County Extension Office is located in Huntsville at 819 Cook Avenue, with Walter B. Harris as the County Extension Coordinator. Dale W. Strong is Chairman of the Madison County Commissioners.

The following chart shows agricultural statistics for Madison County since the area entered the high-technology field – the middle of the 20th century.

	1949	1999	2012
Number of Farms	5,004	973	1,033
Total Farm Land	415,382	210,168	209,253
Percent of County	80	40	40
Average Size, Acres	83	216	203
Cotton, Acres	109,400	44,400	27,850
Cotton, Bales	50,800	51,500	38,300
Cotton Yield, lb/acre	223	557	632
Corn, Acres	56,400	9,800	28,341
Soy Beans, Acres	5,400	27,500	32,244

In 2012 (the latest year for a full report), Madison ranked first in the State in cotton acreage and second in soybeans and corn acreage. In that year, cotton farms in Madison County produced only 75 percent of the number of cotton lint bales produced a decade earlier, possibly indicating that the market demand for American-grown cotton is declining. In terms of acreage, both soy beans and corn exceeded that for cotton. It is also important to note that the average age of the principal operator was 60 years.

The area devoted to farming in Madison County has been continuously shrinking – decreasing more than 50 percent in the first decade of the 21st century. Urban sprawl from Huntsville, Madison (the fastest growing city in Alabama), Decatur, and Athens has taken a lot of the former farmland. Much of the land still in agriculture has small, suburban, and exurban hobby farms, often run by urban commuters. Nevertheless, the amount of cotton lint produced remains substantial – thanks largely to a continuing increase in yield and the increased capability of harvesting equipment.

In 60 years, the Madison County cotton yield increased by a factor of some 2.8 – from 223 to 632 pounds of lint per acre. There are four general methods for increasing yield: more fertilizer, insect control, weed management, and seed genetics. Fertilizer has always been used to augment the natural condition of soil, but it is only applicable to a limit and there have been few increases in effectiveness for many years.

Insect control using aircraft for crop dusting came into being with the advent of boll weevils; it was later used to kill other insects as well as weeds. This is a dangerous and expensive practice, but is still in limited use, particularly on very large farms; there is still at least one service available in Madison County. The chemical glyphosate was discovered to be a highly effective herbicide and was brought to market by Monsanto under the trade named "Roundup" in 1973. It soon became the most effective and most commonly used herbicide worldwide; upon expiration of the original patent, glyphosate is now available under a wide variety of names from many sources.

As noted in the previous chapter, mechanical cotton-harvesting equipment came into widespread use during the last decades of the 20th century, reducing

by a factor of five the man-hours required to produce a cotton crop. This equipment continued to evolve, with highly complex harvesters available from Allis-Chambers, Case-IH, and John Deere by 2012.

The Case-IH Module Express 635 is typical of these. Handling five cotton rows simultaneously, it uses barbed spindles rotating at high speeds to

Modern Cotton Harvester

remove the seed cotton from the plants; the seed cotton is then removed from the spindles by a counter-rotating doffer and blown into a collecting basket. Picking speed is usually between four- and five-MPH. When full, the basket is dumped into a module builder where it is compacted into 'bricks' of seed cotton weighing about 21,000 lb (16 un-ginned bales). These bricks can be left in the field or transported to the gin for ginning. The air-conditioned cab has television monitors, a GPS tracking unit, and a computer data-recording unit.

It was earlier noted that the author of this book was raised on a small farm and made his first bale of cotton when he was nine. After becoming an engineer, he worked two years for J. I. Case Company in designing and field-testing agricultural equipment.

Selected Bibliography / Additional Information for Chapter V

"Alabama Supercomputer Authority",
https://www.asc.edu/aboutasa/historyofASA.shtml

"AMCOM – Warfighter Readiness, Now and in 2025," AMCOM Commander Update, *Army Aviation Magazine*; 2 Feb. 2015.
http://armyaviationmagazine.com/index.php/archive

"AMRDEC – Aviation and Missile Research, Development, and Engineering Center," U.S. Army: http://www.amrdec.army.mil/AMRDEC/

"Army Aviation and Missile Command," *Wikipedia*; http://en.wikipedia.org/wiki/United_States_Army_Aviation_and_Missile_Command

"Army Aviation and Missile Research, Development and Engineering Center," *Wikipedia*; http://en.wikipedia.org/wiki/Aviation_and_Missile_Research,_Development,_and_Engineering_Center

"Army Contracting Command," *Wikipedia*; http://en.wikipedia.org/wiki/Army_Contracting_Command

"Army Materiel Command," *Wikipedia*; http://en.wikipedia.org/wiki/United_States_Army_Materiel_Command

"Industrial Directory," Huntsville / Madison County Chamber of Commerce; Twenty-fifth Edition, 2014

Marshall Space Flight Center, "Core Capabilities and Services", 2014; http://www.nasa.gov/sites/default/files/files/MSFC_Capabilities_09_2014_sm.pdf

"Marshall Space Flight Center," *Wikipedia*; http://en.wikipedia.org/wiki/Marshall_Space_Flight_Center

"MSFC 2014 Update," http://www.nasa.gov/sites/default/files/files/MarshallUpdate_V_6.pdf

"Space Launch System," *Wikipedia*; http://en.wikipedia.org/wiki/Space_Launch_System

U.S. Army Space and Missile Defense Command/Armed Forces Strategic Command, "SMDC/ARSTRAT Technical Center"; http://www.smdc.army.mil/2008/Technical_Center.asp

EPILOGUE

This brief Epilogue summarizes the results to date of the technological evolution of Greater Huntsville. Unlike the usual literary epilogue, it does not bring closure – the technological evolution will continue. Thus, this epilogue provides the author's projection of future technological changes and growth areas.

RESULTS OF TECHNOLOGICAL EVOLUTION

In 1804, the first settlers arrived in the area that became Madison County, and in 1811, Huntsville was the first city incorporated in the territory that became Alabama. At the time this book was being prepared, the 2013 census showed that Huntsville had a population of 186,254; Madison County had 346,892; and the population in the Huntsville Metropolitan area was 435, 737 – the second largest in the State. Technology growth, however, is of greater significance than that of the general population.

Greater Huntsville has one of the largest concentrations of engineers and related professions in the United States, and ranks third as an aerospace and defense manufacturing hub. The Cummings Research Park, with over 300 companies and 30,000 employees, is second in size – first by some measures – in the Nation and third largest in the World. Overall, Greater Huntsville has about 650 technical firms offering products and services in a wide variety of categories. The land on Redstone Arsenal is home to some of the most advanced research and development activities in both NASA and defense agencies. Altogether, NASA, Army and other agencies have some 38,000 federal employees in Greater Huntsville.

An August 2015 article in *Bloomberg BusinessWeek* lists the Huntsville Metropolitan Area as third in the nation for percentage of workers holding jobs in science, technology, engineering, or mathematics (STEM). With 16.7 percent in STEM jobs, Huntsville was only behind San Jose, California, and Framingham, Massachusetts; the national average is 2.9 percent.

FUTURE TECHNOLOGY GROWTH

Previous sections of this book on governmental and industrial activities have included many indications of the future technology status in Greater Huntsville. Beginning in 1941 and the opening of large chemical munitions plants, this has been a "government city," with the economy primarily sustained by Federal expenditures on technical programs. There has been diversification through the years, with non-governmental activities increasing in economic importance – particularly in the telecommunications field. The first years of the 21st century, however, have shown a major upsurge in defense-related work; to a large measure this resulted from BRAC (Base Realignment and Closure) changes.

There is also recognition of the further growing technical capabilities, especially for defense and space work. Raytheon has opened on Redstone Arsenal what they call the "missile factory of the future," centering on extensive application of advanced automation. The Aviation and Missiles Research, Development, and Engineering Center (AMRDEC) is the perhaps the single largest and most technically important activity of the U.S. Army.

There are also expanded activities in foreign intelligence – both in analysis of information and hardware reverse engineering. New activities have been opened on Redstone Arsenal by the Department of Justice. Both of these areas of work will grow with the importance of homeland security and anti-terrorism.

MSFC is responsible for a new heavy-lift rocket, the Space Launch System (SLS). Although this is to be flown in 2017, it is just the beginning of new, long-term activities in deep-space research and travel. Also, MSFC had become the most diversified of NASA centers, with exceptional research capabilities that extend into many avenues of both space- and earth-science work.

An entire book could easily be devoted to just these previously mentioned subjects. The remainder of this Epilogue, however, will be limited to brief notes on two areas: biotechnology and cyber revolution. At the beginning of the 21st century, both of these terms had little common usage – particularly not in activities of Greater Huntsville. Now, only a few years later, both of these terms (or some variations) regularly show up in the local media, technical meetings, and intelligent conversations. It is the author's opinion that in a few years one or both of these areas could give Greater Huntsville another moniker, equivalent to "Rocket City" of the past century. However, it is quickly noted that it will also remain the center for NASA and Army rocket development.

Biotechnology

For thousands of years, people around the world have used processes involving microorganisms to make food products such as bread, beer, and cheese. In a simple form, this was technology based on biology – biotechnology. Although there were some practical and academic developments through the years, it was the 1970s before this field gained any public recognition, led by the discovery of synthetic human insulin. In recent years, biotechnology has provided breakthrough products and techniques to combat diseases, reduce environmental problems, feed the hungry, provide cleaner energy, and improve industrial manufacturing processes.

As described in Chapter V, James Hudson, Lonnie McMillian, and Richard Meyers led in initiating the not-for-profit HudsonAlpha Institute for Biotechnology in 2008. This quickly became the cornerstone of a 152-acre Biotech Campus in West Cummings Research Park. The HudsonAlpha facility includes a Genome Sequencing Center and a Genomic Services Laboratory, as well as housing a number of start-up biotech firms.

In addition to the HudsonAlpha Institute, Huntsville's growing biotechnology community includes: • Applied Genomics • Expression Genetics • EGEN, Inc. • Nektar Therapeutics • New Century Pharmaceuticals • Open Biosystems • Operon Biotechnologies, Inc. • Partnership for Biotechnology Research • SourceCF and • UAHuntsville, which has a respected biotechnology program.

In the *2014 Industrial Directory* from the Huntsville-Madison County Chamber of Commerce, there are 23 firms listing Research and Development in Biotechnology as one of their activities. A number of these are direct affiliates of the HudsonAlpha Institute and occupy space within the facility.

In the past several decades, the field of biology has greatly expanded to incorporate mathematics, statistics, computer science, and areas from engineering to study and process biological data. Sometimes called bioinformatics, this interdisciplinary field has greatly benefited biotechnology advancements, particularly in the fields of genetics and genomics.

Cyber Revolution

Greater Huntsville has one of the nation's largest concentrations of computer scientists and engineers. The *2014 Industrial Directory* shows 99 firms listing Computer Systems Design Services as one of their activities, and 129 firms similarly listing Custom Computer Programming Services.

Within local Federal Government organizations, there are a great many computing specialists. In the Software Engineering Directorate of AMRDEC alone there are about 300 civil service employees and perhaps 3,000 contractors engaged in software development and applications. AMRDEC has one of the few Capability Maturity Model (CMM) Level 4 software engineering facilities in Army organizations.

The largest software firm in Greater Huntsville is Intergraph Corporation. Originally started as a hardware developer, it converted to software and now, owned by a Swedish firm, it provides engineering software products to industrial and governmental activities around the world.

Greater Huntsville also has a major concentration of computer operational facilities. The Alabama Supercomputer Center was the first state-supported facility of its type in the Nation. Three local federal organizations – Marshall Space Flight Center, Missile and Space Defense Center, and Missile and Space Intelligence Center – each have extensive computational capabilities.

Highly specialized electronic systems, subsystems, and components for cyber applications are developed and manufactured by several Huntsville firms. The largest of these is the locally founded ADTRAN, a world leader in developing new data transmission products with extensive lines of products for carrier systems, business networking, and loop access.

In 1946, the American mathematician and philosopher Norbert Wiener originated cybernetics, a type of automatics based on formalization of the notion of feedback; this was an outgrowth of his work during WWII on automatic aiming of anti-aircraft guns. (The author was privileged to have attended Wiener's advanced lectures on cybernetics.) In the 1980s, cybernetics brought in computers and information theory, becoming an adjective that was abbreviated to cyber; this is often used as a prefix.

In recent years, considerable attention has been given to what might be collectively called cyber activities: cyber warfare, cyber protection, cyber security, cyber crime, cyber space, etc. Three-dimensional (3-D) printing – often called additive manufacturing – is also included; this is growing exponentially and is sometimes referred to as the wave of the future in manufacturing. Initial success has recently been shown for 3-D fabrication using biological materials, further linking the cyber revolution and biotechnology.

Cyber Huntsville   Cyber Huntsville, a 501(c)(6) non-profit organization made up of industry, government, and academic institutions, is dedicated to making Huntsville and the Tennessee Valley region a nationally and internationally recognized cyber leader. This is building on Huntsville talent pools and experience in the areas of systems engineering, research and development, modeling and simulation, cyber security, experimentation, and test and evaluation. The organization was initially formed under the leadership of Huntsville's Mayor Tommy Battle in 2010. One of the activities is an annual Cyber Security Conference; the 7th Annual Southeastern Cyber Security Summit, held in June 2015, had some 980 attendees (700 in the previous year).

In the near future, Huntsville is expected to join the small list of "gigabit" locations – cities or small regions having extensive internal fiber-optic networks transmitting at 10^9 bits per second. While several federal agencies and other activities in Greater Huntsville have had for some time dedicated internet connections operating at gigabit rates, it has been announced that a 100-mile long, 10-gigabit circuit will be built around the city, with gigabit links added to customers as needed.

AMRDEC Cyber Campus.   In January 2015, the U.S. Army Aviation and Missile Research, Development and Engineering Center (AMRDEC) acquired two Redstone Arsenal buildings with more than 66,000 ft^2 of space. This has become the AMRDEC Cyber Campus and is expected to be in full operation by 2017.

Initial Cyber Campus

From the conception of Cyber Campus, Wallace E. Kirkpatrick, founder and CEO of DESE Research, has been substantially involved in policy, program, and budget issues at the Congressional level.

Robert H. Goldsmith is the AMRDEC cyber lead. Goldsmith described the operation as follows:

> The primary focus of the AMRDEC Cyber Campus is to provide world-class cybersecurity support to aviation and missile system customers and to conduct cutting-edge research and development in cybersecurity solutions to address their most pressing challenges.
>
> The cyber campus is an expanding home to a cadre of qualified cyber personnel with deep technical expertise in cybersecurity as it relates to hardware, software, firmware, networks, test and evaluation, modeling and simulation, forensics, industrial control systems, supervisory control and data acquisition systems, cyber training and exercise support, and other focus areas.
>
> The campus will support multiple levels of security enabling the cybersecurity team to uncover, dissect, and report on system vulnerabilities to cyber exploits, malicious hardware, software, or firmware, and interface with other government agencies at the highest classification levels.

The Author

Raymond C. Watson, Jr., was born at Anniston, Alabama, in 1926, and raised on a small farm during the Great Depression. After growing a bale of cotton at age nine, he decided that his future wasn't in farming and turned to radio and electrical interests. He entered a special program from the Alabama Polytechnic Institute (now Auburn University) under the Engineering, Science, and Management War Training (see Wikipedia), becoming a non-degreed "instant engineer" in 1942.

Watson's following 18 years included two years of wartime service in the U.S. Navy (attending and teaching in the Electronics Training Program (see Wikipedia); part-time study (finishing his B.S. degree and earning M.S. degrees in engineering and in physics); work in a variety of commercial research and development activities; founding and operating Dixie Service Company (an FCC-recognized engineering consulting firm); and academic positions at Jacksonville State College, the University of Alabama, and the University of Florida. He has been a Registered Professional Engineer since 1958 (Ala. PE No. 3496).

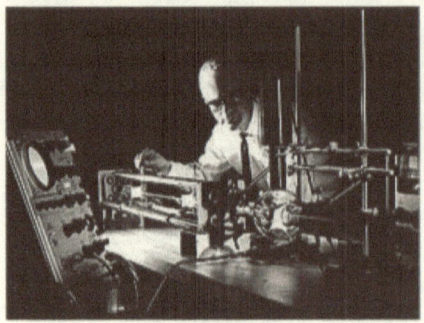

Watson in Laser Laboratory 1966

Watson came to Huntsville in 1960, hired by Milton Cummings to initiate advanced research and development capabilities at Brown Engineering Company (BECO, later Teledyne Brown Engineering, TBE). For the next 42 years, Watson served in a wide variety of technical, management, and consulting positions at BECO / TBE, the last 10 years as the company's Chief Engineer / Chief Scientist, including a period as the Acting President. Continued part-time study resulted in engineering Ph.D. and M.B.A. degrees. He also held teaching and director positions at the University of Alabama in Huntsville, and was later a professor and the President at Southeastern Institute of Technology (see Wikipedia).

Since 2002, and continuing to the present, Watson's work has been primarily through R.C. Watson and Associates; this firm, dating from 1970, provides consulting services to a variety of customers in advanced technologies, strategic and economic analysis, proposal development and review, and program / project management. For the past several years he has especially served as a Staff Consultant to DESE Research. He also served Alabama A&M University in a number of *pro bono* activities, including chairing advisory boards at the department and college levels. Throughout his career, Watson has authored about 450 papers, reports, technical presentations, magazine and encyclopedia articles, and five books.

He is married to Charlotte Bagley Watson, and they have four children, eight grandchildren, and a growing number of great-grandchildren. He is a Christian and a Southern Baptist church member.

Index

Idex – Individuals

Index - Subjects

56

Image Sources

Abbreviations:

P.D. – Public Domain

HMCPL – Huntsville/Madison County Public Library

GNU is a copyleft license for free documentation – rights to copy, redistribute, and modify a work.

AMC – U.S. Army Materiel Command

AMCOM – U.S. Army Missile Command

MDA – U.S. Missile Defense Agency

MSFC – NASA Marshall Space Flight Center

SMDC – U.S. Army Space and Missile Defense Command

---

p. 7, Caption: Early Explorer   Source: Wikimedia Commons   File: De Soto Discovers Mississippi BAH-p20.png   Permission: P.D. (pre-1923)

p. 9, Caption: Early Cherokee House   Source: Wikimedia Commons   File: Reconstruction of a typical Tuskegee Village dwelling   Permission: Released under the GNU Free Documentation License

p. 10, Caption: Territory Early 1800s   Source: Redrawn by author from a more detailed, colored map

p. 11, Caption: Public Land Survey System   Source: Wikimedia Commons   File: Systemic numbering in the Public Land Survey System.gif; from National Atlas   Permission: P.D.

p. 12, Caption: Surveying with Vernier Compass   Source: Colonial Carolina Net; File: Surveying in Colonial Carolina   Permission: P.D.

p. 13, Caption: Madison County Showing the Various Changes in Area Through1887; Prepared by G.W. Jones & Sons, Engineers, in 1934   Source: Huntsville History Collection   Permission: Courtesy of the HMCPL

p. 14, Caption: Theodolite   Source: Maryland Geological Survey Volume Two, by W. B. Clark, 1898. Johns Hopkins University Press.   Immediate source: Google Books   Permission: P.D.

p. 15, Caption: Smooth-Bore Flint-Lock Musket   Source: Reeds Target Shooting Club Web Site (United Kingdom) Technical Diagrams: Flintlock Action   Permission: written permission to use received.

p. 15, Caption: Pocket Compass   Source: Source: Wikimedia Commons File: Kompas Sofia.JPG   Permission: P.D.

p. 15, Caption: Big Spring and Bluff   Source: Part of a painting by James Thomas Neumann; Huntsville History Collection   Permission: Courtesy of the HMCPL

p. 16, Caption: Flatboat Transportation   Source: "Steamboat Times," website; artist: unknown.   Permission: P.D.

p. 17, Caption: Typical Early Log House Interior   Source: Wikimedia Commons; file: Conner-prairie-log-cabin-interior.jpg   Permission: P.D.

p. 18, Caption: Pope's Mansion on The Hill   Source: Huntsville History Collection   Permission: Courtesy of the HMCPL

p. 20, Caption: Blacksmith Shop   Source: Wikimedia Commons; file: Blacksmith shop at Hoover 2002, by Chris Light   Permission: Released under the GNU Free Documentation License.

p. 20, Caption: First Madison County Courthouse; artist C. E. Monroe, Jr.   Source: Huntsville History Collection   Permission: Courtesy of the HMCPL

p. 23, Caption: Battle of Horseshoe Bend   Source: Wikimedia Commons; file: Battle Horseshoe Bend 1914   Permission: P.D.

p. 24, Caption: Present Constitution Village   Source: The George F. Landegger Collection of Alabama Photographs in Carol M. Highsmith's America, Library of Congress   Permission: No Restrictions

p. 24, Caption: First Library Building   Source: Huntsville History Collection   Permission: Courtesy of the HMCPL

p. 25, Caption: First National Bank Building   Source: Huntsville History Collection   Permission: Courtesy of the HMCPL

p. 26, Caption: Huntsville Female College   Source: Wikimedia Commons; file: Huntsville Female College   Permission: P.D.

p. 26, Caption: Cotton Boll   Source: Kathy Clark Artist Website, "Limestone Cotton Boll"   Permission: granted in writing, Kathy Clark

p. 27, Caption: Picking Cotton   Source: Facebook, "Cotton Picking"- Original, Illustrated London News, 1885   Permission: P.D. (pre-1923)

p. 28, Caption: Basic Cotton Gin Mechanisms   Source: "Textile Learner" Website; http://textilelearner.blogspot.com/2014/08/different-parts-of-modern-cotton.html   Permission: granted in writing

p. 28, Caption: Hand-Cranked Cotton Gin   Source: Wikimedia Commons; file: Cotton Gin Harpers (Harpers Weekly 1869)   Permission: P.D. (pre-1923)

p. 28, Caption: Mule-Drawn Gin and Press   Source: Charlotte Mecklenburg Public Library Image Collection, Cotton Gin Photograph 1880   Permission: courtesy of the Robinson-Spangler Carolina Room

p. 29, Caption: Cotton Sellers on Street in Huntsville   Source: Huntsville History Collection   Permission: Courtesy of the HMCPL

p. 30, Caption: Typical Mid-sized Cotton Plantation   Source: "Antebellum Cotton Plantation," U. of Georgia website; http://georgiainfo.galileo.usg.edu/gastudiesimages/Antebellum.jpg   Permission: granted in writing, Edwin L. Jackson

p. 32, Caption: The Bell Factory   Source: Huntsville History Collection   Permission: Courtesy of the HMCPL

p. 36, Caption: Macadam Road Building   Source: Web Site, "The Transportation Revolution", artist, unknown, http://transportationrevolutionproject.weebly.com/resources.html   Permission: P.D. (pre-1923)

p.39, Caption: New Orleans Harbor   Source: Wikipedia, "Flatboat", illustration - Flatboats among the river traffic at New Orleans, 1873; artist: A. Measom Jr.   Permission: P.D. (pre-1923)

p. 40, Caption: Thomas Fearn   Source: Huntsville History Collection   Permission: Courtesy of the HMCPL

p. 42, Caption: Decatur Cotton Port   Source: Encyclopedia of Alabama; File: Decatur Port Scene   Permission: Courtesy of Alabama Department of Archives and History

p. 43, Caption: M&CR Locomotive   Source: Wikipedia, "Memphis and Charleston Railroad," in-text illustration   Permission: P.D.

p. 44, Caption: Hollow Cedar Log Pipes   Source: Project Gutenberg EBook, http://www.gutenberg.org/files/17374/17374-h/17374-h.htm   Permission: No Restrictions

p. 45, Caption: Huntsville Waterworks 1855   Source: Huntsville History Collection   Permission: Courtesy of the HMCPL

p.46, Caption: Gas Street Light   Source: Stars and Stripes, Newspaper Archives, "Lamplighters Linger in Old Munich" http://www.stripes.com/news/lamplighters-linger-in-old-munich-1.23240#   Permission: granted in writing

p. 47, Caption: Simplex Telegraph Circuit   Source: Redrawn from Telecom Milestones Web Site; http://www.telecom-milestones.com/   Permission: P.D.

P. 49, Caption: Union Troops Camped Around Courthouse   Source: Huntsville History Collection
Permission: Courtesy of the HMCPL

p. 51, Caption"|: 1865 Huntsville Map – Drawn by Federal Occupying Troops   Source: Huntsville History
Collection   Permission: Courtesy of the HMCPL

p. 54, Caption: Train Barge   Source: Huntsville History Collection   Permission: Courtesy of the
HMCPL

p. 57, Caption: Watercress Farm   Source: Huntsville History Collection   Permission: Courtesy of the
HMCPL

p. 57, Caption: Lily Flagg   Source: Huntsville History Collection   Permission: Courtesy of the HMCPL

p. 58, Caption: Madison Foundry   Source: Huntsville History Collection   Permission: Courtesy of the
HMCPL

p. 59, Caption: Skinner Carriage Store Downtown   Source: Huntsville History Collection   Permission:
Courtesy of the HMCPL

p. 60, Caption: Typical Small Manufactured Gas Plant   Source: "About Manufactured Gas Plants,"
National Grid / Bay Shore web site   Permission: granted in writing

p. 62, Caption: Street Arc Lamp   Source: Low-Tech Magazine, "Moonlight towers,"
http://www.lowtechmagazine.com/2009/01/moonlight-towers-light-pollution-in-the-1800s.html
Permission: granted in writing

p. 62, Caption: Stock Ticker   Source: Telecom Milestones Web Site; http://www.telecom-milestones.com/
Permission: P.D.

p. 63, Caption: Wall-Mounted Local-Battery Magneto Telephone   Source: Telecom Milestones Web Site;
http://www.telecom-milestones.com/   Permission: P.D.

p. 65, Caption: Spanish-American War Parade for General Wheeler in Downtown Huntsville   Source:
Huntsville History Collection   Permission: Courtesy of the HMCPL

p. 70, Caption: Tracy Pratt   Source: Huntsville History Collection   Permission: Courtesy of the HMCPL

p. 71, Caption: Inside a Cotton Mill   Source: Huntsville History Collection   Permission: Courtesy of the
HMCPL

p. 72, Caption: Lowe Manufacturing Steam Boiler   Source: Huntsville History Collection   Permission:
Courtesy of the HMCPL

p. 72, Caption: Children Spinners and Doffers   Source: Huntsville History Collection   Permission:
Courtesy of the HMCPL

p. 72, Caption: Huntsville Cotton Mill   Source: Huntsville History Collection   Permission: Courtesy of
the HMCPL

p. 73, Caption: Dallas Manufacturing Mills   Source: Huntsville History Collection   Permission: Courtesy
of the HMCPL

p. 73, Caption: Lincoln Mill   Source: Huntsville History Collection   Permission: Courtesy of the
HMCPL

p. 74, Caption: Merrimack Mill   Source: Huntsville History Collection   Permission: Courtesy of the
HMCPL

p. 74, Caption: Lowe Manufacturing   Source: Huntsville History Collection   Permission: Courtesy of the
HMCPL

p.74, Caption: Merrimack Mill Village   Source: Huntsville History Collection   Permission: Courtesy of
the HMCPL

p. 75, Caption: Merrimack Hospital   Source: Huntsville History Collection   Permission: Courtesy of the
HMCPL

p. 75, Caption: Rison School   Source: Huntsville History Collection   Permission: Courtesy of the
HMCPL

p. 75, Caption: Merrimack YMCA   Source: Huntsville History Collection   Permission: Courtesy of the HMCPL

p. 76, Caption: Building Tracks Downtown   Source: Huntsville History Collection   Permission: Courtesy of the HMCPL

p. 76, Caption: Huntsville Streetcar   Source: Huntsville History Collection   Permission: Courtesy of the HMCPL

p. 77, Caption: National Millworkers Strike   Source: Bread and Roses Centennial Web Site; http://breadandrosescentennial.org/   Permission: P.D.

p. 81, Caption: Rural Power   Source: Huntsville History Collection   Permission: Courtesy of the HMCPL

p. 81, Caption: Guntersville Dam   Source: "Tennessee River and Reservoir Update 2011", TVA; http://www.tva.com/email/eRiver/2011/march.html   Permission: P.D.

p. 83, Caption: Graphic of Quick's Airplane   Source: Huntsville Air and Space, by T. Gary Wicks   Permission: granted in writing

p. 84, Caption: First Terminal   Source: Huntsville History Collection   Permission: Courtesy of the HMCPL

p. 86, Caption: Boll Weevil   Source: U.S. Department of Agriculture; Plant Pest and Disease Programs   Permission: P.D.

p. 86, Caption: Crop Dusting (1925)   Source: U.S. Department of Agriculture; Special Collections Research Center at NCSU Libraries   Permission: P.D.

p. 88, Caption: Typical CCC Camp   Source: National Park Service; Photo Gallery, Civilian Conservation Service   Permission: P.D.

p. 89, Caption: Oil Rig in West Huntsville   Source: Huntsville History Collection   Permission: Courtesy of the HMCPL

p. 90, Caption: John Blue Plant   Source: Reconstructed from 1950 advertisement   Permission: P.D.

p. 90, Caption: John Blue G-1000   Source: From John Blue 1975 advertisement   Permission: P.D.

p. 91, Caption: Martin Stove Company   Source: Reconstructed from 2010 realty advertisement   Permission: P.D.

p. 92, Caption: G. W. Jones   Source: Pamphlet on "G. W. Jones & Sons Engineers"; Huntsville History Collection   Permission: Courtesy of the HMCPL

P. 93, Caption: Remember Pearl Harbor   Source: Pearl Harbor History Associates; "Pearl Harbor Attacked"; http://pearlharborattacked.com/   Permission: P.D.

p. 94, Caption: Truman Awarding Bolden   Source: Huntsville History Collection   Permission: Courtesy of the HMCPL

0. 96, No Caption [Newspaper 3 July 1941]   Source: U.S. Army Missile Command (AMCOM) Web Page, "Redstone Arsenal in the 1940's"   Permission: P.D.

p. 98, Caption: Redstone Land Areas   Source: U.S. Army Missile Command (AMCOM) Web Page, "Redstone Arsenal in the 1940's"   Permission: P.D.

p. 98, Caption: Arsenal's First Building   Source: U.S. Army Missile Command (AMCOM) Web Page, "Redstone Arsenal in the 1940's"   Permission: P.D.

p. 99, Caption: Redstone Park Housing   Source: U.S. Army Missile Command (AMCOM) Web Page, "Redstone Arsenal in the 1940's"   Permission: P.D.

p. 100, Caption: Rollo Ditto   Source: U.S. Army Missile Command (AMCOM) Web Page, "Redstone Arsenal in the 1940's"   Permission: P.D.

p. 100, Caption: Squirrel Hill Buildings   Source: U.S. Army Missile Command (AMCOM) Web Page, "Redstone Arsenal in the 1940's"   Permission: P.D.

p. 101, Caption: Chlorine Plant   Source: U.S. Army Missile Command (AMCOM) Web Page, "Redstone Arsenal in the 1940's"   Permission: P.D.

p. 101, Caption: Mustard Gas Shells   Source: U.S. Army Missile Command (AMCOM) Web Page, "Redstone Arsenal in the 1940's"   Permission: P.D.

p. 102, Caption: Incendiary Production   Source: U.S. Army Missile Command (AMCOM) Web Page, "Redstone Arsenal in the 1940's"   Permission: P.D.

p. 102, Caption: Advertisement for Personnel   Source: U.S. Army Missile Command (AMCOM) Web Page, "Redstone Arsenal in the 1940's"   Permission: P.D.

p. 103, Caption: Munitions Warehouse   Source: U.S. Army Missile Command (AMCOM) Web Page, "Redstone Arsenal in the 1940's"   Permission: P.D.

p. 104, Caption: Carroll Hudson   Source: Ordnance Corps Hall of Fame http://www.goordnance.army.mil/hof/1990/1993/hudson.html   Permission: P.D.

p. 104, Caption: Hudson Groundbreaking   Source: U.S. Army Missile Command (AMCOM) Web Page, "Redstone Arsenal in the 1940's"   Permission: P.D.

p. 105, Caption: Mortar Shell Production   Source: U.S. Army Missile Command (AMCOM) Web Page, "Redstone Arsenal in the 1940's"   Permission: P.D.

p. 106, Caption: Weekly Newspaper (Redstone Eagle)   Source: U.S. Army Missile Command (AMCOM) Web Page, "Redstone Arsenal in the 1940's"   Permission: P.D.

p. 107, Caption: Women's Uniforms   Source: U.S. Army Missile Command (AMCOM) Web Page, "Redstone Arsenal in the 1940's"   Permission: P.D.

p. 110, Caption: Typical POW Camp   Source: North Carolina Digital Library; Prisoners of war in North Carolina, http://www.learnnc.org/lp/editions/nchist-worldwar/6047   Permission: Permission: Courtesy of the North Carolina Digital Library

p. 112, Caption: Keller Assembly Line on Redstone Arsenal   Source: Huntsville Rewound, Keller Car; http://huntsvillerewound.com/HSVkellercar.htm   Permission: P.D.

p. 112, Caption: Keller Station Wagon   Source: Huntsville Rewound, Keller Car; http://huntsvillerewound.com/HSVkellercar.htm   Permission: P.D.

p. 113, Caption: Olin Mathieson DDT Plant   Source: Source: U.S. Army Missile Command (AMCOM) Web Page, "Redstone Arsenal in the 1940's"   Permission: P.D.

p. 118, Caption: Wernher von Braun with A-1   Source: ICD Web Site, "Pioniere der Raketentechnik" http://www.weltchronik.de/dch/dch_2786.htm   Permission: P.D.

p. 119, Caption: A-4 (V-2) at Peenemünde   Source: BBC News, "Forgotten cradle of the space age," http://news.bbc.co.uk/2/hi/science/nature/7695425.stm   Permission: P.D.

p. 120, Caption: Robert Goddard   Source: AMRDEC; Celebrating 50 Years of McMorrow Laboratories, http://www.amrdec.army.mil/amrdec/50th/leadership-robert-goddard.html   Permission: P.D.

p. 121, Caption: WAC CORPORAL   Source: Wikipedia, "WAC Corporal"   Permission: Creative Commons Attribution-ShareAlike License   Permission: P.D.

p. 122, Caption: HERMES Missile on Launch Site 33   Source: National Park Service, "Aviation from Sand Dunes to Sonic Booms" http://www.nps.gov/nr/travel/aviation/whi.htm   Permission: P.D.

p. 123, Caption: "Project Paperclip Personnel at Fort Bliss"   Source: Wikipedia, File: Project Paperclip Team at Fort Bliss.jpg, in-text illustration   Permission: P.D.

p. 124, Caption: V-2 Testing Failure at White Sands   Source: White Sands Proving Grounds, Historical Information, Photo Archives (joining two photographs)   Permission: P.D.

p. 125, Caption: BUMPER Missile   Source: NASA/U.S. Army Web Page, File:Bumper8 launch-GPN-2000-000613.jpg   Permission: P.D.

p. 127, Caption: Arsenal for Sale   Source: History.Redstone.Army.Mil; file: Corps of Engineers' sign advertising Huntsville Arsenal for Sale   Permission: P.D.

p. 128, Caption: CORPORAL Missile   Source: Web Site; AMCOM, U.S. Army Historical Information, Photo Archives   Permission: P.D.

p. 128, Caption: Thomas Moore with JETVEST   Source: Web Site; AMCOM, U.S. Army Historical Information, Photo Archives   Permission: P.D.

p. 129, Caption: Thomas Vincent   Source: Web Site; AMCOM, U.S. Army Historical Information, Photo Archives   Permission: P.D.

p. 129, Caption: Holger Toftoy   Source: Web Site; AMCOM, U.S. Army Historical Information, Photo Archives   Permission: P.D.

p. 129, Caption: First Test Stand   Source: Web Site; AMCOM, U.S. Army Historical Information, Photo Archives   Permission: P.D.

p. 130, Caption: Citizenship Ceremony   Source: Web Site; AMCOM, U.S. Army Historical Information, Photo Archives   Permission: P.D.

p. 131, Caption: HONEST JOHN Missile   Source: Web Site; AMCOM, U.S. Army Historical Information, Photo Archives   Permission: P.D.

p. 131, Caption: REDSTONE Rocket   Source: Web Site; AMCOM, U.S. Army Historical Information, Photo Archives   Permission: P.D.

p. 132, Caption: LACROSSE Missile   Source: Web Site; AMCOM, U.S. Army Historical Information, Photo Archives   Permission: P.D.

p. 133, Caption: John Medaris   Source: Web Site; AMCOM, U.S. Army Historical Information, Photo Archives   Permission: P.D.

p. 133, Caption: Redstone Airfield   Source: Web Site; AMCOM, U.S. Army Historical Information, Photo Archives   Permission: P.D.

p. 135, Caption: Comparative Sizes   Source: Derived from 1965 Chrysler Advertisement, America's First Family of Rockets and Missiles   Permission: P.D.

p. 136, Caption: PERSHING Missile   Source: Web Site; AMCOM, U.S. Army Historical Information, Photo Archives   Permission: P.D.

p. 137, Caption: Miss Baker   Source: Web Site; AMCOM, U.S. Army Historical Information, Photo Archives   Permission: P.D.

p. 139, Caption: NIKE HERCULES   Source: Web Site; AMCOM, U.S. Army Historical Information, Photo Archives   Permission: P.D.

p. 139, Caption: Hawk System Radar   Source: Web Site; AMCOM, U.S. Army Historical Information, Photo Archives   Permission: P.D.

p. 140, Caption: Redeye System   Source: Web Site; AMCOM, U.S. Army Historical Information, Photo Archives   Permission: P.D.

p. 142, Caption: DAMP Ship   Source: Wikipedia, imbedded image; The DAMP Project ship, the USAS American Mariner at anchor Clarence Bay, Ascension Island, 1962   Permission: P.D.e

p. 143, Caption: NIKE ZEUS   Source: Web Site; AMCOM, U.S. Army Historical Information, Photo Archives   Permission: P.D.

p. 143, Caption: Zeus System on Kwajalein   Source: Web Site; AMCOM, U.S. Army Historical Information, Photo Archives   Permission: P.D.

p. 144, Caption: Ivey Drewry   Source: Web Site; AMCOM, U.S. Army Historical Information, Photo Archives   Permission: P.D.

p. 145, Caption: SPARTAIN Missile   Source: Web Site; AMCOM, U.S. Army Historical Information, Photo Archives   Permission: P.D.

p. 146, Caption: SPRINT Missile   Source: Web Site; AMCOM, U.S. Army Historical Information, Photo Archives   Permission: P.D.

p. 146, Caption: Missile Site Radar   Source: Web Site; AMCOM, U.S. Army Historical Information, Photo Archives   Permission: P.D.

P. 147, Caption: Julian Davidson   Source: Air Space and Missile Defense Association, Wall of Honor Permission: P.D.

p. 149, Caption: Madkin Antenna   Source: Web Site; AMCOM Photo Archive, Redstone Facilities Permission: P.D.

p. 149, Caption: Carl Duckett   Source: CIA Archives, Carl E. Duckett   Permission: P.D.

p. 150, Caption: Rankin Clinton   Source: Air Space and Missile Defense Association, Wall of Honor Permission: P.D.

p. 150, Caption: U-2 Photograph of Missile Site   Source: Cuban Missile Crisis, Belfer Center for Science and International Affairs, Harvard University   Permission: P.D.

p. 152, Caption: Testing the Antenna   Source: Web Site; AMCOM, U.S. Army Historical Information, Photo Archives   Permission: P.D.

p. 152, Caption: Explorer 1 – America's First Satellite   Source: Web Site; AMCOM, U.S. Army Historical Information, Photo Archives   Permission: P.D.

p. 153, Caption Worried   Source: NASA Photograph, MSFC Public Affairs; Von Braun at Explorer Launch Permission: P.D.

153, Caption: Explorer Leaders   Source: NASA Photograph; Pickering, Van Allen & Von Braun IGY News Conference at National Academy of Sciences   Permission: P.D.

P. 153, Caption: None – Newspaper front page; Jupiter-C Puts Up Moon   Source: NASA MSFC Public Affairs   Permission: P.D.

p. 157, Caption: Horizon Space Dock   Source: Project Horizon: Volume I; Web Site; AMCOM, U.S. Army Historical Information, Photo Archives   Permission: P.D.

p. 157, Caption: Initial Lunar Outpost   Source: Project Horizon: Volume I; Web Site; AMCOM, U.S. Army Historical Information, Photo Archives   Permission: P.D.

p. 158, Caption: Frank McMorrow   Source: AMCOM, U.S. Army Historical Information, Photo Archives Permission: P.D.

p. 158, Caption: McMorrow Laboratories [1964 view]   Source: AMCOM, U.S. Army Historical Information, Photo Archives   Permission: P.D.

p. 158, Caption: John McDaniel   Source: AMCOM, U.S. Army Historical Information, Photo Archives Permission: P.D.

p. 160, Caption: William Davis   Source: Air Space and Missile Defense Association, Wall of Honor Permission: P.D.

p. 160, Caption: Robert Hoye (L) and William McKnight   Source: AMCOM, U.S. Army Historical Information, Photo Archives   Permission: P.D.

p. 161, Caption: LITTLEJOHN Missile   Source: AMCOM, U.S. Army Historical Information, Photo Archives   Permission: P.D.

p. 161, Caption: LAW Missile   Source: AMCOM, U.S. Army Historical Information, Photo Archives Permission: P.D.

p. 161, Caption: DRAGON Missile   Source: AMCOM, U.S. Army Historical Information, Photo Archives Permission: P.D.

p. 162, Caption: SHILLELAGH Missile   Source: AMCOM, U.S. Army Historical Information, Photo Archives   Permission: P.D.

p. 162, Caption: CHAPARRAL Missile   Source: AMCOM, U.S. Army Historical Information, Photo Archives   Permission: P.D.

p. 162, Caption: LANCE Missile   Source: AMCOM, U.S. Army Historical Information, Photo Archives Permission: P.D.

p. 163, Caption: OMMCS Area 1960s   Source: AMCOM, U.S. Army Historical Information, Photo Archives   Permission: P.D.

p. 165, Caption: MSFC Transfer Ceremony   Source: NASA / MSFC Photo   Permission: P.D.

p. 165, Caption: President Eisenhower and Mrs. Marshall at Dedication   Source: NASA / MSFC Photo   Permission: P.D.

p. 166, Caption: L-R: Eberhard Rees, von Braun, and Keith Glennan   Source: NASA / MSFC Photo   Permission: P.D.

p. 169, Caption: Central Office Complex   Source: NASA / MSFC Photo   Permission: P.D.

p. 172, Caption: H-1 Engine   Source: NASA / MSFC Photo   Permission: P.D.

p. 172, Caption: LR-10 Engine   Source: NASA / MSFC Photo   Permission: P.D.

p. 173, Caption: J-2- Engine   Source: NASA / MSFC Photo   Permission: P.D.

p. 174, Caption: F-1 Engine System   Source: Wikipedia, Rocketdyne F-1   Permission: P.D.

p. 175, Caption: Manufacturing Instrument Units   Source: NASA / MSFC Photo   Permission: P.D.

p. 176, Caption: SATURN I in Development   Source: NASA / MSFC Photo   Permission: P.D.

p. 177, Caption: F-1 Static Test   Source: NASA / MSFC Photo   Permission: P.D.

p. 177, Caption: MSFC Michoud Facility   Source: NASA / MSFC Photo   Permission: P.D.

p. 178, Caption: Barges at MSFC Dock   Source: NASA / MSFC Photo   Permission: P.D.

p. 179, Caption: Super Guppy   Source: NASA / MSFC Photo   Permission: P.D.

P. 179. Caption: Launch Complex 39A & B   Source: NASA / MSFC Photo   Permission: P.D.

p. 180, Caption: Saturn Family – Stages and Engines   Source: NASA / MSFC Photo   Permission: P.D.

p. 181, Caption: SATURN IBs in Production   Source: NASA / MSFC Photo   Permission: P.D.

p. 182, Caption: Arthur Rudolph   Source: NASA / MSFC Photo   Permission: P.D.

p. 182, Caption: SATURN V – S1C in Production   Source: NASA / MSFC Photo   Permission: P.D.

p. 183, Caption: Saturn V Apollo Spacecraft Subsystems   Source: NASA / MSFC Photo   Permission: P.D.

p. 184, Caption: Von Braun Driving the BECO MTA   Source & Permission: Photo owned by author

p. 185, Caption: Testing the TBE Soil Penetrometer   Source & Permission: Photo owned by author

p. 189, Caption: First Step on the Moon   Source: NASA / MSFC Photo   Permission: P.D.

p. 190, Caption: First Lunar Excursion   Source: NASA / MSFC Photo   Permission: P.D.

p. 192, Caption: Pegasus Experiment   Source: NASA / MSFC Photo   Permission: P.D.

p. 193, Caption: Apollo 17 ALSEP   Source: NASA / MSFC Photo   Permission: P.D.

p. 194, Caption: Ernst Stuhlinger   Source: NASA / MSFC Photo   Permission: P.D.

p. 197, Caption: Josiah Gorgas Research Laboratory   Source: Redstone Army Military History   Permission: P.D.

p. 198, Caption: First 34-Inch Solid Rocket Motor   Source: Thiokol Capabilities Brochure 1955   Permission: P.D.

p. 198, Caption: Thiokol Area 1959   Source: AMCOM Photo Archive, Redstone Facilities   Permission: P.D.

p. 200, Caption: First Building   Source & Permission: Photo owned by author

p. 201, Caption: HIC Building   Source & Permission: Photo owned by author

p. 201, Caption: Milton Cummings and Wernher von Braun at BECO   Source & Permission: Photo owned by author

p. 201, Caption: 1963 Complex in Research Park   Source & Permission: Photo owned by author

p. 202, Caption: William Giardini   Source & Permission: Photo owned by author

p. 203, Caption: IBM 7040   Source & Permission: Photo owned by author

p. 203, Caption: Joseph Moquin   Source & Permission: Photo owned by author

p. 203, Caption: Raymond Watson   Source & Permission: Photo owned by author

p. 206, Caption: Teledyne Wah Chang Huntsville   Source & Permission: Photo owned by author

p. 208, Caption: SCI 8600 S Mem. Parkway   Source: Real Estate Advertisement   Permission: P.D.

p. 211, Caption: Engineering Design Before Computers   Source & Permission: Photo owned by author

p. 211, Caption: Boeing – Wynn Drive   Source: Real Estate Advertisement   Permission: P.D.

p. 212, Caption: Wyle Laboratories 1962   Source: Wyle Brochure 1963   Permission: P.D.

p. 214, Caption: Lockheed 1963   Source: Lockheed Brochure 1965   Permission: P.D.

p. 217, Caption: PPG Huntsville   Source: Huntsville Industrial Brochure 1970   Permission: P.D.

p. 220, Caption: Dedication of CRP   Source & Permission: Photo owned by author

p. 221, Caption: Edward Buckbee   Source: Space and Rocket Center Brochure   Permission: P.D.

p. 221: Caption: U.S. Space & Rocket Center   Source: Space and Rocket Center Brochure   Permission: P.D.

p. 221, Caption: Space & Socket Center Interior   Source: Space and Rocket Center Brochure   Permission: P.D.

p. 223, Caption: Morton Hall   Source: UAH Catalog   Permission: P.D.

p. 225, Caption: Alabama A&M University   Source: AAMU Catalog   Permission: P.D.

p. 226, Caption: Planned Layout for Space City USA   Source: Huntsville Rewound Web Site, Space City USA Brochure   Permission: P.D.

p. 236, Caption: Working on Skylab in NBS   Source: NASA / MSFC Photo   Permission: P.D.

p. 236, Caption: Components of Skylab and ATM   Source: NASA / MSFC Photo   Permission: P.D.

p. 237, Caption: Skylab with Missing Solar Panel   Source: NASA / MSFC Photo   Permission: P.D.

p. 238, Caption: At MSFC for Ground Testing   Source: NASA / MSFC Photo   Permission: P.D.

p. 239, Caption: Solid Rocket Booster   Source: NASA / MSFC Photo   Permission: P.D.

p. 239, Caption: Shuttle with External Fuel Tank and Solid Rocket Boosters   Source: NASA / MSFC Photo   Permission: P.D.

 p. 241, Caption: TBE's Experiment Bridge   Source: NASA / MSFC Photo   Permission: P.D.

p. 242, Caption: Spacelab in Shuttle Cargo Bay   Source: NASA / MSFC Photo   Permission: P.D.

p. 243, Caption: Unity Module Under Construction   Source: NASA / MSFC Photo   Permission: P.D.

 p. 245, Caption: Components of Hubble Space Telescope   Source: NASA / MSFC Photo   Permission: P.D.

 p. 246, Caption: Components of XAXF – Chandra X-ray Observatory   Source: NASA / MSFC Photo   Permission: P.D.

 p. 248, Caption: HEAO Configuration   Source: NASA / MSFC Photo   Permission: P.D.

p. 248, Caption: HEAO-A in Construction   Source: NASA / MSFC Photo   Permission: P.D.

P. 249, Caption: LAGEOS-1   Source: NASA / MSFC Photo   Permission: P.D.

p. 250, Caption: Gravity Probe B   Source: NASA / MSFC Photo   Permission: P.D.

 p. 251, Caption: GP-B Dewar   Source: NASA / MSFC Photo   Permission: P.D.

p. 252, Caption: Optical Transient Detector   Source: NASA / MSFC Photo   Permission: P.D.

p. 253, Caption: Flight Robotics Laboratory   Source: NASA / MSFC Photo   Permission: P.D.

p. 255, Caption: Dual Sprint Fired From Meck   Source: U.S. Army SMDC Photo Gallery   Permission: P.D.

p. 255, Caption: Mickelsen Complex   Source: U.S. Army SMDC Photo Gallery   Permission: P.D.

p. 257, Caption: Kinetic Kill Vehicle used in the HOE   Source: Lockheed Horizons magazine Webb Page Permission: P.D.

p. 258, Caption: President Reagan Announces the SDI   Source: U.S. Missile Defense Agency, Media Resources   Permission: P.D.

p. 258, Caption: Early SDI Space-Based Concepts   Source: Wikipedia, Strategic defense Initiative Permission: P.D.

p. 260, Caption: HEDI SDI Missile   Source: U.S. Missile Defense Agency, Media Resources Permission: P.D.

p. 260, Caption: HELSTF   Source: U.S. Army White Sands Test Center, High Energy Laser Systems Test Facility   Permission: P.D.

p. 261, Caption: ERINT Hit   Source: U.S. Missile Defense Agency, Media Resources   Permission: P.D.

p. 263, Caption: William McCorkle   Source: U.S. Missile Defense Agency, Media Resources Permission: P.D.

p. 264, Caption: STINGER Missile   Source: Raytheon Products Photo   Permission: P.D.

p. 265, Caption: ATACSS Missile   Source: Vought Products Photo   Permission: P.D.

p.265, Caption: HELLFIRE System   Source: Lockheed Martin Products Photo   Permission: P.D.

p. 265, Caption: HYDRA-70 Rocket Pod   Source: U.S. Missile Defense Agency, Media Resources Permission: P.D.

p. 266, Caption: JAVELIN System   Source: U.S. Missile Defense Agency, Media Resources Permission: P.D.

p. 267, Caption: PATRIOT System   Source: Wikipedia, MIM-104 Patriot   Permission: P.D.

p. 267, Caption: AN-MPQ Radar   Source: RadarTutorial – AN-MPQ Raytheon Photo   Permission: P.D.

p. 267, Caption: Test Center with Tank Target   U.S. Army Redstone Test Center – History   Permission: P.D.

p. 270, Caption: Apache   Source: Consortium of Defense Analysts Web Page   Permission: P.D.

p. 270, Caption: Black Hawk   Source: Sikorsky Aircraft Products Photo   Permission: P.D.

p. 271, Caption: Chinook   Source: U.S. Army Military Web Page "Howitzer slung under a Chinook helicopter"   Permission: P.D.

p. 271, Caption: Kiowa   Source: Bell Helicopter Products Photo   Permission: P.D.

p. 272, Caption: Comanche   Source: Army Technology Web Page "RAH-66 Comanche"   Permission: P.D.

p. 272, Caption: Shadow UAV   Source: Defense Industry Daily Web Page "Laser Designators for RQ-7 Shadow UAVs"   Permission: P.D.

P. 273, Caption: Sparkman and Heflin Centers   Source: Civil Air Patrol (CAP)-119 Web Page (Aug 21 2014)   Permission: P.D.

p. 275, Caption: Hypervelocity Gun Building   Source: Courtesy of UAHuntsville Aerospace Physics Center   Permission: P.D.

p. 276, Caption: Large Two-Stage Light-Gas Gun   Source: Courtesy of UAHuntsville Aerospace Physics Center   Permission: P.D.

p. 277, Caption: Large Optics & X-Ray Test Facility   Source: NASA / MSFC Photo   Permission: P.D.

p. 278, Caption: Testing in Instrument Chamber   Source: NASA / MSFC Photo   Permission: P.D.

p. 279, Caption: Alabama Supercomputer Facility   Source: Alabama Supercomputer Authority Web Page   Permission: P.D.

p. 286, Caption: TBE Campus in 1900s   Source: Teledyne Brown Engineering Capabilities Brochure   Permission: P.D.

p. 288, Caption: Olin King   Source: SCI Web Site   Permission: P.D.

p. 288, Caption: Eugene Sapp   Source: SCI Web Site   Permission: P.D.

p. 288, Caption: SCI Headquarters   Source: Real Estate Advertisement   Permission: P.D.

p. 301, Caption: Roy Nichols   Source: Randolph School Web Page (5-19-20110   permission: P.D.

p. 301, Caption: Chris Horgan   Source: PeopleTec Web Site – Board of Directors   Permission: P.D.

p.307, Caption: Frank Collazo   Source: Business Alabama Web Page, April 2011   Permission: In writing

p. 309, Caption: Wallace Kirkpatrick   Source: Air Space and Missile Defense Association, Wall of Honor   Permission: P.D.

p. 312, Caption: HMCRA Engine 8933   Source: Hawkins Short Lines Web Site, Huntsville & Madison County Rail Authority; Bob Baudendistel for HMCR photograph   Permission: P.D.

p. 314, Caption: John Stallworth   Source: PROJECTXYZ, Inc. Board of Advisors Web Site   Permission: P.D.

p. 316, Caption: Mark Smith and ADTRAN Campus   Source: Huntsville Times File Photo   Permission: ??????

P. 338, Caption: Franklin Street Offices   Source: Huntsville History Collection, Brochure of former G.W. Jones Engineering   Permission: Courtesy of the HMCPL

p. 341, Caption: SIT Facility   Source & Permission: Photo owned by author

p. 342, Caption: Typical SIT Class Settings   Source & Permission: Photo owned by author

p. 354, Caption: Patrick Scheuermann   Source: MSFC Multimedia Gallery   Permission: P.D.

P. 355, Caption: Christopher Singer   Source: MSFC Multimedia Gallery   Permission: P.D.

p. 356, Caption: Friction-Stir Welding Facility   MSFC, Core Capabilities and Services   Permission: P.D.

p. 357, Caption: Propulsion research Laboratory   MSFC, Core Capabilities and Services   Permission: P.D.

p. 358, Caption: Test of Sub-Scale SLS Booster   MSFC, Core Capabilities and Services   Permission: P.D.

p. 358, Caption: Huntsville Operations Support Center   MSFC, Core Capabilities and Services   Permission: P.D.

p. 359, Caption: Jody Singer   Source: MSFC Multimedia Gallery   Permission: P.D.

p. 360, Caption: Stephen Cash   Source: MSFC Multimedia Gallery   Permission: P.D.

p. 360, Caption: Todd May   Source: MSFC Multimedia Gallery   Permission: P.D.

p. 361, Caption: SLS Development Schedule   MSFC, Core Capabilities and Services   Permission: P.D.

p. 361, Caption: Daniel Schumacher   Source: MSFC Multimedia Gallery   Permission: P.D.

p. 362, Caption: James Spann   Source: MSFC Multimedia Gallery   Permission: P.D.

p. 364, Caption: Mighty Eagle Free Flight   Source: MSFC, Core Capabilities and Services   Permission: P.D.

p. 365, Caption: Roy Malone   Source: MSFC Multimedia Gallery   Permission: P.D.

p. 366. Caption: Completed ISS   Source: MSFC, Core Capabilities and Services   Permission: P.D.

p. 368, Caption: NSSTC Building   Source: UAHuntsville, Research Laboratory   Permission: P.D.

p. 369, Caption: Testing JWST Segments   Source: MSFC, Core Capabilities and Services   Permission: P.D.

p. 372, Caption: Von Braun Complex   Source: SMDC Photo Gallery   Permission: P.D.

p. 372, Caption: TEDAC Laboratory   Source: FBI — Photo Gallery: Inside TEDAC   Permission: P.D.

p. 376, Caption: Myra Gray   Source: AMCOM Photo Gallery: Myra Gray   Permission: P.D.

p. 378, Caption: AMRDEC Main Complex   Source: AMCOM Photo Gallery: AMRDEC   Permission: P.D.

p. 378, Caption: James Lackey   Source: AMCOM Photo Gallery: AMRDEC   Permission: P.D.

p. 379, Caption: Specializations in Tactical Missiles   Source: AMCOM Photo Gallery: AMRDEC   Permission: P.D.

p. 379, Caption: William Craig   Source: AMCOM Photo Gallery: AMRDEC   Permission: P.D.

p. 380, Caption: David Byrd   Source: AMCOM Photo Gallery: Test Center   Permission: P.D.

p. 380, Caption: Rotary Wing Center Hangar Facility   Source: AMCOM Photo Gallery: Test Center   Permission: P.D.

p. 381, Caption: Mark Clark   Source: AMCOM Photo Gallery: Intelligence Center   Permission: P.D.

p. 383, Caption: LTG David Mann   SMDC Photo Gallery: General Mann   Permission: P.D.

p. 383, Caption: Julie Schumacher   SMDC Photo Gallery: Schumacher   Permission: P.D.

p. 383, Caption: Larry Burger   SMDC Photo Gallery: Burger   Permission: P.D.

p. 384, Caption: GEN Ann Dunwoody   AMC Photo Gallery, General Dunwoody   Permission: P.D.

p. 384, Caption: GEN Dennix Via   AMC Photo Gallery, General Via   Permission: P.D.

p. 385, Patrick O'Neill   AMC Photo Gallery, O'Neill   Permission: P.D.

p. 386, Caption: Logistics Information Warehouse   AMC Photo Gallery, LOGSA   Permission: P.D.

p. 387, Caption: MDA Central Facility in Von Braun Complex   MDA Photo Gallery, VBC   Permission: P.D.

p. 388, Caption: VAD James Syring   Source: MDA Photo Gallery, Syring   Permission: P.D.

p. 392, Caption: HEL Beam Director   Source: SMDC Photo Gallery   Permission: P.D.

p. 392, Caption: HEL Mobile Demonstrator   Source: SMDC Photo Gallery   Permission: P.D.

p. 394, Caption: FBI Hazardous Devices School   Source: FBI Photo Gallery   Permission: P.D.

p. 394, Caption: Explosives Training   Source: ATF Photo Gallery   Permission: P.D.

p. 394, Caption: FBI TEDAC Laboratory   Source: FBI Photo Gallery   Permission: P.D.

p. 395, Caption: Greater Huntsville Research and Industrial Parks   Source: Huntsville-Madison County Chamber of Commerce, Industrial Directory   Permission: P.D.

p. 400, Caption: Janice Hess   Source: Teledyne Brown Engineering Web Page   Permission: P.D.

p. 400, Caption: Stephen Kuffner   Source: Teledyne Brown Engineering Web Page   Permission: P.D.

p. 401, Caption: Roy Rice   Source: Teledyne Brown Engineering Web Page   Permission: P.D.

p. 401, Caption: Anthony Jones   Source: Boeing Web Page   Permission: P.D.

p. 402, Caption: Boeing Complex   Source: Boeing Web Page   Permission: P.D.

p. 402, Caption: Kevin Campbell   Source: Northrop-Grumman Web Site   Permission: P.D.

p. 403, Caption: John Holly   Source: Lockheed Martin Web Page   Permission: P.D.

p. 403, Caption: Intergraph Headquarters   Source: Intergraph Web Page   Permission: P.D.

p. 404, Caption: John Gully   Source: SAIC Web Page   Permission: P.D.

p. 405, Caption: Marc Bendickson   Source: Dynetics Web Page   Permission: P.D.

p. 405, Caption: Dynetics Campus   Source: Dynetics Web Page   Permission: P.D.

p. 405, Caption: Corporate Officers   Source: Dynetics Web Page   Permission: P.D.

p. 406, Caption: Wyle CAS Group Facility   Source: Wyle CAS Web Page   Permission: P.D.

p. 406, Caption: Sanmina-SCI Factory   Source: Sanmina-SCI Web Page   Permission: P.D.

p. 407, Caption: Qualitest Pharmaceuticals Complex   Source: Qualitest Web Page   Permission: P.D.

p. 408, Caption: ADTRAN Corporate Campus   Source: ADTRAN Web Page   Permission: P.D.

p. 408, Caption: CINRAM Facility   Source: CINRAM Web Page   Permission: P.D.

p. 409, Caption: TMMAL Complex   Source: TMMAL Web Page   Permission: P.D.

p. 413, Caption: Factory-of-the-Future   Source Raytheon Huntsville Web Page   Permission: P.D.

p. 415, Caption: EFT-1 Booster for Delta IV Heavy in Decatur   Source: United Launch Alliance Web Page   Permission: P.D.

p. 424, Caption: Jim Hudson   Source: HudsonAlpha Web Page   Permission: P.D.

p. 424, Caption: Lonnie McMillian   Source: HudsonAlpha Web Page   Permission: P.D.

p. 424, Caption: Richard Meyers   Source: HudsonAlpha Web Page   Permission: P.D.

p. 424, Caption: HudsonAlpha Institute   Source: HudsonAlpha Web Page   Permission: P.D.

p. 425, Caption: Bryan Dodson   Source: Phoenix Industries Web Page   Permission: P.D.

p. 425, Caption: Sewing Flags at Phoenix Industries   Source: Phoenix Industries Web Page   Permission: P.D.

p. 426, Caption: HU Dispatch Center   Source: Huntsville Utilities Web Page   Permission: P.D.

p. 426, Caption: Huntsville Hospital Central Campus   Source: Huntsville Hospital Web Page   Permission: P.D.

p. 427, Caption: Clinical Laboratory   Source: Huntsville Hospital Web Page   Permission: P.D.

p. 427, Caption: Central Hall of Davidson Center   Source: U.S. Space & Rocket Center Web Page   Permission: P.D.

p. 427, Caption: Deborah Barnhart   Source: U.S. Space & Rocket Center Web Page   Permission: P.D.

p. 428, Caption: Central Control   Source: Alabama Supercomputer Center Web Page   Permission: P.D.

p. 429, Caption: Ray Vaughn   Source: University of Alabama in Huntsville Web Site   Permission: P.D.

p. 430, Caption: Chance Glenn   Source: Alabama A&M University Web Site   Permision: P.D.

p. 432, Caption: TVREC at Belle Mina   Source: Tennessee Valley Research and Extension Center Web Page   Permission: P.D.

p. 434, Caption: Modern Cotton Harvester   Source: Web Site advertisement for John Deere 9930   Permission: P.D.

p. 438, No Caption – Cyber Graphis   Source: Integrated by author from several graphics

p. 439, Caption: Initial Cyber Campus   Source: U.S. Army Press Release   Permission: P.D.

p. 453, Caption: Watson in Laser Research Laboratory, 1964   Source & Permission: Photo owned by author

Front cover: Pictures from interior of book

The author, Raymond C. Watson, Jr., Ph.D., P.E., has been an engineer since 1942, and has been intimately involved with many of the technical activities of Greater Huntsville since 1960. His overall career combined a broad variety of industrial and academic positions, and he is still fully engaged as a consultant and technical historian.

His recent books are Solving the Radar Crisis (Trafford 2007) and Radar Origins Worldwide (Trafford 2009).

www.ingramcontent.com/pod-product-compliance
Lightning Source LLC
Chambersburg PA
CBHW020719180526
45163CB00001B/36